Classical Biological Control of *Bemisia tabaci* in the United States - A Review of Interagency Research and Implementation

Progress in Biological Control

Volume 4

Published:

Volume 1
H.M.T. Hokkanen and A.E. Hajek (eds.):
Environmental Impacts of Microbial Insecticides – Need and Methods for Risk Assessment. 2004
ISBN 978-1-4020-0813-9

Volume 2
J. Eilenberg and H.M.T. Hokkanen (eds.):
An Ecological and Societal Approach to Biological Control. 2007
ISBN 978-1-4020-4320-8

Volume 3
J. Brodeur and G. Boivin (eds.):
Trophic and Guild Interactions in Biological Control. 2006
ISBN 978-1-4020-4766-4

Forthcoming:

Integration of Insect-Resistant GM Crops within IPM Programs
Edited by J. Romeis, A.M. Shelton and G.G. Kennedy

Use of Microbes for Control and Eradication of Invasive Arthropods
Edited by A.E. Hajek, M. O'Callaghan and T. Glare

Ecological & Evolutionary Relationships among Entomphagous Arthropods and Non-prey Foods
By J. Lundgren

Biocontrol-based Integrated Management of Oilseed Rape Pests
Edited by I.H. Williams and H.M.T. Hokkanen

Biological Control of Plant-Parasitic Nematodes: Building Coherence between Microbial Ecology and Molecular Mechanisms
Edited by Y. Spiegel and K. Davies

Juli Gould • Kim Hoelmer • John Goolsby
Editors

Classical Biological Control of *Bemisia tabaci* in the United States - A Review of Interagency Research and Implementation

Springer

Juli Gould
USDA-APHIS
MA
USA

Kim Hoelmer
USDA-ARS
Newark
DE
USA

John Goolsby
USDA-ARS
Weslaco
TX
USA

ISBN 978-90-481-7708-0 e-ISBN 978-1-4020-6740-2

Cover illustrations: A female *Eretmocerus* (an undescribed species native to Florida) feeds on fluids exuding from a *Bemisia tabaci* nymph that has been pierced by the wasp's ovipositor (top picture) and oviposits underneath a *B. tabaci* nymph (bottom picture).
Photograph credits: Mike Rose

Printed on acid-free paper

9 8 7 6 5 4 3 2 1

springer.com

Dedication

This book is dedicated to Mike Rose (1945–2004), an author of Chapter 5, preeminent biological control specialist and inspirational mentor to many of the book's authors. Mike Rose began his career at the University of California, Riverside, in the early 1960s working with Paul DeBach on biological control of whitefly, scale and mealybug pests of citrus. His career continued at Texas A&M University, and even in nominal retirement in Montana he remained very active as a biological control consultant. Mike's expertise in biological control of whitefly and in the aphelinid genus *Eretmocerus* made him a natural leader and proponent of the biological control program for *Bemisia tabaci* in the USA. Mike's influence can be seen in many aspects of the research reported in this book, especially in the taxonomy, quarantine evaluation, and postrelease evaluation of the natural enemies.

Progress in Biological Control

Series Preface

Biological control of pests, weeds, and plant and animal diseases utilising their natural antagonists is a well-established and rapidly evolving field of science. Despite its stunning successes world-wide and a steadily growing number of applications, biological control has remained grossly underexploited. Its untapped potential, however, represents the best hope to providing lasting, environmentally sound, and socially acceptable pest management. Such techniques are urgently needed for the control of an increasing number of problem pests affecting agriculture and forestry, and to suppress invasive organisms which threaten natural habitats and global biodiversity.

Based on the positive features of biological control, such as its target specificity and the lack of negative impacts on humans, it is the prime candidate in the search for reducing dependency on chemical pesticides. Replacement of chemical control by biological control – even partially as in many IPM programs – has important positive but so far neglected socio-economic, humanitarian, environmental and ethical implications. Change from chemical to biological control substantially contributes to the conservation of natural resources, and results in a considerable reduction of environmental pollution. It eliminates human exposure to toxic pesticides, improves sustainability of production systems, and enhances biodiversity. Public demand for finding solutions based on biological control is the main driving force in the increasing utilisation of natural enemies for controlling noxious organisms.

This book series is intended to accelerate these developments through exploring the progress made within the various aspects of biological control, and via documenting these advances to the benefit of fellow scientists, students, public officials, policy-makers, and the public at large. Each of the books in this series is expected to provide a comprehensive, authoritative synthesis of the topic, likely to stand the test of time.

Heikki M.T. Hokkanen, Series Editor

Editors Preface

This book reviews interagency research and development of classical (importation) biological control of *Bemisia tabaci* (biotype B) conducted in the USA from 1992-2002. The successful discovery, evaluation, release, and establishment of at least five exotic *B. tabaci* natural enemies in rapid response to the devastating infestations in the USA represents a landmark in interagency cooperation and coordination of multiple disciplines. The review covers all key aspects of the classical biocontrol program, beginning with foreign exploration and quarantine culture, through development of mass rearing methodology, laboratory and field evaluation for efficacy, to field releases, integration with other management approaches, and monitoring for establishment and potential non-target impacts. The importance of morphological and molecular taxonomy to the success of the program is also emphasized. The book's contributors include 28 USDA, state department of agriculture, and university scientists who participated in various aspects of the project.

Bemisia tabaci continues to be a pest of major concern in many parts of the world, especially since the recent spread of the Q biotype, so the publication of a review of the biological control program for the B biotype is especially timely. We anticipate that our review of the natural enemies that were evaluated and which have established in the USA will benefit researchers and IPM practitioners in other nations affected by *B. tabaci.* This book will also serve as a useful reference for scientists in the USA conducting research on the Q biotype of *B. tabaci.* It will complement other recent works on *Bemisia* that deal more broadly with a wide range of subject areas and consequently must treat importation biological control in much less detail. Although the book's theme is *B. tabaci*, the organization and conduct of the project serves as a useful model for programs directed at biological control of other whitefly species, as well as biocontrol programs for other pests. This book should also support and encourage classical biological control inputs into other integrated pest management systems.

We would like to acknowledge Deborah Winograd (USDA-APHIS-PPQ, Center for Plant Health Science and Technology) for her assistance in reviewing the book chapters for grammar, consistency, and reference citations.

Juli Gould
Kim Hoelmer
John Goolsby

Contents

Contributing Authors

Earl R. Andress
USDA-APHIS Pink Bollworm Rearing Facility
3645 E. Chipman Rd., Phoenix AZ 85040 USA

James Brown
California Department of Food & Agriculture, Biological Control Program
3288 Meadowview Rd., Sacramento, California 95832 USA

Albino B. Chavarria
USDA-APHIS-PPQ, Center for Plant Health Science and Technology Laboratory
22675 N. Moorefield Rd., Edinburg, TX, 78541-9398 USA

Matthew A. Ciomperlik
USDA-APHIS-PPQ, Center for Plant Health Science and Technology Laboratory
22675 N. Moorefield Rd., Edinburg, TX, 78541-9398 USA

Nick Colletto
USDA-APHIS Pink Bollworm Rearing Facility
3645 E. Chipman Rd., Phoenix, AZ 85040 USA

Robert M. Faust
USDA-ARS, National Program Staff (retired)
5601 Sunnyside Ave, Beltsville, MD 20705-5103 USA

John A. Goolsby
USDA-ARS, Kika de la Garza Subtropical Agricultural Research Center
Beneficial Insects Research Unit
2413 East Highway 83, Weslaco, Texas 78596 USA

Juli Gould
USDA-APHIS-PPQ, Center for Plant Health Science and Technology Laboratory
Building 1398, Otis ANGB, MA 02542 USA

Thomas J. Henneberry
USDA-ARS Arid Land Agricultural Research Center (retired)
21881 North Cardon Lane, Maricopa, AZ 85238 USA

John M. Heraty
Department of Entomology
University of California, Riverside, CA 92521 USA

Kim A. Hoelmer
USDA-ARS, Beneficial Insect Introduction Research Unit
501 South Chapel St., Newark, DE 19713-3814 USA

Alan A. Kirk
USDA-ARS, European Biological Control Laboratory
Campus International de Baillarguet, CS 90013, Montferrier-sur-Lez
34988 St. Gély du Fesc Cedex, France

Lawrence A. Lacey
USDA-ARS, Yakima Agricultural Research Center
5230 Konnowac Pass Road, Wapato, WA 98951 USA

Benjamin C. Legaspi, Jr.
Florida Public Service Commission
2540 Shumard Oak Blvd. Tallahassee, FL 32399 USA

Jesusa C. Legaspi
USDA-ARS CMAVE – Florida A&M University, Center for Biological Control
6383 Mahan Drive, Tallahassee, Florida 32308 USA

Paul Merten
USDA-ARS, Arid Land Agricultural Research Center
21881 North Cardon Lane, Maricopa, AZ 85238 USA

Eric T. Natwick
ANR UCCE, University of California, Desert Research & Extension Center
1050 East Holton Road, Holtville, CA 92250-9615 USA

Charles H. Pickett
California Department of Food & Agriculture, Biological Control Program
3288 Meadowview Rd., Sacramento, California 95832 USA

Andrew Polaszek
Entomology Department, Natural History Museum
Cromwell Road, London, SW7 5BD U.K.

Mark Quinn
Department of Crop and Soil Sciences, Washington State University
P.O. Box 646420, Pullman, WA 99164-6420 USA

William J. Roltsch
California Department of Food & Agriculture, Biological Control Program
3288 Meadowview Rd., Sacramento, California 95832 USA

Mike Rose*
Department of Entomology, Montana State University (*deceased)
Bozeman, MT 59717 USA

Raul A. Ruiz-Arce
USDA-APHIS-PPQ, Center for Plant Health Science and Technology Laboratory
22675 N. Moorefield Rd., Edinburg, TX, 78541-9398 USA

Michael E. Schauff
USDA-ARS, Plant Sciences Institute, BARC West, Bldg 003, Room 231
10300 Baltimore Ave., Beltsville, MD 20705 USA

David J. Schuster
University of Florida, Institute of Food & Agricultural Sciences
Gulf Coast Research & Education Center
14625 CR 672, Wimauma, FL 33598 USA

Gregory S. Simmons
USDA-APHIS PPQ, Center for Plant Health Science and Technology Laboratory
3645 East Wier Ave., Phoenix, AZ 85040 USA

Don C. Vacek
USDA-APHIS, PPQ, Center for Plant Health Science and Technology Laboratory
22675 N. Moorefield Rd, Edinburg, TX, 78541-9398 USA

Diane Waldner
USDA-APHIS PPQ, Center for Plant Health Science and Technology Laboratory
3645 East Wier Ave., Phoenix, AZ 85040 USA

Stephen P. Wraight
USDA-ARS, Plant, Soil, and Nutrition Laboratory
Tower Rd., Ithaca, NY 14853 USA

Gregory Zolnerowich
Department of Entomology, Kansas Sate University
Manhattan, KS 66506 USA

Chapter 1
Introduction

Thomas J. Henneberry[1] and Robert M. Faust[2]

Bemisia tabaci (Gennadius) biotype B (= *B. argentifolii* Bellows and Perring) was described in 1889 as a tobacco pest in Greece and named *Aleyrodes tabaci*, the tobacco whitefly (Gennadius 1889). Numerous synonymies (Russell 1957, 1975) and nomenclatural issues (Brown et al. 1995) have occurred since its first description. Perring (2001) indicates that the existence of a species complex is reaching acceptance by the scientific community. The complex has many biotypes and two described extant, cryptic species. Improved transportation technology and increased frequency of international transport of plant material has contributed to the extension of the geographical range of the *B. tabaci* complex. At present, it is globally distributed and occurs on all continents except Antarctica (Martin 1999; Martin et al. 2000). Losses from the species complex in worldwide agricultural systems have been extensive. Table 1.1, modified and updated from Oliveira et al. (2001), Cock (1986, 1993), and Ioannou (1997) shows the international scope of *B. tabaci* as an economic pest. Its emergence as a major threat in agricultural production systems has been characterized by outbreaks in many parts of the world (Gerling and Henneberry 2001). In the 1980s and early 1990s, infestations in the USA were particularly damaging.

1.1 Brief History of *B. tabaci* and its Economic Impact in the USA

The first *B. tabaci* collected in the New World was found in 1894 in the USA on sweet potato and described as *Aleyrodes inconspicua* Quaintance and given the name sweetpotato whitefly (Quaintance 1900). Except for its role as a vector of cotton leaf crumple in the late 1950s and early 1960s (van Schaik et al. 1962), *B. tabaci* was not recognized as an economic pest in the USA. However, the serious

[1] USDA-ARS (retired), Arid Land Agricultural Research Center, 21881 North Cardon Lane, Maricopa, AZ 85238

[2] USDA-ARS-NPL (retired), 5601 Sunnyside Ave, Beltsville, MD 20705-5103

J. Gould et al. (eds.), *Classical Biological Control of* Bemisia tabaci *in the United States.*

Table 1.1 Some reports of *Bemisia tabaci* economic pest status: Crops involved, monetary losses, or percentages of crop-loss estimates.

Commodity	Time frame	Geographical areas	Losses: US$ (millions)	Losses: % of crop	Sources
Aubergine	1970	Egypt	"Extensive" crop loss	–	Herakly and El-Ezz (1970)
Cassava	1970s	Africa, India	–	30–80	Sauti (1982); Phillips (1973); Lopez-Avila and Cock (1986)
	1988	Ivory Coast	–	40	Fargette et al. (1988)
	1990	Africa	2,000	50	Fauquet and Fargette (1990)
Cotton	1920s–1930s	Northern India (now Pakistan)	"Extensive" crop loss	–	Misra and Lamba (1929)
	1920s–1984	India, Sudan, Iran, El Salvador, Mexico, Brazil, Turkey, Israel, Thailand, Ethiopia	–	41–81	Basu (1995)
	1960s	USA, California, Imperial Valley (leaf crumple)	–	50	van Schaik et al. (1962)
	1965	Sudan Gezira	–	20–30	Mound (1965)
	1980s	Worldwide (sticky cotton)	30–35	–	Strolz (1992)
	1981	USA (Arizona, California)	>50[a]	–	Duffus and Flock (1992)
	1982	Sudan (sticky cotton)	15	–	Khalifa (1980); Khalifa and Gameel (1982); Khalifa (1983); Lopez-Avila and Cock (1986)
	1981–1983	India	–	1–31	Sukhija et al. (1986)
	1992	USA, Arizona (sticky cotton)	10.3[b]	–	Wade and Tronstad (1993)
	1995	Australia	?		DeBarro (1995)
Lettuce	1981	USA, California, Arizona	–	50–75	Duffus et al. (1986)
Sugar beets	1981	USA, California, Arizona	–	20–30	Duffus et al. (1986)
Mung bean	1970s	India	–	Up to 74	Cock (1986)

Tomato	1973	India	–	20–95	Sastry and Singh (1973)
	1991	USA, Florida	125	–	Schuster (1992)
Sunflower	1985	India	–	9.2	Balasubramanian and Chelliah (1985)
Okra	1989	India	–	54	Chaudhary and Dadheech (1989)
Cotton, melons squash, watermelons, sesame, lettuce, vegetables	1991	Rio Grande Valley, Texas	100		Kirk et al. (2001); Riley and Sparks (1993)
	1992	Imperial Valley, California	>100		
	1991, 1992	Mexico, Mexicali Valley	33[a]	–	Medina-Esparza and Leon-Paul (1994)
	1991, 1992	USA (Arizona, California, Texas, Florida)	200–500	–	Perring et al. (1993)
	1991–1995	USA, California, Imperial Valley	>100/year[b]	–	Birdsall et al. (1995)
	1995	Mexico, Sonora	[c]	–	Silva-Sanchez (1997)
Cotton	1994–1998	USA (Arizona, California, Texas) (stickiness)	154	–	Ellsworth et al. (1999)
Cowpea	1980s–present	Nigeria	–	Up to 90	Cock (1986)
Black bean, mung bean, soybean	1991	India	300	–	Varma et al. (1991)
Soybeans	1992–1994	Mexico, Sonora	41[d]	–	American Soybean Association (2000)
Tomato, okra, cotton tobacco, melon	1990s	Central American area, Cuba, Barbados, Costa Rica, Dominican Republic, El Salvador, Haiti, Honduras, Guatemala, Jamaica, Montserrat, Nicaragua, Santa Lucia	"Extensive" crop loss	–	Hilje (1996); Vazquez (1999)

(continued)

Table 1.1 (continued)

Commodity	Time frame	Geographical areas	Losses US$ (millions)	Losses % of crop	Sources
Melon, tomato pepper	1998–1999	Guatemala	–	30–50	Dávila (1999)
Beans, tomatoes, cotton, melons, watermelons, okra, cabbage and other crops	1995–present	South America, Brazil, many areas losses reported from Argentina, Columbia, Paraguay, Bolivia	>5,000	–	Lima et al. (2000); Viscarret et al. (2000); Quintero et al. (1998); Morales and Anderson (2001)
Poinsettias, tomatoes	1990s–present	Mediterranean Basin, Italy, Southern France	"Extensive" crop loss	–	Traboulsi (1994); Ioannou (1997)
Citrus	1990s–present	Azerbaijan, Georgia	"Extensive" crop loss	–	Dantsing and Shenderovska (1988); Traboulsi (1994)
Vegetables, ornamentals, citrus, cotton, olives, pears, melons, watermelons	1990s–present	Near East, Algeria, Bahrain, Cyprus, Egypt, Islamic Republic of Iran, Iraq, Jordan, Kuwait, Lebanon, Libya, Malta, Morocco, Saudi Arabia, Somalia, Sudan, Tunisia, Turkey, United Arab Emirates	–	10–90 5–80 in field culture; 0–60 in protected culture	Traboulsi (1994), Bedford et al. (1994); Ioannou (1997); Lopez-Avila and Cock (1986)
Cotton	1990s–present	Pakistan	1,100	37 2–25	Robinson and Taylor (1996); ICAC (1994); Traboulsi (1994)
Multiple crops	1990s–present	China, Taiwan, Yunnan, found in 10 additional provinces by 1995	"Extensive" crop loss	?	Rumei (1996)
Greenhouse production	1991–present	New Zealand, Spain, worldwide	"Extensive" crop loss		Oetting and Buntin (1996); Price et al. (1986); Martin (1989)
	1992	USA	23		Barr and Drees (1992)

[a]Cotton 39,415 ha to 653 ha.
[b]7.4 cents discount/454 g = 10.3 million loss in 1992 in Arizona.
[c]Cotton acreage reduced 65%.
[d]Soybeans 89,000–124,000 ha to 25,000 ha.

nature of the *B. tabaci* problem and the potential for serious impact on agricultural communities in the USA and northern Mexico became dramatically evident in the 1980s and early 1990s. Outbreaks in California and Arizona in 1981, presumably *B. tabaci* biotype A, were followed by heavy infestations on poinsettias and appearance of silverleaf symptoms on squash (Price et al. 1986; Maynard and Cantliffe 1989) by a new biotype in Florida in 1986. Reproductive, host plant, allozyme and other differences resulted in designation of the new pest as *Bemisia tabaci* biotype B (Costa and Brown 1990) and subsequently a new species *B. argentifolii* Bellows and Perring was described (Perring et al. 1993). As mentioned earlier in this paper, the taxonomic definition of the *B. tabaci* complex remains open for discussion and *B. tabaci* biotype B (= *B. argentifolii*) will be referred to in this book as *B. tabaci*.

In Arizona, California, Texas, and Florida, economic losses from *B. tabaci* in 1991 and 1992 were estimated to range from $200 to $500 million (US dollars) (Perring 1996). In Imperial Valley, California, between 1991 and 1995, over $100 million were lost annually (Birdsall et al. 1995). In Arizona, California and Texas, cotton growers spent $154 million (Ellsworth et al. 1999) during 1994–1998 to control sweetpotato whitefly and prevent cotton lint stickiness. Gonzalez et al. (1992) estimated that for every million dollars of primary *B. tabaci*-induced crop loss in a multi-commodity-growing agricultural community, there was an estimated $1.2 million loss of personal income as well as the elimination of 42 jobs. *Bemisia tabaci* infestations in the US greenhouse and ornamental production also caused estimated losses in millions of dollars (Barr and Drees 1992). Losses to the tomato industry in Florida in 1991 were reported to exceed $125 million (Schuster 1992). Similar crop and financial losses occurred in adjacent agricultural areas in northern Mexico (Medina-Esparza and Leon-Paul 1994; Silva-Sanchez 1997; American Soybean Association 2000).

These unacceptable *B. tabaci*-caused financial, social, and environmental losses highlighted the need for a nationally coordinated effort to provide long- and short-term solutions to the problem. The *B. tabaci* outbreaks were unexplained but clearly suggested biological and host plant preference differences compared to previously encountered *B. tabaci* populations. Immediate and aggressive attention was required to address the issues arising from the unprecedented outbreaks of the new type of *B. tabaci*.

1.2 National Research and Action Plan for Development of Management and Control Methodology for the Sweetpotato Whitefly

In October, 1991, a sweetpotato whitefly Ad Hoc Working Group meeting was held in Atlanta, Georgia, to initiate planning for a coordinated research effort on *B. tabaci* (USDA 1992a). Twenty-six participants, representing USDA-ARS, USDA-APHIS and USDA-CSREES, state experiment stations, several universities and various commodity groups were in attendance. The need for high-priority research was agreed upon, and plans were made to organize a comprehensive working conference. Subject area coordinators from various agencies and institutions were identified to aid in the development of the conference. With support from the Secretary of Agriculture's Office, a group of 40 individuals representing several state universities, USDA-ARS, USDA-APHIS, USDA-CSREES, and commodity groups, met in Reno, NV in December 1991 (USDA 1992b) to further coordinate these activities. A draft of a coordinated, cooperative research and action plan was reviewed, and priority areas were highlighted for immediate action and assembly into a formal written document. The plan was finalized and accepted at a meeting of more than 200 participants in Houston, Texas in February 1992 (USDA 1992c).

At the national level, the USDA Sweetpotato Whitefly Research, Education and Implementation Coordinating Group (two members from ARS, two members from APHIS, two members from CSREES, and one member from a state agricultural experiment station) was formed in 1992 to coordinate the interagency activities related to the plan. The coordinating group and partner state agricultural experiment stations ensured a unified effort for the program, and provided an annual review to exchange research information, plan cooperative work, and evaluate research progress.

The high-priority research areas set forth for the 1992–1997 national plan were: (1) ecology, population dynamics, and dispersal; (2) fundamental research on behavior, biochemistry, biotypes, morphology, physiology, systematics, virus diseases, and virus vector interactions; (3) chemical control, biorationals, and pesticide application technology; (4) biological control; (5) crop management systems and host plant resistance; and (6) integrated techniques, approaches, and philosophies. Mandated annual reviews were held to review programs, priorities, consider new research thrusts and exchange information.

The need for research continuity, continuing high levels of communication, technology transfer, and coordination resulted in development of a second 5-year plan. The Silverleaf Whitefly (*Bemisia argentifolii* Bellows and Perring) Research, Action, and Technology Transfer Plan was finalized at the annual review meeting at San Diego, California in January 1997 (Henneberry et al. 1997). The high-priority research areas were: (A) biology, ecology, and population dynamics; (B) viruses, epidemiology, and virus-vector interactions; (C) chemical control, biopesticides, resistance management, and application methods; (D) natural enemy ecology and biological control; (E) host plant resistance, physiological disorders, and host-plant interactions; and (F) integrated and area-wide pest management approaches, and crop management systems. The last meeting for the second 5-year plan occurred in February 2002 at San Diego, California.

1.3 The Role of Biological Control

Biological control was identified as a high-priority research area in the US national research and action plans. Developing long-term integrated *B. tabaci* population management, with a strong natural enemy component, in lieu of individual farmers focusing on local infestations, was a mandate developed in the formative phase of the research and action plan. The positive role of natural enemy interactions in *B. tabaci* populations and their potential as control agents have been recognized by numerous authors (Mound and Halsey 1978; Greathead and Bennett 1981; Cock 1986, 1993; Gerling 1990, 1996; Gerling and Heneberry 2001). The complexity of nomenclature issues for *B. tabaci* and its natural enemies, agroecosystem and geographic variability and the lack of essential biological and ecological information have made evaluations of the impact of natural enemies on *B. tabaci* populations a formidable challenge to biological control workers worldwide.

Although high *B. tabaci* nymph parasitism (70–80%) often occurs in southern California cotton, adequate control of *B. tabaci* has not been obtained with native parasitoids (Gerling 1967; Natwick and Zalom 1984; Bellows and Arakawa 1988; Hoelmer 1996; Gerling and Naranjo 1998). Similar results have been reported from Israel (Gerling et al. 1980; Gerling 1986). In the USA (Nuessly 1990) and Israel (Gerling 1996) the results of introductions of new parasitoid species in 1985–1987 were disappointing. In contrast, reports from the Sudan (Abdelrahman and Munir 1989), Syria (Stam and Elmosa 1990), and Egypt (Hafez et al. 1979) indicate effective parasitoid regulation of *B. tabaci* populations in diverse cropping ecosystems when no insecticides were used (Hafez et al. 1979; Abdel-Fattah et al. 1986; Abdel-Gawaad et al. 1990). There are many possible explanations for the differences in biological control efficacy: *B. tabaci* host range; multiple cropping systems, providing year-round host biomass; lack of information on natural enemy-*B. tabaci*-host interactions; geographical variability; and different crop production inputs. Insecticides have also frequently been identified as the cause of suppression of natural enemies, resulting in *B. tabaci* outbreaks (Eveleens 1983). Resistance of *B. tabaci* to insecticides, in combination with hormoligosis (increased reproduction of resistant strains), has been suggested as contributing to outbreaks (Dittrich et al. 1990). Under laboratory and greenhouse conditions, highly toxic effects of insecticides on several parasitoid species have been reported, but species responses vary and generalizations appear to be risky (see Hoelmer 1996 for review). In the field, Hoelmer (1996) suggested that insecticide impact on some parasitoids may not be as severe as under controlled laboratory conditions. Alternate approaches such as manipulating timing and placement of insecticides and the use of selective and new chemicals offer potential for integrating chemical and biological control. This possibility was strengthened considerably for *B. tabaci* with the development of the insect growth regulators (IGRs), such as buprofezin and pyriproxyfen, for control on cotton and imidacloprid for control on melons. Natural enemy conservation was found to be much improved with IGR use in cotton (Naranjo 2001). Ellsworth and Martinez-Carillo (2001) found that the combination of natural enemy conservation and IGR use increased *B. tabaci* mortality by more than 50% compared to conventional chemistry because of direct mortality by the IGRs plus increased predation. Soil applications of imidacloprid on melons were also found to be environmentally compatible and broke the host continuity by reducing dispersal from melons to cotton in the spring and cotton to melons in the fall (Palumbo et al. 2001).

The precise combinations of biotic and abiotic factors that trigger *B. tabaci* outbreaks remain unknown, but the large number of natural enemies species recorded attacking *B. tabaci* and the high level of observed activity, leading to effective *B. tabaci* control in some areas, strongly supported the need to exploit their usefulness (Greathead and Bennett 1981; Onillon 1990; Hoelmer 1996; Gerling and Kravchenko 1996). *Bemisia tabaci* natural enemy records have been cataloged by several authors (Greathead and Bennett 1981; Lopez-Avilla 1986; Lopez-Avilla and Cock 1986; Gerling 1990). A summary of the most recent *Encarsia* spp. status was included in a 1993 natural enemy update (Cock 1993). Additional reviews of *B. tabaci* fungal entomopathogens (Lacey et al. 1996), *Eretmocerus* spp. (Rose et al. 1996), and *B. tabaci* predators (Nordlund and Legaspi 1996) further informed the effort to locate new *B. tabaci* biological control agents. New introductions from this broad base of biological material to complement existing natural enemies, development of mass-rearing and release augmentation, and conservation approaches were considered important components for long-term *B. tabaci* management systems (Cock 1986; Gerling 1990; Onillon 1990; Cock 1993). Cock (1986, 1993) suggested that workers in *B. tabaci* infested areas lacking specific natural enemies noted as beneficial in other areas should consider introduction of these effective natural enemies into their areas. This strategy was considered a particularly promising way to strengthen *B. tabaci* biological control by providing new natural enemies to supplement indigenous species. The approach was further supported by the overall *B. tabaci* research management effort to develop ecologically oriented technology to conserve natural enemy resources and provide a receptive environment for augmentation and new introductions.

Thus, foreign explorations for natural enemies were initiated within the framework of the national plans in the early 1990s by the USDA-ARS European Biological Control Laboratory, Montpellier, France (Kirk et al. 1996). Areas selected for initial natural enemy exploration were in Greece, Spain and the Indian subcontinent. These areas were chosen because their climate and crop-productions systems were similar to those areas in the USA with problem *B. tabaci* populations. It was expected that if new natural enemies were identified they could easily adapt after introduction into similar US ecosystems (Kirk et al. 2001). Explorations were focused on the Indian subcontinent because the area has been suggested as the point of origin for *B. tabaci* (Brown et al. 1995). From the worldwide explorations in 28 countries, 55 parasitoid cultures were established. Numerous isolates of the fungal pathogen, *Paecilomyces fumosoroseus* (Wize), were collected from five countries. Of these, a large number of strains were isolated that have been reported as having good *B. tabaci* insecticidal activity. Field studies have been promising (Kirk et al. 2001), but additional research will be required to develop these materials to effectively control *B. tabaci* populations.

After field collection, all exotic parasitoids were shipped to the USDA-APHIS Quarantine Facility in Mission, Texas for further study (Goolsby et al. 1996, 1998). Several native parasitoids were also evaluated in comparison with exotic species, and *Eret. eremicus* has been widely used for augmentative release, especially in greenhouse crops. Parasitoid species that showed high fecundity on major commercially cultivated field crops in quarantine studies were further evaluated in field cages in the Imperial Valley, California and the Rio Grande Valley, Texas. The parasitoids that performed best under field conditions were then mass reared for release programs (Goolsby et al. 1998, 1999).

Large-scale exotic parasitoid augmentation controlled *B. tabaci* in melons, and releases were found compatible with a commonly used systemic insecticide (imidacloprid) (Goolsby and Ciomperlik 1999; Simmons et al. 1998). Area-wide parasitoid release programs to reduce *B. tabaci* overwintering populations in central California (Pickett et al. 1999) also proved particularly promising. Three of the imported and released parasitoid species, *Eretmocerus emiratus* Zolnerowich and Rose, *Eretmocerus* sp. nr *emiratus*, and *Encarsia sophia* (Girault and Dodd), have been established in agricultural ecosystems in California and Arizona (Hoelmer and Kirk 1999; Gould et al. 1998; Goolsby et al. 2005, chapters 12–14) and two additional species (*Eret. mundus* Mercet and *Eret. hayati* Zolnerowich and Rose) in Texas (Goolsby et al. 1998, chapter 11). Continuing long-term monitoring will be essential to determine the spread of these species into *B. tabaci* habitats and to quantify their impact on *B. tabaci* population dynamics.

The successful exploration, screening, evaluation, importation and establishment of at least five exotic *B. tabaci* natural enemies in rapid response to the devastating infestations occurring in the USA in the late 1980s and early 1990s is a landmark in interagency, multiple discipline coordination and cooperation. The integration of exotic biological control components into highly effective *B. tabaci* management programs has been achieved. Key contributing factors to this achievement were the efforts of many scientists who developed multifaceted *B. tabaci* management strategy using (1) non-*B. tabaci* preferred cultivars, (2) spatial and temporal considerations in sequential crop systems, (3) intensive sampling and monitoring of *B. tabaci* populations, (4) chemical control focused on natural enemy conservation, action thresholds, alternating chemistry, new chemistry, and resistance monitoring, (5) optimum crop yield goals, allowing for early harvests and destruction of crop residues, and (6) active education and extension outreach to provide timely communication of new developments and guidelines for implementation of new technology (Henneberry et al. 1998). In this volume, the various authors will present the detailed documentations of natural enemy exploration, introduction, and evaluation efforts that will serve as a guide to support and encourage classical biological control inputs into other integrated pest management systems.

References

Abdelrahman, A.A. and B. Munir. 1989. Sudanese experience in integrated pest management of cotton. Insect Science Applications 10: 787–794.

Abdel-Fattah, M.J., A. Hendi, and A. El-Said. 1986. Ecological studies on parasites of the cotton whitefly *Bemisia tabaci* (Genn.) in Egypt. Bulletin of the Entomological Society of Egypt Economic Series 14: 95–105.

Abdel-Gawaad, A.A., A.M. El Sayed, F.F. Shalaby, and M.R. Abo-El-Ghar. 1990. Natural enemies of *Bemisia tabaci* Genn. and their role in suppressing the population density of the pest. Agricultural Research Review 68: 185–195.

American Soybean Association. 2000. Latin American Market Reports, Mexico Intelligence Report, http://www.ag.uiuc.edu/~asala/english/market/MEMay00.htm).

Balasubramanian, G. and S. Chelliah. 1985. Chemical control of pests of sunflower. Pesticides 19 (4): 21–22.

Barr, C.L. and B.M. Drees. 1992. The poinsettia strain of sweetpotato whitefly. Texas Nursery 23: 8–12.

Basu, A.N. 1995. *Bemisia tabaci* (Gennadius): Crop Pest and Principal Whitefly Vector of Plant Viruses, Westview Press, New Delhi.

Bedford, I.D., R.W. Briddon, P. Jones, N. Alkaff, and P.G. Markham. 1994. Differentiation of three whitefly-transmitted geminiviruses from the Republic of Yemen. European Journal of Plant Pathology 100: 243–257.

Bellows, T.S., Jr. and K. Arakawa. 1988. Dynamics of pre-imaginal populations of *Bemisia tabaci* (Homoptera: Aleyrodidae) in Southern California cotton. Environmental Entomology 17: 483–487.

Birdsall, S.L., D. Ritter, and P.L. Cason. 1995. Economic impact of the silverleaf whitefly in Imperial Valley, CA. In T.J. Henneberry, N.C. Toscano, R.M. Faust, and J.R. Coppedge, eds., 1995 Supplement to the 5-Year National Research and Action Plan. Third Annual Review, San Diego, California, January 28–31, 1995. US Department of Agriculture, Agricultural Research Service 1995–2, p.162.

Brown, J.K., D.R. Frohlich, and R.C. Rossell. 1995. The sweet potato or silverleaf whiteflies: Biotypes of *Bemisia tabaci* or a species complex? Annual Review of Entomology 40: 511–534.

Chaudhary, H.R. and L.N. Dadheech. 1989. Incidence of insects attacking okra and the avoidable losses caused by them. Annals of Arid Zone 28(3–4): 305–307.

Cock, M.J.W. 1986. *Bemisia tabaci* – A literature survey on the cotton whitefly with an annotated bibliography. FAO/CAB International Institute of Biological Control, Ascot, UK.

Cock, M.J.W. 1993. *Bemisia tabaci*: an update 1986–1992 on the cotton whitefly with an annotated bibliography. International Institute of Biological Control. Ascot, UK.

Costa, H.S. and J.K. Brown. 1990. Variability in biological characteristics, isozyme patterns and virus transmission among populations of *Bemisia tabaci*. Genn. in Arizona. Phytopathology 80: 888.

Dantsing, E.M. and L.P. Shenderovska. 1988. Cotton whitefly. Zashchita Rastenii 12: 40.

Dávila, A.G.H. 1999. La mosca blanca (Homoptera: Aleyrodidae) en Guatemala. In VII Taller Latinoamericano y del Caribe Sobre Moscas-Blancas y Geminivirus, pp. 125–126. IPA, Recife.

DeBarro, P.J. 1995. *Bemisia tabaci* biotype B: a review of its biology, distribution and control. Second Edition, Commonwealth Scientific and Industrial Research Organization, Canberra, Australia. Division of Entomology Technical Paper 36, 1–58.

Dittrich, V., S. Uk, and H. Ernst. 1990. Chemical control and insecticide resistance of whiteflies. In D. Gerling, ed., Whiteflies: Their Bionomics, Pest Status and Management, pp. 263–286. Intercept, Andover, UK.

Duffus, J.E. and R.A. Flock. 1992. Whitefly-transmitted disease complex of the desert Southwest. California Agriculture 36: 4–6.

Duffus, J.E., R.C. Larsen, and H.Y. Liu. 1986. Lettuce infectious yellows virus – a new type of whitefly transmitted virus. Phytopathology 76: 97–100.

Ellsworth, P.C. and J. Martinez-Carillo. 2001. IPM for *Bemisia tabaci*: a case study from North America. Crop Protection 20: 853–869.

Ellsworth, P.C., R. Tronstad, J. Leser, P.B. Goodell, L.D. Godfrey, T.J. Henneberry, D. Hendrix, D. Brushwood, S.E. Naranjo, S. Castle, and R.L. Nichols. 1999. Sticky cotton sources and solutions. IPM Series, No. 13. The University of Arizona Cooperative Extension. Publication AZ1156. Tucson.

Eveleens, K.G. 1983. Cotton-insect control in the Sudan Gezira: analysis of a crisis. Crop Protection 2: 273–287.

Fauquet, C. and D. Fargette. 1990. African cassava mosaic virus: Etiology, epidemiology, and control. Plant Disease 74(6): 404–411.

Fargette, D., C. Fauquet, and J.C. Thouvenel. 1988. Yield losses induced by African cassava mosaic virus in relation to the mode and the date of infection. Tropical Pest Management 34(1): 89–91.

Gennadius, P. 1889. Disease of tobacco plantations in the Trikonia. The aleurodid of tobacco. Ellenike Georgia 5: 1–3.

Gerling, D. 1967. Bionomics of the whitefly parasite complex associated with cotton in southern California (Homoptera: Aleurodidae: Hymenoptera: Aphelinidae). Annals of the Entomological Society of America 60: 1306–1321.

Gerling, D. 1990. Natural enemies of whiteflies: Predators and parasitoids. In D. Gerling, ed., Whiteflies: Their Bionomics, Pest Status and Management, pp.147–185. Intercept, Andover, UK.

Gerling, D. 1996. Status of *Bemisia tabaci* in the Mediterranean countries: opportunities for biological control. Biological Control 6: 11–22.

Gerling, D. 1996. Natural enemies of *Bemisia tabaci,* biological characteristics and potential as biological control agents: a review. Agriculture Ecosystems and Environment 17: 99–110.

Gerling, D. and T.J. Henneberry. 2001. The status of *Bemisia* as a cotton pest: Trends and future possibilities VII. In F.M. Gillham, ed., Proceedings of the Worldwide Cotton Research Conference 2, New Frontiers in Cotton Research, pp. 639–43. Athens, Greece, September 6–12, 1998, International Cotton Advisory Committee, Washington, DC.

Gerling, D. and V. Kravchenko. 1996. Pest management of *Bemisia* out of doors. In D. Gerling and R.T. Mayer, eds., *Bemisia* 1995 Taxonomy, Biology, Damage, Control and Management, pp. 667–680. Intercept, Andover, UK.

Gerling, D. and S.E. Naranjo. 1998. The effect of insecticide treatments in cotton fields on the levels of parasitism of *Bemisia tabaci* (Gennadius). Biological Control 12: 33–41.

Gerling, D., U. Motro, and R. Horowitz. 1980. Dynamics of *Bemisia tabaci* (Gennadius) (Homoptera: Aleyrodidae) attacking cotton in the coastal plain of Israel. Bulletin of Entomological Research 70: 213–219.

Gerling, D., O. Alomar, and J. Arno. 2001. Biological control of *Bemisia tabaci* using predators and parasitoids. Crop Protection 20: 779–799.

Gonzalez, R.A., G.E. Goldman, E.T. Natwick, H.R. Rosenberg, J.I. Grieshop, S.R. Sutter, T. Funakoshi, and S. Davila-Garcia. 1992. Whitefly invasion in Imperial Valley costs growers, workers millions in losses. California Agriculture 46(5): 7–8.

Goolsby, J.A. and M.A. Ciomperlik. 1999. Development of parasitoid inoculated seedling transplants for augmentative biological control of silverleaf whitefly (Homoptera: Aleyrodidae). Florida Entomologist 82: 532–545.

Goolsby, J.A., J.C. Legaspi, and B.C.Legaspi, Jr. 1996. Quarantine evaluation of exotic parasites of the sweetpotato whitefly, *Bemisia tabaci* (Gennadius). Southwestern Entomologist 21: 13–21.

Goolsby, J.A., M.A. Ciomperlik, B.C. Legaspi, Jr., J.C. Legaspi, and L.E. Wendel. 1998. Laboratory and field evaluation of exotic parasitoids of *Bemisia tabaci* (Biotype "B") in the Lower Rio Grande valley of Texas. Biological Control 12: 127–135.

Goolsby, J.A., Ciomperlik, M.A., Kirk, A.A., Jones, W.A., Legaspi, B.C., Ruiz, R.A., Vacek, D.C., Wendel, L.E. 2000. Predictive and empirical evaluation for parasitoids of *Bemisia tabaci* (Biotype "B"), based on morphological and molecular systematics, pp. 347–358. In Austin, A. and Dowton, M. eds., Hymenoptera: Evolution, Biodiversity, and Biological Control. 4th International Hymenopterists Conference. CSIRO Publishing, Collingwood, Victoria, Australia.

Goolsby, J.A., P. De Barro, A.A. Kirk, R. Sutherst, M. Ciomperlik, P. Ellsworth, J. Gould, K. Hoelmer, S. Naranjo, M. Rose, W. Roltsch, R. Ruiz, C. Pickett, and D. Vacek. 2005. Post-release evaluation of the biological control of *Bemisia tabaci* biotype "B" in the USA and the development of predictive tools to guide introductions for other countries. Biological Control 32: 70–77.

Gould, J., D. Waldner, N. Colleto, L. Antilla, and R. Santangelo. 1998. Release of exotic parasitoids for establishment in Arizona. In T.J. Henneberry, N.C. Toscano, T.M. Perring, and R.M. Faust, eds., Silverleaf Whitefly: 1998 Supplement to the 5-Year National Research and Action Plan, Charleston, South Carolina, February 3–5, 1998. US Department of Agriculture, Agricultural Research Service 1998–01, p. 60.

Greathead, D.J. and F.D. Bennett. 1981. Possibilities for the use of biotic agents in the control of whitefly, *Bemisia tabaci*. Biocontrol News and Information 2: 7–13.

Hafez, M., K.T. Awadallah, M.F.S. Tawfik, and A.A. Sarhan. 1979. Impact of the parasite *Eretmocerus mundus* Mercet on population of the cotton whitefly, *Bemisia tabaci* (Genn.) in Egypt. Bulletin of the Entomological Society of Egypt 62: 23–32.

Henneberry, T.J., N.C. Toscano, and S.J. Castle. 1998. *Bemisia* spp. (Homoptera: Aleyrodidae) in the United States: history, pest status, and management. Recent Developments in Entomolology 2: 151–161.

Henneberry, T.J., N.C. Toscano, T.M. Perring, and R.M. Faust, eds., 1997. Silverleaf Whitefly, 1997 Supplement to the Five-year National Research and Action Plan: Progress Review, Technology Transfer, and New Research and Action Plan (1997–2001), 5th Annual Review, US Department of Agriculture, Agricultural Research Service 1997–02.

Herakly, F.A. and A.A. El-Ezz. 1970. The cotton whitefly, *Bemisia tabaci* Genn., infesting cucurbits, in U.A.R. Agricultural Research Review 48: 110–118.

Hilje, L. 1996. Introducci6n. In L. Hilje, ed., Metodologias para el Estudio y Manejo de Moscas Blancas y Geminivirus, pp. vii–xv. CATIE, Turrialba, Costa Rica.

Hoelmer, K.A. 1996. Whitefly parasitoids: Can they control field populations of *Bemisia*? In D. Gerling and R. Mayer, eds., *Bemisia* 1995: Taxonomy, Biology, Damage Control and Management, pp. 451–476. Intercept, Andover, Hampshire, UK.

Hoelmer, K.A. and A.A. Kirk. 1999. An overview of natural enemy explorations and evaluations for *Bemisia* in the US. IOBC/WPRS Bulletin 22: 109–112.

International Cotton Advisory Committee. 1994. Leaf curl virus in Pakistan. The ICAC Recorder 7: 6–10 (September).

Ioannou, N. 1997. Integrated pest management strategies for the whitefly virus complex in the Near East. In N. Ioannou, ed., Management of the whitefly virus complex. Proceedings of the FAO Workshop on Management of the Whitefly-virus Complex in Vegetable and Cotton Production in the Near East. Larnaca, Cyprus October 2–6, 1995. FAO Plant Production and Protection Paper 143.

Khalifa, H. 1980. Cotton stickiness. International Textile Bulletin, World Edition. Spinning 2: 80.

Khalifa, H. 1983. Variation of cotton stickiness and methods of sampling. Textil Praxis International 38: 1187–1189.

Khalifa, H. and O.I. Gameel. 1982. Control of cotton stickiness through breeding cultivars resistant to whitefly (*Bemisia tabaci* (Genn.)) infestation: improvement of oil-seed industrial crops by induced mutations, pp. 181–186, IAEA, Vienna.

Kirk, A.A., L.A Lacey, and J.A. Goolsby. 1996. A systematic approach to foreign exploration for natural enemies of *Bemisia tabaci* and some current results. In D. Gerling and R.T. Mayer, eds., *Bemisia* 1995. Taxonomy, Biology, Damage, Control and Management, pp 530–533. Intercept, Andover, UK.

Kirk, A.A., L.A Lacey, and J.A. Goolsby. 2001. Classical biological control of *Bemisia* and successful integration of management strategies in the United States. In K.F. Harris, O.P. Smith, and J.E. Duffus, eds., Virus-Insect-Plant Interactions, pp. 309–329. Academic Press, New York.

Lacey, L.A., J.J. Fransen, and R. Carruthers. 1996. Global distribution of naturally occurring fungi of *Bemisia,* their biologies and use as biological control agents. In D. Gerling and R.T. Mayer, eds., *Bemisia* 1995. Taxonomy, Biology, Damage, Control and Management, pp. 401–433. Intercept, Andover, UK.

Lima, L.H. C., M.C. Moretzohn, and M.R.V. Oliveira. 2000. Survey of *Bemisia tabaci* (Gennadius) (Hemiptera: Aleyrodidae) biotypes in Brazil using RAPD markers. Genetics and Molecular Biology 23: 1–5.

Lopez-Avila, A. 1986. Natural enemies. In M.J.W. Cock, ed., *Bemisia tabaci* – A Literature Survey on the Cotton Whitefly with an Annotated Bibliography, pp. 27–35. CAB International Institute of Biological Control, Wallingford, UK/FAO, Rome.

Lopez-Avila, A. and M.J.W. Cock. 1986. Economic damage. In M.J.W. Cock, ed., *Bemisia tabaci* – A Literature Survey on the Cotton Whitefly with an Annotated Bibliography, pp. 51–54. CAB International Institute of Biological Control, Wallingford, UK/FAO, Rome.

Martin, J.H. 1999. The whitefly fauna of Australia (Sternorrhyncha: Aleyrodidae): a taxonomic account and identification guide. Commonwealth Scientific and Industrial Research Organization Entomology Technical Paper No. 38, Canberra.

Martin, N.A. 1989. Greenhouse tomatoes. A survey of pest and disease control. DSIR Plant Protection Report 1: 1–42.

Martin, J. H., D. Mifsud, and C. Rapisarda. 2000. The whiteflies (Hemiptera: Aleyrodidae) of Europe and the Mediterranean Basin. Bulletin of Entomological Research 90: 407–448.

Maynard, D.N. and D.J. Cantliffe. 1989. Squash silverleaf and tomato irregular ripening: new vegetable disorders in Florida. University of Florida, IFAS, Florida Cooperative Extension Service, Vegetable Crops Fact Sheet VC-37.

Medina-Esparza, J.J. and R.L. Leon-Paul. 1994. Evaluation of pesticides for the control of white-fly on cotton. INIFAP-CIRNO-CEMEXI, Mexicali Valley-1993. In Proceedings of the International Pest Work Committee, Mazatlan, Mexico, pp. 50–55. California Department of Food and Agriculture, Sacramento, CA.

Misra, C.S. and S.K. Lamba. 1929. The cotton whitefly (*Bemisia gossypiperda* n.sp.). Bulletin of the Agricultural Research Institute of Pusa 196: 1–7.

Morales, F.J. and P.K. Anderson. 2001. The emergence and dissemination of whitefly-transmitted geminiviruses in Latin America. Archives of Virology 146: 415–441.

Mound, L.A. 1965. Effect of whitefly (*Bemisia tabaci*) on cotton in the Sudan Gezira. Empire Cotton Growing Review 42: 290–294.

Mound, L.A. and S.H. Halsey. 1978. Whitefly of the World: A Systematic Catalogue of the Aleyrodidae (Homoptera) with Host Plant and Natural Enemy Data. Wiley, New York.

Naranjo, S.E. 2001. Conservation and evaluation of natural enemies in IPM systems for *Bemisia tabaci*. Crop Protection 20: 835–852.

Natwick, E.T. and F.G. Zalom. 1984. Surveying sweetpotato whitefly in the Imperial Valley. California Agriculture 38: 11.

Nordlund, D.A. and J.C. Legaspi. 1996. Whitefly predators and their potential for use in biological control. In D. Gerling and R.T. Mayer, eds., *Bemisia* 1995 Taxonomy, Biology, Damage, Control and Management, pp. 499–513. Intercept, Andover, UK.

Nuessly, G.S. 1990. Biological control of the sweetpotato whitefly in California. In R.K. Yokomi, K.R. Narayanan, and D. Schuster, eds., Proceedings, Workshop on Sweetpotato Whitefly Mediated Vegetable Disorders in Florida, Feb. 1–2, 1990. Homestead, FL. Institute of Food and Agricultural Science, University of Florida, Gainesville. pp. 81–82.

Oetting, R.D. and G.D. Buntin. 1996. *Bemisia* damage expression in commercial greenhouse production. In D. Gerling and R.T. Mayer, eds., *Bemisia:* 1995 Taxonomy, Biology, Damage, Control and Management, pp. 201–208. Intercept, Andover, UK.

Oliveira, M.R.V., T.J. Henneberry, and P. Anderson. 2001. History, current status, and collaborative research projects for *Bemisia tabaci*. Crop Protection 20: 709–723.

Onillon, J.C. 1990. The use of natural enemies for the biological control of whiteflies. In D. Gerling, ed., Whiteflies: Their Bionomics, Pest Status and Management, pp. 287–313. Intercept, Andover, UK.

Palumbo, J.C., A.R. Horowitz, and N. Prabhaker. 2001. Insecticidal control and resistance management for *Bemisia tabaci*. Crop Protection 20: 739–765.

Perring, T.M. 1996. Biological differences of two species of *Bemisia* that contribute to adaptive advantages. In D. Gerling and R.T. Mayer, eds., *Bemisia* 1995 Taxonomy, Biology, Damage, Control and Management, pp. 3–16, Intercept, Andover, UK.

Perring, T.M. 2001. The *Bemisia tabaci* species complex. Crop Protection 20: 725–737.

Perring, T.M., D.A. Cooper, R.J. Rodriguez, C.A. Farrar, and T.S. Bellows. 1993. Identification of a whitefly species by genomic and behavioral studies. Science 259: 74–77.

Phillips, T.P. 1973. Cassava utilization and potential markets, pp. 107–108. International Development Research Centre, Ottawa.

Pickett, C.H., G.S. Simmons, J.A. Goolsby, and D. Overholt. 1999. Fall releases of parasites into citrus. In T.J. Henneberry and R.M. Faust, eds., Silverleaf Whitefly: 1999 Supplement to the 5-Year National Research, Action and Technology Transfer Plan, 1997–2001: Second Annual Review of the Second 5-Year Plan, US Department of Agriculture, Agricultural Research Service Publication 1999–01, p. 83.

Price, J.F., D. Schuster, and D. Short. 1986. Managing sweetpotato whitefly. Greenhouse Grower 55–57 (December).

Quaintance, A.L. 1900. Contribution towards a monograph of the American Aleurodidae. US Department of Agriculture, Technical Services, Bureau of Entomology 8: 9–64.

Quintero, C., C. Cardona, D. Ramirez, and N. Jimenez. 1998. Primer registro del biotipo B de *Bemisia tabaci* (Homoptera: Aleyrodidae) en Colombia. Revista Colombiana de Entomologia 24: 23–28.

Riley, D.G. and A.N. Sparks, Jr. 1993. Management of the sweetpotato whitefly in the Lower Rio Grande Valley of Texas. Texas A & M Extension Bulletin B-5082.

Robinson, J.R.C. and M.J. Taylor. 1996. Economic case studies of whitefly management. In D. Gerling, R. T. Mayer, eds., *Bemisia*: 1995 Taxonomy, Biology, Damage, Control and Management, pp. 225–236. Intercept, Andover, UK.

Rose, M., G. Zolnerowich, and M.S. Hunter. 1996. Systematics, *Eretmocerus* and biological control. In D. Gerling and R.T. Mayer, eds., *Bemisia*: 1995 Taxonomy, Biology, Damage, Control and Management, pp. 477–497. Intercept, Andover, UK.

Rumei, X. 1996. The occurrence and distribution of *Bemisia* in China. In D. Gerling, R.T. Mayer, eds., *Bemisia*: 1995 Taxonomy, Biology, Damage, Control and Management, pp. 125–131. Intercept, Andover, UK.

Russell, L.M. 1957. Synonyms of *Bemisia tabaci* (Gennadius) (Homoptera: Aleyrodidae). Bulletin of the Brooklyn Entomological Society 52: 122–123.

Russell, L.M. 1975. Collection records of *Bemisia tabaci* (Gennadius) in the United States (Hemiptera: Homoptera: Aleyrodidae). Cooperative Economic Insect Report 25: 229–230.

Sastry, K.S.M. and S.J. Singh. 1973. Assessment of losses in tomato by tomato leaf curl virus. Indian Journal of Mycology & Plant Pathology 3(1): 50–54.

Sauti, R.F.W. 1982. Malavi in root crops in Eastern Africa. In Proceedings of Workshop, Kigali, Rwanda, November 23–27, 1980, pp. 104–106. International Development Research Centre, Ottawa.

Schuster, D.J. 1992. Newsletter of Work Group on *Bemisia tabaci,* Newsletter No. 5, January.

Silva-Sanchez, C. 1997. Behavior of the silverleaf whitefly (*Bemisia argentifolii* B&P) in the cotton crop on the coast of Hermosillo, Sonora. In Proceedings of the International Cotton Pest Work Committee, p. 34. California Department of Food and Agriculture, Sacramento, CA.

Simmons, G.S., K.A. Hoelmer, R.S. Staten, T. Boratynski, and E. Natwick. 1998. Biological control of silverleaf whitefly infesting cantaloupe with large-scale releases of exotic parasitoids in the Imperial Valley of California. In T.J. Henneberry, N.C. Toscano, T.M. Perring, and R.M. Faust, eds., Silverleaf Whitefly: 1998 Supplement to the 5-Year National Research and Action Plan, US Department of Agriculture, Agricultural Research Service Publication 1998–01, p. 84.

Stam, P.A. and H. Elmosa. 1990. The role of predators and parasites in controlling populations of *Earias insulana, Heliothis armigera* and *Bemisia tabaci* on cotton in the Syrian Republic. Entomophaga 35: 315–327.

Strolz, H. 1992. Cotton contamination. A global view from the spinning industry. In Proceedings of the 21st International Cotton Conference, Bremen Faserinstitut, Bremen, pp. 11–19.

Sukhija, H.S., N.S. Butter, and J. Singh. 1986. Determination of the economic threshold of white-fly, *Bemisia tabaci* Genn., on American cotton in the Punjab. Tropical Pest Management 32(2): 34–136, 190, 194. [Spanish & French summaries]

Traboulsi, R. 1994. *Bemisia tabaci*: A report on its pest status with particular reference to the Near East. FAO Plant Protection Bulletin 42: 33–58.

US Department of Agriculture 1992a. Appendix F, 1992 Sweetpotato Whitefly Ad Hoc Working Group Meeting, October 24–25, 1991, Atlanta, GA. In R.M. Faust, ed., Conference Report and 5-year National Research and Action Plan for the Development of Management Control Methodology for the Sweetpotato Whitefly, US Department of Agriculture, Agricultural Research Service, ARS-107, pp. 132–134.

US Department of Agriculture 1992b. Appendix G, Sweetpotato Whitefly Ad Hoc Working Group Meeting, December 12–13, 1991, Reno, NV. Summary/research and action plan priorities/current activities/research gaps. In R.M. Faust, ed., Conference Report and 5-year National Research and Action Plan for the Development of Management Control Methodology for the Sweetpotato Whitefly, US Department of Agriculture, Agricultural Research Service, ARS-107, pp. 135–162.

US Department of Agriculture 1992c. Appendix B, Inter-agency Workshop for the Development of Management and Control Methodologies for the Sweetpotato Whitefly, February 18–21, 1992, Houston, TX. In R.M. Faust, ed., Conference Report and 5-year National Research and Action Plan for the Development of Management Control Methodology for the Sweetpotato Whitefly, US Department of Agriculture, Agricultural Research Service, ARS-107, pp. 67–80.

Van Schaik, P.H., D.C. Erwin, and M.J. Garber. 1962. Effects of time of symptom expression of the leaf-crumple virus on yield and quality of fiber of cotton. Crop Science 2: 275–277.

Varma, D., A. K. Dhar, and B. Mandal. 1991. MYMV – the virus, the vector and their control in India. In Consultative group meeting for MYMV, July 2–6, AVRDC, Taiwan.

Vazquez, L.L. 1999. Mosca blanca-geminivirus en el Caribe: Estado actual y perspectivas. VII Taller Latinoamericano y del Caribe Sobre Moscas-Blancas y Geminivirus, pp. 45–58. IPA, Recife.

Viscarret, M.M., E.N. Botto, and A. Polaszek. 2000. Whiteflies (Hemiptera: Aleyrodidae) of economic importance and their natural enemies (Hymenoptera: Aphelinidae, Signiphoridae) in Argentina. Revista Chilena de Entomologia 26: 5–11.

Wade, J.C. and R. Tronstad. 1993. Regional price differentials. Cotton Management Economic Notes 2: 1–2.

Chapter 2
Foreign Exploration for Insect Natural Enemies of *Bemisia* for Use in Biological Control in the USA: A Successful Program

Alan A. Kirk[1], Lawrence A. Lacey[2], and John A. Goolsby[3]

Abstract European Biological Control Laboratory scientists (USDA-ARS) and collaborators sent 130 shipments of *Bemisia* species and natural enemies from 28 countries to the Mission Biological Control Laboratory (MBCL) in Mission Texas. More than 235 collections resulted in 13 species of parasitoids and several predators for evaluation in the USA. Climate modeling software was used to focus on collecting areas with climates similar to Arizona, California and Texas. Field crops, glasshouse crops and weeds were searched and many host plant species yielded parasitized *Bemisia*. Field parasitism by *Bemisia* parasitoids was shown to be 39–44% in Spain and 0–67% in Thailand. Taxonomists identified *Bemisia* biotypes, parasitoids and predators; geneticists characterized *Bemisia* and natural enemy species. This information was used for evaluation, release, and experimentation.

2.1 Introduction

Bemisia tabaci (Gennadius) has been recorded as collected from over 900 plant species in 74 families (Cock 1986, 1993; Mound and Halsey 1978). Taxonomically *B. tabaci* is now regarded as a species complex, and in 1994 a new species, *B. argentifolii* (Bellows and Perring), known as silverleaf whitefly, was described for the form known as "biotype B" (Bellows et al. 1994). The name *B. tabaci* will be used here to avoid confusion and because natural enemies were obtained from various biotypes of *B. tabaci*. Outbreaks of *B. tabaci* biotype B in Arizona, California, Florida, and Texas caused estimated crop losses in excess of $500

[1] USDA-ARS, European Biological Control Laboratory, Campus International de Baillarguet, CS 90013, Montferrier sur Lez, 34988, St. Gély du Fesc CEDEX, France

[2] USDA-ARS, Yakima Agricultural Research Center, 5230 Konnowac Pass Road, Wapato, WA 98951

[3] USDA-ARS, Kika de la Garza Subtropical Agricultural Research Center, Beneficial Insects Research Unit, 2413 East Highway 83, Weslaco, TX 78596

J. Gould et al. (eds.), *Classical Biological Control of* Bemisia tabaci *in the United States.*

million in 1992 (Faust and Coppedge 1995). First recorded in the USA in Florida in 1894 and in Texas and Georgia in the late 1940s, *B. tabaci* was not known as a serious pest until 1986–1988 in Florida (Schuster et al. 1990), and late 1989 in Texas and California (Brown et al. 1991). Before the late 1980s *Bemisia* was unknown on cole crops that are grown as a winter crop in the southern USA. After its accidental introduction, presumably from the Middle East, *B. tabaci* biotype B was found attacking cole crops in the agricultural region of southern California and was recognized as a new problem when several visible disorders of cruciferous crops appeared at the same time (Perring et al. 1991). In southern Texas severe damage to cabbage was also noted (Elsey and Farnham 1994). In addition to sustaining considerable direct damage, cole crops acted as reservoir plants for overwintering whitefly populations that moved onto melons in spring when the crucifers were harvested.

Recent genetic evidence points to an expansion in range of *B. tabaci* from an ancestral Mediterranean home throughout the world (Brown et al. 2000). Without a doubt, increased transportation of ornamental plants as seedlings and full-grown plants has led to this global spread. As a rule, natural enemies do not travel with their host and in the case of *Bemisia* extraordinary attempts at obtaining clean plants by applying pesticides for export would have eliminated the natural enemies at the source. *Bemisia tabaci*, however, because of its comprehensive resistance to pesticides would have traveled with its host plant.

A series of planning meetings to develop coordinated research and management plans for *B. tabaci* led to a 5-year national research and action plan for development of management and control methodology for *Bemisia* (Faust 1992, Chapter 1). The plan identified six areas of priority research, including biocontrol. The diverse landscapes and agricultural systems present worldwide suggested a potential for foreign exploration of many suitable habitats for whiteflies and natural enemies. The USDA-ARS European Biological Control Laboratory (EBCL) in Montpellier, France conducted exploration for *Bemisia* and its natural enemies throughout the world from 1991 to 1998 for importation and evaluation in the USA (Kirk et al. 1993; Lacey et al. 1993; Kirk and Lacey 1996). In addition to the main effort by EBCL some collections were made by collaborators and exported to the USA (Legaspi et al. 1996; Goolsby et al. 1998).

The potential of aphelinid (Hymenoptera: Aphelinidae) insect parasitoids as biocontrol agents of *Bemisia* was considered to be very high. They are widespread and their ability to find and attack whitefly nymphs is well documented (Cock 1986, 1993). The hymenopterous parasitoids obtained through foreign exploration were identified by morphological taxonomy, or by a characteristic identifying pattern using the RAPD-PCR molecular technique (Chapter 6), an important tool in maintaining the quality of parasitoid species colonies in quarantine. Colonies of these parasitoids were established, evaluation experiments were performed, and selected natural enemies were then released into the field. Whiteflies from source collections were also identified using morphological characters, and molecular characterization was accomplished using a DNA fragment of the mitochondrial cytochrome oxidase I (mtCOI) gene (Kirk et al. 2000).

2.2 Foreign Collections

Discussion at annual *Bemisia* workgroup meetings emphasized the need for parasitoids that were climatically adapted to the varied agroecosystems where *B. tabaci* was a pest. Based on predictions using CLIMEX climate matching software (Sutherst et al. 1999; Skarratt et al. 1995) of foreign locations closely matching agricultural production zones of the USA impacted by *B. tabaci*, *Bemisia* parasitoids were located in the Multan area of Pakistan, an area climatically similar to subtropical lower Rio Grande Valley, Texas (Fig. 2.1a); in Ethiopia and the United Arab Emirates, which are climatically similar to desert climates of Arizona and the Imperial Valley in southern California (Fig. 2.1b); and the region of Murcia in southeast Spain, which is climatically similar to the San Joaquin Valley, California (Fig. 2.1c). Other areas such as Thailand (Fig. 2.1d), Malaysia, and Indonesia were also intensively surveyed, but their climates were not predicted to be close matches

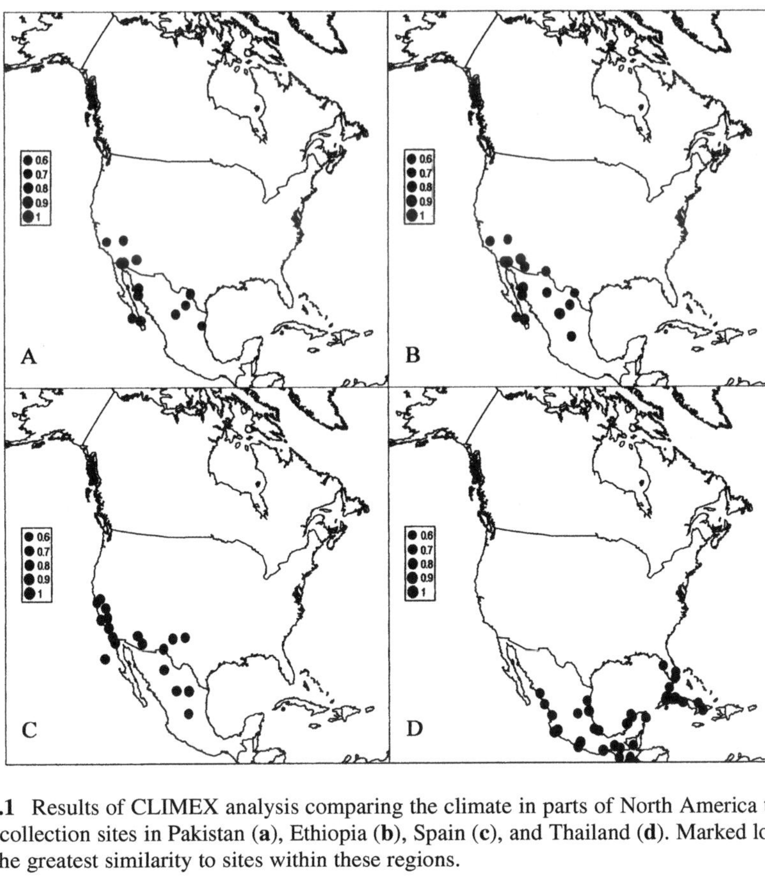

Fig. 2.1 Results of CLIMEX analysis comparing the climate in parts of North America to parasitoid collection sites in Pakistan (**a**), Ethiopia (**b**), Spain (**c**), and Thailand (**d**). Marked locations have the greatest similarity to sites within these regions.

by CLIMEX to California, Arizona or Texas, but similar to subtropical Florida. None of the parasitoids collected in these latter areas became established in the USA (Goolsby et al. 2005).

The first collections of *Eretmocerus mundus* Mercet were made in June, November and December 1991 from *Bemisia*-infested cotton and weeds near Murcia, southeast Spain. Tens of thousands of parasitized *B. tabaci* nymphs were sent to the APHIS quarantine at Mission, Texas as a result of these collections (Lacey et al. 1993; Kirk and Lacey 1996). Descendants of these parasitoids were released and subsequently became established in the San Joaquin Valley, California (Chapter 14). Collecting efforts were then concentrated on the Indian subcontinent, which was believed at that time to be the center of origin of *B. tabaci* based on the apparent high diversity of its natural enemies.

The objectives of foreign explorers were to deliver large quantities of parasitized whitefly nymphs on leaves to quarantines in the USA as quickly as possible (usually 2–4 days, occasionally much longer). Field collections were made by pulling infested leaves from plants, placing them directly into paper bags and storing them at refrigerator temperatures. Waxed paper bags absorbed condensation and enabled partial drying of the leaves to prevent rotting. The leaf material was kept cool using refrigeration or insulated containers with coolant gel packs to delay development of parasitoids until shipment. Refinements of this basic system led to a very high emergence rate (50–75% of parasitized late instar whitefly) in the receiving quarantine facility.

Studies were carried out to elucidate optimal temperatures for storage of collected material. Lacey et al. (1999) showed that the eggs and pupae of *B. tabaci* and pupae of the parasitoid *Enc. formosa* survived after storage at 8°C for several weeks. Shipments were made using air freight companies providing tracking of the packages whenever possible; other mail services were less reliable. Where no suitable means of shipment was found, material was hand carried to EBCL in Montpellier and shipped to the USA from there. All samples were collected and stored separately, based upon host plant and locality. Table 2.1 gives the complete list of parasitic Hymenoptera imported into the USA for biological control of *B. tabaci*. Only a few predators were collected and sent to the USA: *Clitostethus arcuatus* Rossi (Coleoptera: Coccinellidae) from Mazaron, Spain, *Serangium parcesetosum* Sicard (Coleoptera: Coccinellidae) from Padappai, India, and *Acletoxenus formosus* (Loew) (Diptera: Drosophilidae) from Crete and the Canary Islands (Kirk et al. 1993; Lacey et al. 1993). *Macrolophus caliginosus* (Wagner) (Hemiptera: Miridae) was collected at Stomion, Crete but was not sent to the quarantine facility at Mission.

Twenty eight countries were visited at least once and often several times and hundreds of thousands of miles were traveled, which resulted in about 130 shipments of natural enemies sent to the APHIS-MBCL quarantine facility. During these explorations, *Bemisia* was collected from more than 50 host species, including 17 crops, 13 ornamentals, and 20 weeds. Two hundred and thirty five accessions resulted in 55 parasitoid cultures and 13 distinct species. Of these, thirty-eight exotic plus two native parasitoid populations were evaluated by USDA-APHIS in laboratory and field assays for efficacy (Chapters 7–9). In a separate effort, 15 exotic parasitoid species were introduced into quarantine in Gainesville, Florida

Table 2.1 Parasitic Hymenoptera imported into the USA and evaluated for biological control of *Bemisia tabaci* (biotype "B"), from 1992 to 1998. (All specimens were collected from *Bemisia tabaci* complex unless otherwise noted).

Species	MBCL accession code	MBCL DNA pattern	Collection locality	Collector[a]	Date	Identifier	Host plant	Biology
***Encarsia* species**								
Enc. bimaculata Heraty and Polaszek	M92018	EN-1	India, Parbhani	G. Butler	1/92	Woolley and Schauff		Autoparasitoid
Enc. bimaculata	M93010	EN-1	India, Parbhani	G. Butler	1/92	Woolley and Schauff		Autoparasitoid
Enc. bimaculata	M95023	EN-5	Thailand, Doi Suthep	Carruthers and Legaspi	5/95	Heraty	Unknown woody plant	Autoparasitoid
Enc. formosa Gahan[c]	M92017	EN-2	Greece, Angelohori	J. Kashefi	1/92	Woolley and Schauff	Bean	Uniparental
Enc. formosa	M92030	EN-2	Egypt, Nile Delta	Kirk and Lacey	1/92	Schauff	Lantana	Uniparental
Enc. lutea (Masi)	M93064	EN-10	Cyprus, Mazotos	Kirk and Lacey	1/93	Woolley and Johnson	Lantana, Soybean	Autoparasitoid
Enc. lutea	M94107	EN-10	Israel, Givat Haim	Kirk and Lacey	10/94	Woolley and Johnson	Cotton	Autoparasitoid
Enc. lutea	M94115	EN-10	Israel, Ein Gedi, Dead Sea	Kirk and Lacey	10/94	Woolley and Johnson	Lantana	Autoparasitoid
Enc. lutea	M94129	EN-10	Spain, Mazarron Casas Nuevas	Kirk and Lacey	11/94	Woolley and Johnson	*Ipomea* sp.	Autoparasitoid
Enc. lutea	M96044	EN-10	Sicily, Ragusa	Kirk and Campobasso	9/96	Johnson	Solanaceous weed	Autoparasitoid
Enc. nr. *pergandiella*	M94055	EN-15	Brazil, Sete Lagoas	Rose	2/94	Rose and Woolley	Poinsettia, Soybean	Uniparental
Enc. nr. *hispida*	M94056	EN-16	Brazil, Sete Lagoas	Rose	2/94	Rose and Woolley	Poinsettia, Soybean	Uniparental
Enc. sophia (Girault and Dodd)	M93003	EN-7	Spain, Murcia	Kirk and Lacey	1/93	Woolley and Schauff	Lantana	Autoparasitoid

(continued)

Table 2.1 (continued)

Species	MBCL accession code	MBCL DNA pattern	Collection locality	Collector[a]	Date	Identifier	Host plant	Biology
Enc. sophia	M94017	EN-3	Taiwan, Shan-Hua	Legaspi, Carruthers, Poprawski	3/94	Woolley and Johnson	Soybean, Tomato	Autoparasitoid
Enc. sophia	M94019	EN-4	Taiwan, Shan-Hua	Legaspi, Carruthers, Poprawski	3/94	Woolley and Johnson	Soybean, Tomato	Autoparasitoid
Enc. sophia	M94041	EN-5	Thailand, Chiang Mai	Kirk and Lacey	3/94	Woolley and Johnson	Poinsettia	Autoparasitoid
Enc. sophia	M94047	EN-5	Malaysia, Kuala Lumpur	Kirk and Lacey	3/94	Woolley and Johnson	*Mussaenda* sp.	Autoparasitoid
Enc. sophia	M95107	EN-5	Pakistan, Multan	Kirk and Lacey	11/95	Goolsby	Cotton	Autoparasitoid
Enc. sophia	M96065	EN-5	Pakistan, Jalari	Kirk	10/96	Goolsby	Cotton	Autoparasitoid
Enc. sophia[c]	M94014	EN-11	Philippines, Benguet	Legaspi, Carruthers, Poprawski	3/94	Woolley and Johnson	White potato	Autoparasitoid
Enc. sophia	M94016	EN-11	Taiwan, Shan-Hua	Legaspi, Carruthers, Poprawski	3/94	Woolley and Johnson	Poinsettia	Autoparasitoid
Encarsia sp.	M94024	EN-6	Thailand, Kampang Saen	Kirk and Lacey	3/94	Woolley and Johnson	Snakeweed	Autoparasitoid
Enc. sp. (*parvella* group)	M95001	EN-18	Dominican Republic, Azua	Ciomperlik	1/95	Schauff	Tomato	Autoparasitoid
***Eretmocerus* spp.**								
Eret. emiratus Zolnerowich and Rose	M95104	ERET-12	United Arab Emirates	Porter, Romadon	11/95	Rose and Zolnerowich	Okra	Biparental
Eret. sp. nr. *furuhashii*[b]	M95026	ERET-11	Taiwan, Chiuju	Kirk	5/94	Goolsby	Cabbage	Biparental

Eret. sp. nr. *furuhashii*	M95098	ERET-11	Taiwan, Tainan	Talekar and Jones	10/95	Rose and Zolnerowich	Tomato	Biparental
Eret. hayati Zolnerowich and Rose	M93005	ERET-2	India, Thirumala	Kirk and Lacey	1/93	Rose and Zolnerowich		Biparental
Eret. hayati	M95012	ERET-10	Pakistan, Multan	Kirk, Lacey and Akey	4/95	Rose and Zolnerowich	Mulberry	Biparental
Eret. hayati	M95105	ERET-10	Pakistan, Multan	Kirk and Lacey	9/95	Rose and Zolnerowich	Eggplant	Biparental
Eret. hayati[b]	M96064	ERET-10	Pakistan, Jalari	Kirk	10/96	Goolsby	Cotton	Biparental
Eret. melanoscutus Zolnerowich and Rose	M94036	ERET-3	Thailand, Chiang Mai	Kirk and Lacey	3/94	Rose and Zolnerowich	*Chromolaena odorata*	Biparental
Eret. melanoscutus	M94040	ERET-3	Thailand, Kampang Saen	Kirk and Lacey	3/94	Rose and Zolnerowich	Cotton	Biparental
Eret. melanoscutus[b]	M94023	ERET-8	Thailand, Sai Noi Klong Ha Roi	Kirk and Lacey	3/94	Rose and Zolnerowich	Eggplant, melon	Biaparental/ uniparental
Eret. melanoscutus	M95097	ERET-3	Taiwan, Tainan	Talekar and Jones	10/95	Rose and Zolnerowich	Tomato	Biparental
Eret. mundus Mercet	M92014	ERET-1	Spain, Murcia	Kirk, Chen, Sobhian	1/92	Schauff	Cotton	Biparental
Eret. mundus	M92019	ERET-1	India, Padappai	Kirk and Lacey	1/92	Rose and Zolnerowich	Eggplant	Biparental
Eret. mundus	M92027	ERET-1	Egypt, Cairo	Kirk and Lacey	1/92	Rose and Zolnerowich	Lantana	Biparental
Eret. mundus[b]	M93004	ERET-1	Spain, Murcia	Kirk and Lacey	1/93	Woolley and Schauff	*Sonchus*	Biparental
Eret. mundus	M93058	ERET-1	Taiwan, Tainan	Moomaw	12/93	Rose and Zolnerowich	Tomato	Biparental
Eret. mundus[b]	M94085	ERET-1	Italy, Frascati	Kirk and Campobasso	9/94	Rose and Zolnerowich	Hibiscus	Biparental

(continued)

Table 2.1 (continued)

Species	MBCL accession code	MBCL DNA pattern	Collection locality	Collector[a]	Date	Identifier	Host plant	Biology
Eret. mundus	M94092	ERET-1	Italy, Castel Gondolfo	Kirk and Campobasso	9/94	Rose and Zolnerowich	*Ipomea* sp.	Biparental
Eret. mundus	M94097	ERET-1	Italy, Testa Di Lespe	Kirk and Campobasso	9/94	Rose and Zolnerowich	Eggplant	Biparental
Eret. mundus	M94103	ERET-1	Israel, Gat	Kirk and Lacey	10/94	Rose and Zolnerowich	Kohlrabi	Biparental
Eret. mundus	M94105	ERET-1	Israel, Gat	Kirk and Lacey	10/94	Rose and Zolnerowich	*Sonchus* sp.	Biparental
Eret. mundus	M94120	ERET-1	Israel, Golan Ma'Aleh Gamla	Kirk and Lacey	10/94	Rose and Zolnerowich	Melons	Biparental
Eret. mundus	M94124	ERET-1	Israel, Negev Desert	Kirk and Lacey	10/94	Rose and Zolnerowich	Cucumber	Biparental
Eret. mundus	M94125	ERET-1	Israel, Golan Kibbutz	Kirk and Lacey	10/94	Rose and Zolnerowich	*Euphorbia* spp.	Biparental
Eret. mundus	M96028	ERET-1	Sicily, Santa Groce	Kirk and Campobasso	9/96	Goolsby	Eggplant	Biparental
Eret. mundus	M97046	ERET-1	Cyprus, Nicosia	Kirk	7/97	Goolsby	Lantana	Biparental
Eretmocerus sp. nr. *emiratus*	M96076	ERET-13	Ethiopia, Melka Werer	Gerling, Terefe	11/96	Goolsby	Cotton	Biparental

[a] Affiliations of collectors besides the chapter authors or others mentioned in text include: D. Akey and G. Butler (USDA-ARS Phoenix, AZ), G. Campobosso (USDA-ARS, Rome, Italy), K. Chen (USDA-ARS, Montferrier, France), M. Ciomperlik (USDA-APHIS, Mission, TX), W. Jones (USDA-ARS, Weslaco, TX), J. Kashefi (USDA-ARS, Thessaloniki, Greece), J. and B. Legaspi (USDA-ARS, Weslaco, TX), C. Moomaw (Texas A& M University), T. Poprawski (USDA-ARS, Weslaco, TX), R. Sobhian (USDA-ARS, Montferrier, France), N. Talekar (AVRDC, Shanhua, Taiwan).

[b] Not evaluated; all other species evaluated at MBCL quarantine and/or in field.

[c] Host *Trialeurodes vaporariorum* or *Trialeurodes* sp.

between 1990 and 1995 by University and state entomologists. Eight of these received release approval and were released in Florida during this period, including *Amitus bennetti* Viggiani and Evans *ex* Puerto Rico, *Eretmocerus rui* Zolnerowich and Rose (*ex* Hong Kong), *Eretmocerus* sp. *ex* Sudan, *Eretmocerus mundus* Mercet and *Encarsia lutea* (Masi) *ex* Israel, *Encarsia bimaculata* Heraty and Polaszek *ex* India, and an *Eretmocerus* sp. and *Encarsia* sp. *ex* Guatemala (Nguyen 1996; Nguyen and Bennett 1995).

A classic example of the fate of a collection during shipment was that from Multan, Pakistan in 1997. A new species of *Eretmocerus*, eventually named *Eret. hayati* Zolnerowich and Rose, was collected from *Bemisia* on a lone mulberry tree growing in an open latrine near Multan. After long delays in Karachi, Pakistan the shipment finally reached the APHIS-MBCL quarantine facility 3 weeks after collection. During this time the leaves inside the bag had completely decomposed and if any parasitized whiteflies had been present they were now immersed in green slime. From the decomposed leaf material, 20 late instar whiteflies were collected, washed in distilled water and placed in a humidified chamber (~70% RH) for emergence of parasitoids. From this lot, 12 females and a few males of *Eret. hayati* emerged. Each individual was fed with honey, mated and placed on a hibiscus leaf with 2nd instar whitefly nymphs. Despite the extreme stress during transit the parental females produced a large number of progeny and the colony quickly prospered. This species was among those permitted for release following evaluations and *Eret. hayati* is now dominant in the Lower Rio Grande Valley of Texas, which is climatically similar to the Indus River valley surrounding Multan, Pakistan. This was an example for which climatic adaptation may have been an important factor in the establishment and success of the imported whitefly parasitoids.

Parasitoids were also needed that could withstand the harsh conditions of the Imperial Valley of California. It was hypothesized that parasitoids from the arid desert climate of the Persian Gulf region might be suitable for the Imperial Valley. Pending hostilities prevented Kirk and Lacey (1996) from traveling to this area to explore, so Ed Porter of the USDA-Foreign Agricultural Service at the US Embassy in Dubai was contacted to obtain a shipment of parasitized *B. tabaci* from the United Arab Emirates (UAE). A packet of information was sent to Porter which described the appearance of *B. tabaci* nymphs on the underside of leaves and which host plants were likely to have infestations. Porter and M. Romadan of the US Embassy in Dubai drove south of Dubai towards the Straits of Hormuz where they searched in a vegetable garden. They located *Bemisia* nymphs on okra plants; infested leaves were wrapped in paper towels, placed in waxed paper bags, packaged and shipped back to MBCL, Mission, Texas. An *Eretmocerus* species emerged and was identified as unique compared to known whitefly parasitoid RAPD-PCR banding patterns; it was later described as *Eret. emiratus* by Zolnerowich and Rose (1998).

At a whitefly meeting in 1995, J. Goolsby proposed to Professor D. Gerling of the University of Tel Aviv that *B. tabaci* biotype B could have entered Israel from Ethiopia at the time of the airlift of Ethiopian Jews during the Ethiopian famine in the mid-1980s. If Ethiopia was the area of origin of the *B. tabaci* it would be a good place to explore for natural enemies. Dr. Gerling arranged through Israeli colleagues working in Ethiopia to make a collection of cotton leaves infested with *B. tabaci* and send them to MBCL. In November of 1996 a small envelope in transit for nearly a month arrived at MBCL containing the cotton leaves and parasitized whitefly nymphs. Fifty *Eretmocerus* adults emerged in quarantine and were identified by RAPD-PCR as a species near *Eret. emiratus*. Although these had a unique PCR banding pattern, Rose and Zolnerowich (personal communication) were unable to find clear morphological characteristics to distinguish it from *Eret. emiratus*. The Ethiopian *Eretmocerus* subsequently became established in Arizona, Texas and California. Little is known about the climate of Melka Werer where the parasitoid was originally collected, which is a high, tropical desert unlike any of the locations in the USA where it has established. It was speculated that the distribution of the parasitoid probably extends down to the Red Sea near Djibouti and Eritrea, where the climate is hot, arid desert similar to the Persian Gulf and to the drier parts of the southwestern USA. The discovery, collection and shipment of *Eret.* sp. nr. *emiratus* from Melka Werer, Ethiopia (M96076) was accomplished through the collaborative efforts of Dr. Gerling, APHIS, and Terefe of the Melka Werer Cotton Institute.

The collection of *Eret.* sp. nr. *emiratus* from Ethiopia highlights the international collaboration that was the hallmark of the US *B. tabaci* program. Numerous people contributed to the many successful shipments of insects. A comparison can be made here between the widespread presence of cooling devices in even remote areas of the world today and the efficiency of airline freight even from remote airfields, to the long boat trips and risky on-board rearing of natural enemies carried out by earlier foreign explorers looking for natural enemies of *Levuana iridescens* B. Baker (Lepidoptera: Zygaenidae) in Fiji. A trip from Singapore to Suva, Fiji took 30 days by steamer in 1925 (Tothill et al. 1930) a trip that takes 12.5h by plane via Brisbane, Australia today. Successful shipments of short-lived *Bemisia* and its aphelinid parasitoids would have been very difficult under such conditions!

2.3 Characterizing and Identifying *Bemisia* and Parasitoid Specimens

Microscope slides of adult parasitoid specimens and whitefly nymphal case remains were prepared using the techniques described by Noyes (1982) and Martin (1987). Brown et al. (1995a, b), Brown et al. (2000) and Kirk et al. (2000) describe the characterization and distribution of *Bemisia* biotypes worldwide. RAPD-PCR patterns were generated from target DNA fragments of individual hymenopteran parasitoids (Kirk et al. 2000, Chapter 6), which were previously identified morphologically. The ability to rapidly generate a molecular fingerprint allowed new cultures to be characterized well in advance of the assignment of a taxonomic identification, which sometimes required new species descriptions. This made it possible to detect and remedy culture contamination, and allowed researchers in California, Arizona, and Texas to track the presence and establishment of exotic parasitoids in the field. The voucher specimens are held by the identifiers referred to in Table 2.1, and the RAPD-PCR gel pattern images stored on computer discs are held at MBCL. These were cross referenced and a high degree of correlation was found between distinctive RAPD patterns and final species descriptions. Table 2.1 lists the insect parasitoids collected, their taxonomic identifications and RAPD-PCR patterns.

The combined use of classical taxonomic and molecular methods established the identities of *Bemisia* and its natural enemies and has aided in achieving accurate estimates of the relationships between biogeographic lineages of whiteflies and natural enemies. Areas containing the greatest diversity in the populations of both the host and parasitoid are thought to be the location in which these insects arose. This information suggests that a region that includes the Near East, the Arabian Peninsula and the Horn of Africa is the center of origin of the *Bemisia tabaci* species complex.

2.4 Whitefly Populations and Apparent Field Parasitism

As can be imagined, very little time was available during a foreign exploration trip to study in any detail the impact of natural enemies on *Bemisia.* However in two cases, leaves were conserved and kept for examination at EBCL in Montpellier after the conclusion of travel. Examples of apparent field parasitism shown by parasitoids in Spain and Thailand are given here. These collections are "snapshots" of the situation in the field and do not allow for replication, which is difficult to accomplish during foreign exploration.

2.4.1 Spain 1991 and 1992

Collections were made in Murcia, southeastern Spain, an intensively farmed area with a dry Mediterranean climate. The 1991 collections were made from senescing cotton leaves on which pesticide spraying had ended 2 weeks prior to collection. One leaf was taken from each of 10 cotton plants in Murcia. The following year, in November, one leaf from each of ten *Sonchus oleraceae* (common sowthistle) plants growing under the cotton was collected; the cotton was still being sprayed at the time of collection. All whitefly pupae in five 1 cm^2 areas across each leaf were counted. Parasitized nymphs were recognized by the presence of an aphelinid larva or pupa, the presence of meconia, color (nymphs parasitized by *Eret. mundus* were citron yellow in color and those by *Encarsia* spp. black or grey), and the typical circular hole in the *B. tabaci* nymph made by emerging aphelinid adults. Populations (n = 50) of whitefly nymphs on cotton leaf samples were 12 nymphs/cm^2 (±1.61) and on *Sonchus*, 13 nymphs/cm^2 (±1.21). The percentage parasitism by *Eret. mundus* of whitefly nymphs on cotton was 39% (±4.24) and on *Sonchus* 44% (±4.9), despite the spraying. *Eretmocerus mundus* from cultures established from this collection were later found to be tolerant of several insecticides when evaluated in the laboratory in Texas (Jones et al. 1996).

2.4.2 Thailand 1994

In 1994, collections were made as described above from 11 host plant species in 19 localities throughout Thailand in localities that ranged from tropical areas to subtropical areas having heavy seasonal rains and a distinct dry season. Six *Bemisia* parasitoids emerged. The parasitized nymphs were identified morphologically and by RAPD-PCR, and all whitefly eggs and nymphs were sampled and counted. Populations of whitefly nymphs were generally sparse on weeds and field crops (0–1 nymph/cm^2 leaf). Numbers of whitefly eggs, an indicator of current whitefly activity, ranged from 0 to 12 per cm^2. Percentage parasitism of *Bemisia* on these various hosts varied from 65% on *Physalis minima* L. to 42% on *Xanthium* to 14% on *Sida* sp. and 0% on *Polygonum* sp.

2.5 Discussion and Conclusions

Foreign exploration for classical biological control programs traditionally focuses on the known or presumed center of origin of the pest, which is assumed to be the area of greatest diversity of parasitoids. In the case of *B. tabaci* this appeared to be the Indian subcontinent, based on literature (Cock 1986; Lopez-Avila 1986) available at the start of the US *Bemisia* program. Recent molecular characterization of *B. tabaci* populations collected during the foreign explorations for natural enemies shows that this complex probably evolved in the region encompassing the UAE and Ethiopia (Frohlich et al. 1999). Goolsby et al. (2005) present evidence that natural enemies obtained from homologous climates selected using CLIMEX climate matching software (Sutherst et al. 1999) became established and spread more rapidly than natural enemies collected from less well matched areas, eventually becoming dominant on *B. tabaci*. The most successfully established parasitoids were those from the UAE/Ethiopia area which had a climate most similar to the affected areas in the USA with the exception of *Eret. hayati* from the Indian subcontinent. To our knowledge, this is the first example in which molecular methods have been employed to identify and/or track both the host insect and its natural enemies and have shown a plausible correlation. Clearly, knowledge of the origin of a target pest can be exceedingly useful in identifying promising sites of collection for natural enemies, which in many programs is done retrospectively. The genetic characterization and phylogeny of the pest and natural enemies should be a priority of classical biocontrol programs in which foreign exploration provides the material for those studies.

In the *B. tabaci* project, strong links were made between the USDA-ARS European Biological Control Laboratory (EBCL), The University of Arizona (genetic characterization of *Bemisia* populations collected by EBCL); USDA-APHIS, Mission, Texas (rearing, evaluation, release of natural enemies, characterization of *Bemisia* parasitoids and application for release permits); APHIS and California Department of Food & Agriculture in Brawley, California (surveys for local parasitoids, EBCL data on collection habitats for evaluation, releases of parasitoids collected by EBCL); USDA-ARS in Weslaco, Texas (foreign exploration in Southeast Asia, and insecticide tolerance of parasitoids collected by EBCL); and many local cooperators (help in collecting, shipping and obtaining foreign permits when necessary). Without such linkages a classical biological control program cannot reach fruition. All of these elements were in place for the *Bemisia* program, in addition to the annual *Bemisia* workgroup meetings where ideas, results and plans were discussed. The involvement of the foreign explorers did not stop at collecting *B. tabaci* and its natural enemies, but expanded into close collaboration with workers in the fields of taxonomy, quarantine, release, and implementation.

Acknowledgements Numerous administrators, scientific colleagues, farmers, their workers and often village youngsters, who cannot all be named here, contributed in the friendliest way to the foreign explorations. Colleagues in the USA maintained the natural enemy colonies, working with them to add to our knowledge, and supported the objectives of area-wide biological control-based integrated management of *Bemisia*. We thank them all. We are grateful to the following people at MBCL for their contributions: A. Chavarria, D. Vasquez, and E. Lopez for production rearing; R. Sheetz for quarantine rearing; D. Vacek and R.A. Ruiz, for genetic fingerprinting. Whitefly specimens were sent to S. Nakahara at the Systematic Entomology Center, Beltsville, MD and R. Gill, California Department of Food and Agriculture, Sacramento, California. All *Encarsia* species were sent to M. Schauff at the Systematic Entomology Center, Beltsville, MD, J. Heraty at University of California, Riverside, and J. B.Woolley at Texas A&M University, College Station, Texas; the *Eretmocerus* species were sent to M. Rose and G. Zolnerowich at Texas A&M University, College Station, Texas; Mme G. Berrebi of EBCL, Montpellier, France prepared the whitefly and parasitoid slides. We are especially grateful to Dr Alfredo Lacasa, C.R.I. A., La Albarca, Murcia, Spain, for generously opening his Center to Drs. Kirk, Lacey, Sobhian and Chen; EBCL, Montpellier for assistance with field collections; Amporn Winotai, Pimpun Sommartya and Sirichai Divasiri (NBCRC, Bangkok), for their help in Thailand; and Ed Porter and Romadan for collecting natural enemies in the UAE. We acknowledge the expert assistance of Ms. Ivone Torres-Jerez (UA Center, Tucson, Arizona) with PCR and cloning of reference and field collections of *B. tabaci*. Terefe, an Ethiopian collaborator, was reported as missing following the civil war in Ethiopia; his contribution is gratefully acknowledged here.

References

Bellows, T.S., Jr., T.M. Perring, R.J. Gill, and D.H. Headrick. 1994. Description of a species of *Bemisia* (Homoptera: Aleyrodidae). Annals of the Entomological Society of America 87: 195–206.

Brown, J.K., H.S. Costa, and F. Laemmlen. 1991. First incidence of whitefly-associated squash silverleaf (SSL) of *Cucurbita*, and of white streaking (WSt) disorder of cole crops in Arizona and California. Plant Diseases. 76: 426.

Brown, J.K., D.R. Frohlich, and R.C. Rosell. 1995a. The sweetpotato or silverleaf whiteflies: biotypes of *Bemisia tabaci* or a species complex? Annual Review of Entomology 40: 511–534.

Brown, J.K., S.A. Coats, I.D. Bedford, P.G. Markham, J. Bird, and D.R. Frohlich. 1995b. Characterization and distribution of esterase electromorphs in the whitefly, *Bemisia tabaci* (Genn.) (Homoptera: Aleyrodidae). Biochemical Genetics 33: 205–214.

Brown, J.K., T.M. Perring, A.D. Cooper, I.D. Bedford, and P.G. Markham. 2000. Genetic analysis of *Bemisia* (Homoptera: Aleyrodidae) populations by isoelectric focusing electrophoresis. Biochemical Genetics 38: 13–25.

Cock, M.J.W. ed., 1986. *Bemisia tabaci*, A Literature Survey on the Cotton Whitefly with an Annotated Bibliography. CAB International Institute of Biological Control, Ascot, UK.

Cock, M.J.W. ed., 1993. *Bemisia tabaci*, an update 1986–1992 on the Cotton Whitefly with an Annotated Bibliography. CAB International Institute of Biological Control, Ascot, UK.

Elsey, K.D. and M.W. Farnham. 1994. Response of *Brassica oleracea* L. to *Bemisia tabaci* (Gennadius). HortScience 29: 814–817.

Faust, R.M. 1992. Conference report and 5-year national research and action plan for development of management and control methodology for the sweetpotato whitefly. US Department of Agriculture, Agricultural Research Service, ARS-107. Washington, DC.

Faust, R.M. and J.R. Coppedge. 1995. Foreword. In T.J. Henneberry, N.C. Toscano, R.M. Faust, and J.R. Coppedge, eds., 1995 Supplement to the 5-year National Research and Action Plan-Third Annual Review, San Diego, CA, January 28–31, 1995. USDA, ARS, 1995–2.

Frohlich, D., I. Torres-Jerez, I.D. Bedford, P.G. Markham, and J.K. Brown. 1999. A phylogeographic analysis of the *Bemisia tabaci* species complex based on mitochondrial DNA markers. Molecular Ecology 8: 1593–1602.

Goolsby, J.A., M.A. Ciomperlik, B.C. Legaspi, Jr., J.C. Legaspi and L.E. Wendel, 1998. Laboratory and field evaluation of exotic parasitoids of *Bemisia tabaci* (Biotype 'B') in the Lower Rio Grande Valley of Texas. Biological Control 12: 127–135.

Goolsby, J.A., P. De Barro, A.A. Kirk, R. Sutherst, M. Ciomperlik, P. Ellsworth, J. Gould, K. Hoelmer, S. Naranjo, M. Rose, W. Roltsch, R. Ruiz, C. Pickett, and D. Vacek. 2005. Post-release evaluation of the biological control of *Bemisia tabaci* biotype "B" in the USA and the development of predictive tools to guide introductions for other countries. Biological Control 32: 70–77.

Jones, W.A., D.F. Wolfenbarger, and A.A. Kirk. 1996. Response of adult parasitoids of the sweet-potato whitefly *Bemisia tabaci* (Hom: Aleyrodidae) to leaf residues of selected cotton insecticides. Entomophaga 40: 1–9.

Kirk, A.A. and L.A. Lacey. 1996. A systematic approach to foreign exploration for natural enemies of *Bemisia*. In D. Gerling and R. Mayer, eds., *Bemisia* 1995: Taxonomy, Biology, Damage, Control and Management, pp. 531–533. Intercept, Andover, UK.

Kirk, A.A., L.A. Lacey, N. Roditakis, and J.K. Brown. 1993. The status of *Bemisia tabaci* (Hom: Aleyrodidae), *Trialeurodes vaporariorum* (Hom: Aleyrodidae) and their natural enemies in Crete. Entomophaga 31: 183–190.

Kirk, A.A., L.A Lacey, J.K. Brown, M.A. Ciomperlik, J.A. Goolsby, D.C. Vacek, L.E. Wendel, and B. Napompeth. 2000. Variation in the *Bemisia tabaci* s. l. species complex (Hemiptera: Aleyrodidae) and its natural enemies leading to successful biological control of *Bemisia* biotype B in the USA. Bulletin of Entomological Research 90: 317–327.

Lacey, L.A., A.A. Kirk, and R.D. Hennessey. 1993. Foreign exploration for natural enemies of *Bemisia tabaci* and implementation in integrated control programs in the United States. Proceedings Third International Conference on Pests in Agriculture, Montpellier, France, 7–9 December 1993, pp. 351–360. Association Nationale de Protection des Plantes, Paris.

Lacey, L.A., L. Millar, A.A. Kirk, and T.M. Perring. 1999. Effect of storage temperature and duration on survival of eggs and nymphs of *Bemisia argentifolii* (Homoptera: Aphelinidae) and pupae of the whitefly parasitoid *Encarsia formosa* (Hymenoptera: Aphelinidae). Annals of the Entomological Society of America 92: 430–434.

Legaspi, J.C., B.C. Legaspi Jr., R.I. Carruthers, J. Goolsby, W.A. Jones, A.A. Kirk, C. Moomaw, T.J. Poprawski, R.A. Ruiz, N.S. and D. Vacek. 1996. Foreign exploration for natural enemies of *Bemisia tabaci* from southeast Asia. Subtropical Plant Science 48: 43–48.

Lopez-Avila, A. 1986. Taxonomy and biology. In M.J.W. Cock, ed., *Bemisia tabaci*, A Literature Survey on the Cotton Whitefly with an Annotated Bibliography, pp. 3–11. CAB International Institute of Biological Control, Ascot, UK.

Martin, J.H. 1987. An identification guide to common whitefly pest species of the world (Homoptera, Aleyrodidae). Tropical Pest Management 33: 298–322.

Mound, L.A. and S.H. Halsey. 1978. Whitefly of the World: a Systematic Catalogue of the Aleyrodidae (Homoptera) with Host Plant and Natural Enemy Data. British Museum (Natural History). Wiley, Chichester, London.

Nguyen, R. 1996. Importation and field release of parasites against silverleaf whitefly, *Bemisia argentifolii* (Bellows and Perring) in Florida from 1990–1995. In Henneberry, T.J., N.C. Toscano, R.M. Faust, and J.R. Coppedge, eds., Silverleaf Whitefly: 1996 Supplement to the Five Year National Research and Action Plan – 4th Annual Review, February 4–6, 1996, San Antonio, Texas. US Department of Agriculture, Agricultural Research Service, ARS-1996–01, p. 134.

Nguyen, R. and Bennett, F.D. 1995. Importation and field release of parasites against silverleaf whitefly, *Bemisia argentifolii* (Bellows and Perring) in Florida from 1990–1994. Proc. Florida State Horticultural Society 108: 43–47.

Noyes, J.S. 1982. Collecting and preserving chalcid wasps (Hymenoptera:Chalcidoidea). Journal of Natural History 16: 315–334.

Perring, T.M., A. Cooper, D.J. Kazmer, C. Shields, and J. Shields. 1991. New strain of sweetpotato whitefly invades California vegetables. California Agriculture 45 (Nov.-Dec.): 10–12.

Schuster, D.J., T.F. Mueller, J.B. Kring, and J.F. Price. 1990. Relationship of the sweetpotato whitefly to a new fruit disorder of tomato in Florida. HortScience 25: 1618–1620.

Skarratt, D.B., R.W. Sutherst, and G.F. Maywald. 1995. CLIMEX for windows Version 1, CSIRO, CRC, Brisbane.

Sutherst, R.W., G.F. Maywald, T.Yonow, and P.M. Stevens. 1999. CLIMEX. Predicting the effects of climate on plants and animals. CD-ROM and User guide. CSIRO Publishing, Collingwood.

Tothill, J.D., T.H.C. Taylor, and R.W. Pazine. 1930. The coconut moth in Fiji: a history of its control by parasites. Imperial Bureau of Entomology, London.

Zolnerowich, G. and M. Rose. 1998. *Eretmocerus* Haldeman (Hymenoptera: Aphelinidae) imported and released in the United States for control of *Bemisia* (*tabaci* complex) (Homoptera: Aleyrodidae). Proceedings of the Entomological Society of Washington 100: 311–323.

Chapter 3
Entomopathogenic Fungi for Control of *Bemisia tabaci* Biotype B: Foreign Exploration, Research and Implementation

Lawrence A. Lacey[1], Stephen P. Wraight[2], and Alan A. Kirk[3]

Abstract Prior to the global outbreak of *Bemisia tabaci* type B in the early 1990s, very little attention was paid to the potential of fungal pathogens for control of this whitefly. A massive foreign exploration effort was mounted in 1991 by the USDA Agricultural Research Service to collect and develop fungi and other natural enemies of whiteflies. From 1990 until 1996, dozens of trips were made by scientists from the USDA European Biological Control Laboratory (EBCL) in Montpellier, France specifically for the purpose of finding natural enemies of whiteflies. The countries visited by EBCL ranged from European countries bordering the Mediterranean (Spain, France, Italy, Greece, Cyprus), to the Middle East (Israel, Egypt), Western Asia (Pakistan, India, Nepal), Southeast Asia (Thailand, Malaysia) and Latin America (Argentina and Brazil). Other exploratory efforts for entomopathogenic fungi within the USA and in other foreign countries (Philippines, Indonesia, Taiwan) were also undertaken by USDA-ARS personnel in Weslaco, Texas. The most prevalent fungus attacking *B. tabaci* in the field was *Paecilomyces fumosoroseus*. Fungi from whiteflies or from other insects with good activity against the silverleaf whitefly include *Paecilomyces* spp., *Lecanicillium lecanii* (=*Verticillium*), *Beauveria bassiana* and *Aschersonia* spp. Hundreds of isolates of these fungi were collected and shipped to the USA. Subsequent research involved screening of isolates for activity, study of factors that limited or enhanced their activity, and evaluation of candidate fungi in field and glasshouse crops. The literature on research for the development and implementation of these fungi is reviewed with recommendations for future avenues of research and development.

[1]USDA-ARS, Yakima Agricultural Research Center, 5230 Konnowac Pass Road, Wapato, WA 98951

[2]USDA-ARS, Plant, Soil, and Nutrition Laboratory, Tower Rd., Ithaca, NY 14853

[3]USDA-ARS, European Biological Control Laboratory, Campus International de Baillarguet, CS 90013, Montferrier sur Lez, 34988, St. Gély du Fesc CEDEX, France

J. Gould et al. (eds.), *Classical Biological Control of* Bemisia tabaci *in the United States.*

3.1 Introduction

A huge variety of insect pathogens have been reported from virtually every insect group. Many of these provide natural regulation of insect populations and some have been commercially produced as microbial insecticides (Lacey et al. 2001; Kaya and Lacey 2007). For insects with piercing and sucking mouth parts, such as whiteflies and aphids, entomopathogenic fungi offer the most promise as naturally occurring and applied insect pathogens (Latgé and Papierok 1988; Lacey et al. 1996; Wraight et al. 2007). Prior to the global outbreak of *Bemisia tabaci* (Genn.) biotype B (Homoptera: Aleyrodidae) (= *B. argentifolii* Perring and Bellows) in the early 1990s, very little attention was paid to the potential of entomopathogenic fungi for control of *Bemisia*. However, a substantial amount of basic and applied research had been conducted on the potential of entomopathogenic fungi for the control of the greenhouse whitefly, *Trialeurodes vaporariorum* (Westwood) (Hall 1981) and to a lesser extent other whitefly species (Fransen 1990b). Since 1990 the urgent need for alternatives to conventional insecticides for the management and control of *Bemisia* resulted in a tremendous effort to search for and develop fungi and other natural enemies. In this chapter we present a synopsis of the diverse work that went into the development and implementation of entomopathogenic fungi for whitefly control.

The fungi that have received the majority of attention for *B. tabaci* control are the Deuteromycetes because of their prevalence, possibility for production on artificial media, ease of application, and relatively long shelf lives. Deuteromycetes collected from whiteflies or other insects with activity against whiteflies include *Paecilomyces* spp., *Lecanicillium* (= *Verticillium*)[1] *lecanii, Beauveria bassiana* and *Aschersonia* spp. Several reviews on the utilization of fungi for control of *Bemisia* (Osborne and Landa 1992; Smith 1993; Lacey et al. 1996; Chen and Feng 1999; Faria and Wraight 2001) cover the majority of the literature published before 2000. General techniques used for isolation, identification, culture and preservation of these and other fungi are provided by Goettel and Inglis (1997), Lacey and Brooks (1997), and Humber (1997a, b). Their safety to beneficial and other nontarget organisms is presented by Goettel et al. (1990) and Vestergaard et al. (2003).

[1] A revision by Zare and Gams (2001) based on molecular and morphological studies has resulted in the placement of the common insect pathogenic *Verticillium* species in the new genus, *Lecanicillium*. Their revision assigns different species names for some of the isolates in the species complex previously and collectively referred to as *Verticillium lecanii*. For example, the whitefly active species used in the Koppert product Mycotal is *Lecanicillium muscarium* while the aphid active isolate used in the Koppert product Vertalec is *Lecanicillium longisporum*.

3.2 Foreign Exploration

During the period of most intensive foreign exploration for natural enemies of *Bemisia* (1990–1996), dozens of trips were made by scientists from the USDA European Biological Control Laboratory (EBCL) in Montpellier, France specifically for the purpose of finding natural enemies of whiteflies (Lacey et al. 1993, 1996; Kirk and Lacey 1996; Poprawski and Lacey 2000; Kirk et al. 2001). The countries visited by EBCL ranged from European countries bordering the Mediterranean (Spain, France, Italy, Greece, Cyprus), to the Middle East (Israel, Egypt), Western Asia (Pakistan, India, Nepal), Southeast Asia (Thailand, Malaysia) and Latin America (Argentina and Brazil). Other exploratory efforts for entomopathogenic fungi within the USA and to other foreign countries (Philippines, Indonesia and Taiwan) were also undertaken by USDA Agricultural Research Service (ARS) personnel in Weslaco, Texas. The principal repository for collected isolates is the USDA Agricultural Research Service Entomopathogenic Fungal Culture collection (ARSEF) housed in Ithaca, New York (Humber 2002). A summary of collections from *Bemisia* made by USDA personnel and others is presented in Table 3.1.

3.2.1 Paecilomyces fumosoroseus

Paecilomyces fumosoroseus has a diverse range of insect hosts (Samson 1974; Smith 1993; Sterk et al. 1996) and has also been isolated from soil and other habitats (Tigano-Milani et al. 1992; Sterk et al. 1996). Exploration in the tropics and subtropics consistently revealed *P. fumosoroseus* to be the predominant species of entomopathogenic fungus in field populations of *Bemisia* species (Lacey et al. 1993, 1996; Humber 2002). During certain humid seasons, massive epizootics were observed as the predominant factor controlling populations of *Bemisia* in Pakistan, India and Nepal (Lacey et al. 1993, 1996). *Paecilomyces fumosoroseus* and *Beauveria bassiana* have been the two fungal species studied most extensively with respect to efficacy against *Bemisia*, factors influencing activity, and materials and methods for improving production.

3.2.1.1 Screening of Isolates

Several bioassay methods were developed during the 1990s to assess the activity of *P. fumosoroseus* and other fungi, as well as the biotic and abiotic factors that influence their activity. The most effective methods that provided a realistic measurement of fungal activity employed excised or rooted host plant leaves (Vidal et al. 1997b; Negasi et al. 1998; Wraight et al. 1998; Lacey et al. 1999). Several dozen isolates of *P. fumosoroseus* have been evaluated in laboratory studies against whitefly nymphs by Osborne et al. (1990a, b), Mellín Rosas and Garza Gonzalez (1993),

Table 3.1 Natural occurrence of entomopathogenic fungi in *Bemisia* spp.

	Location	Reference
Hypocreales		
Aschersonia aleyrodis	Taiwan	Yen and Tsai (1969)
	USA	Berger (1921)
Aschersonia andropogonis	Taiwan	Yen and Tsai (1969)
Aschersonia goldiana	Brazil	Lourenção et al. (1999)
	Taiwan	Yen and Tsai (1969)
Beauveria bassiana	Israel	Ben-Ze'ev et al. (1994)
Paecilomyces farinosus	Greece	Kirk et al. (1993)
	India	Nene (1973); Balakrishnan and Nene (1980)
Paecilomyces sp.	Martinique	Ryckewaert and Alauzet (2002)
Paecilomyces fumosoroseus	Brazil	Sosa-Gomez et al. (1997); Rancel *cited in* Humber (2002)
	Colombia	Garza Gonzalez (unpub. Report); López-Ávila et al. (2001)
	Cuba	Castineiras (1995)
	Ecuador	López-Ávila et al. (2001)
	India	Rao et al. (1989); Rao and Reddy (1992); Lacey et al. (1993)
	Indonesia	Humber (2002)
	Japan	S. Kurogi (personal communication)
	Mexico	Humber (2002); Garza Gonzalez (1993); Gonzalez (1993)
	Nepal	Lacey et al. (1993)
	Pakistan	Lacey et al. (1993)
	Philippines	T. Poprawski and R. Carruthers (personal communication)
	Trinidad	Hall et al. (1994a)
	Venezuela	R. Hall (personal communication)
	USA	Humber (2002); Carruthers et al. (1993); Osborne et al. (1990b); R. Humber (personal communication)
Lecanicillium spp.	Denmark	Steenberg (Humber 2002)
Lecanicillium lecanii	Brazil	Lourenção et al. (2001)
	Colombia	Drummond et al. (1987); López-Ávila et al. (2001)
	Denmark	Steenberg (Humber 2002)
	Ecuador	López-Ávila et al. (2001)
	Israel	Ben Ze'ev et al. (1994)
	Japan	S. Kurogi, (personal communication)
	Mexico	Nier et al. (1991)
	Spain	Lacey et al. (1993)
	Venezuela	R. Hall (personal communication)
Entomophthorales		
Conidiobolus spp.	Israel	Ben Ze'ev (1993); Gindin and Ben Ze'ev (1994a, b)
	USA	R. Carruthers (personal communication)
Entomophthora sp.	USA	R. Carruthers (personal communication)
Zoophthora radicans	Chad	Silvie and Papierok (1991)
	Israel	Ben-Ze'ev et al. (1988)
Unidentified spp.	Brazil	Sosa-Gomez et al. (1997)
	USA	S. Wraight (unpublished observation)

Vidal et al. (1997b), Negasi et al. (1998), Wraight et al. (1998), Herrera et al. (1999), and others. A broad range of activity has been observed. The most virulent isolates produced 50% mortality in *B. tabaci* biotype B nymphs at 50–150 spores/cm^2. Some of the isolates that originated from hosts other than whiteflies are as active against *Bemisia* as isolates from whiteflies (Vidal et al. 1997b). One of the most virulent isolates studied by Osborne et al. (1990b) is the active component of PFR-97 (20% Water Disperable Granules), a commercially produced microbial insecticide. Efficacy under greenhouse and field conditions is presented in Table 3.2 and will be discussed in Section 3.3.

3.2.1.2 Factors Affecting Persistence, Growth and Insecticidal Activity

Biotic and abiotic factors that have been studied for their influence on the activity of *P. fumosoroseus* against whiteflies include the effect of temperature, solar radiation, effect of original insect host, host plant, method of application, type of propagule, effect of exogenous nutrients on germination, and exposure to certain allelochemicals.

3.2.1.3 Effects of Abiotic Factors

Temperature has a profound effect on growth and sporulation of *P. fumosoroseus*. Hall et al. (1994a) observed optimal growth at 28°C, but the temperature optimum for sporulation was 24°C. Vidal et al. (1997a) reported a wide range of temperatures tolerated by the fungus with distinct optimal temperatures for isolates originating from different geographical areas having diverse meteorological conditions. Optimal growth rates were observed between 20 and 30°C. Higher temperatures (30–40°C) were more limiting than lower ones (8–11°C). Similar observations were made by Fargues et al. (1992) for several non-whitefly isolates of *P. fumosoroseus* and other Hypocreales. Temperature and drying method has a profound effect on survival of spores, especially blastospores (Jackson et al. 1997; Cliquet and Jackson 1997). Blastospores stored at 4°C survived significantly longer than those stored at 28°C.

The effect of humidity on activity of *P. fumosoroseus* against the *Bemisia tabaci* complex has not been extensively studied, but Wraight et al. (2000) observed that the fungus was able to infect third-instar whitefly nymphs on excised hibiscus leaves incubated at relative humidities as low as 25–30%. This suggests the existence of a high humidity boundary layer surrounding the nymphs. Research on the fungus with other insects has revealed a need for high humidity to promote germination and infection (Poprawski et al. 1985; Mohamed and Varma 1992; Lacey et al. 1997; Mesquita et al. 1999). Conidia of *P. fumosoroseus* are rapidly killed by solar radiation, particularly UV-B (Fargues et al. 1996, 1997b; Smits et al. 1996, 1997).

Table 3.2 Control of *Bemisia* spp. using entomopathogenic fungi in greenhouse and field crops.

Fungal species	Crop	Efficacy/reduction	Reference
Paecilomyces fumosoroseus	Cabbage, cucumber, tomato (greenhouse)	82–88% (nymphs)	Vidal et al. (1998b)
	Cantaloupe, cucumber, (field)	>90% (nymphs) Ineffective (adults)	Wraight et al. (2000)
	Cotton (field)	Ineffective (nymphs, adults)	Wraight et al. (1996)
	Cotton (field)	75–78% (nymphs)	Akey and Henneberry (1998)
	Hibiscus, poinsettia (greenhouse)	Good control (nymphs, adults)	Osborne and Landa (1994)
	Poinsettia (greenhouse)	Significant control (nymphs)	Labanowski and Soika (1999)
	Tomato (greenhouse)	Effective control (nymphs)	Ohta et al. (1998)
Paecilomyces farinosus	Tomato, pepper (field)	Significant control (IPM)[a]	Ruiz-Vega and Aquino-Bolaños (1999)
Paecilomyces spp.	Tomato, pepper (field)	Significant control (IPM)[a]	Ruiz and Medina, (2001)
Beauveria bassiana	Beans (field, greenhouse)	57–87% (nymphs, IPM)[a]	Alves et al. (2001)
	Cantaloupe (field)	64% (nymphs)	Jaronski and Lord (1996)
	Cantaloupe (field)	71–74% (large nymphs)	Jaronski and Lord (1996)
	Cantaloupe (field)	68–79% (adults)	Liu et al. (1999)
	Cantaloupe, cucumber Squash (field)	>90% (nymphs) Ineffective (adults)	Wraight et al. (2000)
	Cantaloupe (field)	77–82% (nymphs) 64–74% (adults)	Orozco-Santos et al. (2000)
	Cantaloupe (field)	35–63% (nymphs)	Liu and Meister (2001)
	Cucumber (greenhouse)	100% (nymphs)	Zaki (1998)
	Collards (field)	68% (nymphs)	Poprawski (1999)
	Cotton (field)	64% (large nymphs)	Akey and Henneberry (1994)
	Cotton (field)	Ineffective (nymphs, adults)	Wraight et al. (1996)
	Cotton (field)	78–88% (nymphs)	Akey and Henneberry (1998)
	Cotton (field, 1992)	Ineffective (nymphs, adults)	Liu et al. (1999)
	Cotton (field, 1993)	Significant reduction (nymphs, adults)	Liu et al. (1999)
	Cotton (field, 1995)	Significant reduction (adults)	Liu et al. (1999)
	Hibiscus (greenhouse)	62–84% (nymphs)	Liu et al. (1999)
	Poinsettia (greenhouse)	58% (nymphs)	Costa (1998)

(continued)

Table 3.2 (continued)

Fungal species	Crop	Efficacy/reduction	Reference
	Poinsettia (greenhouse)	Highly significant reduction (nymphs, adults)	Olson and Oetting (1999)
	Tomato, eggplant (field)	41–50% infection (pupae)	Liu et al. (1999)
	Tomato (greenhouse)	Effective control (nymphs)	Ota et al. (1999)
	Tomato (field)	short-term reduction of yellow leaf curl	Rushtapakornchai and Petchwichit (1996)
Lecanicillium lecanii	Tomato (greenhouse)	Good control (nymphs)	Saito (1992, 1993)
Aschersonia spp.	Poinsettia (greenhouse)	Up to 70% (nymphs)[b]	Meekes et al. (2002)
Aschersonia aleyrodis	Cucumber (greenhouse)	Good control (nymphs)[a]	Steinberg and Prag (1994)
Metarhizium anisopliae	Eggplant	30–92% (nymphs)	Batta (2003)

[a] Indicated efficacy is from a combination of control agents/practices; efficacy of the fungus alone was not determined.

[b] Results from screening of 44 *Aschersonia* isolates against *Bemisia tabaci* nymphs on potted plants.

3.2.1.4 Effect of Whitefly Stage and Age

Osborne et al. (1990b) reported that all stages of *B. tabaci,* including eggs, are susceptible to *P. fumosoroseus*. However, Hall et al. (1994a) reported that first instar nymphs are refractory to the fungus. It should be noted that when epizootics are observed in the field, infected adults are often more numerous than infected nymphs (Carruthers et al. 1993) indicating a greater sensitivity to the fungus in adults. However, this is apparently the case only under humid conditions as applications of the fungus under normal field conditions do not induce epizootics among adult populations (Wraight et al. 2000). Eggs are the least susceptible of whitefly stages to fungi. Relatively low, but significant levels of infection were reported after exposure of eggs to blastospores of two *P. fumosoroseus* isolates (Lacey et al. 1999).

3.2.1.5 Host Plant Effects

Poprawski et al. (2000) and Poprawski and Jones (2001) reported significant differences in the susceptibility of *B. tabaci* and *T. vaporariorum* to *P. fumosoroseus* and *B. bassiana* reared on different host plant species. The two fungi were more active against *B. tabaci* when the whitefly was reared on melon than on cotton. Germination of conidia was strongly inhibited on the cuticle of nymphs fed on cotton (<12% germination), but was > 95% on nymphs raised on melon (Poprawski and Jones 2001). *Trialeurodes vaporariorum* nymphs reared on cucumber and treated with either *P. fumosoroseus* or *B. bassiana* were significantly more susceptible than those reared on tomato (Bolckmans et al. 1995; Poprawski et al. 2000). Poprawski et al. (2000) proposed that the differences in susceptibility could be due to antimicrobial compounds produced by the host plant and sequestered in the nymphs. Lacey and Mercadier (1998) observed inhibition of germination of conidia and blastospores of *P. fumosoroseus* exposed to several allelochemicals (tomatine, solanine, xanthotoxin, tannic acid and camptothecin). Mycelial growth was most affected by camptothecin, tomatine, and xanthotoxin. Vega et al. (1997) observed delayed germination of blastospores when they were exposed to catechol, salicyclic acid and tannic acid. However, no significant effect of host plant (tomato, cabbage, cucumber) on insecticidal activity of *P. fumosoroseus* against *B. argentifolii* was reported by Vidal et al. (1998b) in greenhouse trials where relative humidity was 70–100% with over 12 h per day at 100% RH. Research on the effect of tomato leaf boundary layer climate on fungi used for whitefly control in a dry greenhouse was presented by Boulard et al. (2002). They noted that tomato leaves in a dry greenhouse are associated with nearly moisture-saturated conditions required for fungal conidium germination for only a few hours per day. The complex issue of boundary layer effects on spore germination under a variety of host plant and environmental conditions warrants additional investigation in order to better comprehend the factors that govern the success or failure of entomopathogenic fungi in the field and to provide suggestions for improving activity through formulation and other avenues.

3.2.1.6 Effect of Propagule

Freshly produced blastospores were significantly more infectious for whitefly nymphs than were conidia of the same isolates (Lacey et al. 1999). However, blastospores can rapidly lose their infectivity after drying, depending upon growth medium (Jackson et al. 1997), method of drying (Cliquet and Jackson 1997) and adjuvants (Sandoval-Coronado et al. 2001). James (2001) reported significantly greater infectivity of conidia of both *B. bassiana* and *P. fumosoroseus* that were germinated prior to application against *B. tabaci* nymphs. Conidia of *P. fumosoroseus* and other Hypocreales formed early in colony growth on artificial media tend to germinate faster than those formed later (Hall et al. 1994b). Improvements in formulation of conidia have led to successes with other fungi for control of insects in less humid habitats (Bateman et al. 1993; Burges 1998). Smith (1997) reviews the options for formulation of *P. fumosoroseus* for improved whitefly control.

3.2.1.7 Effect of Growth Medium

P. fumosoroseus readily grows on simple artificial media, such as Sabouraud dextrose agar with yeast extract (SDAY) and forms conidia at 24°C in 10–14 days.

Peterkin and Hall (1994) produced conidia on semi-artificial media (parboiled rice) that germinated more rapidly than those produced on SDAY. Increased production of conidia was reported by Sakamoto et al. (1985) when cultures were incubated at 25°C for 4 days under white light followed by 4 days of darkness. Studies conducted by Jackson et al. (1997) and Vidal et al. (1998a) investigated a variety of media for their effect on production and stability of blastospores. Production was greatest on MS medium containing 80g glucose L^{-1} and 13.2g Casamino acids L^{-1}. De la Torre and Cardenas-Cota (1996) investigated the effects of light, temperature, yeast extract, and carbon to nitrogen (C:N) ratio on conidia and blastospore production in submerged culture. Cycles of 12h light and 12h dark, incubation at 37° C for 12h prior to incubation at 30° C, and a C:N ratio of 25 produced the highest number of spores. James (2001) observed marked effects of exogenous nutrients on spore germination; sugars stimulated only 11% germination whereas treatment with yeast extract or peptone resulted in 95–100% germination.

3.2.1.8 Genetic and Molecular Studies

The results of PCR analysis of dozens of isolates of *P. fumosoroseus* are reported by Tigano-Milani et al. (1995), Cantone and Vandenberg (1998), Azevedo et al. (2000a), and Fargues et al. (2002). Tigano-Milani et al. (1995) analyzed the genetic variability among 27 isolates of *P. fumosoroseus*, 15 of which were isolated from *B. tabaci*. The level of divergence among isolates observed by these authors suggests that the various strains represent a species aggregate. Fargues et al. (2002) reported on the genetic variability among *P. fumosoroseus* isolates as a function of geographical and host insect origins. Azevedo et al. (2000b) identified isolates with double stranded RNA, but they exhibited no differences in virulence from isolates with single stranded RNA. Cantone and Vandenberg (1998) considered that DNA fingerprinting with RAPD-PCR could be useful in tracking the fate of fungi released into the environment as biological control agents. Genetic transformations of *P. fumosoroseus* and the use of fluorescent protein for investigations of the fungus were reported by Cantone and Vandenberg (1999a, b).

3.2.2 Lecanicillium[1] *(= Verticillium) lecanii*

Lecanicillium lecanii is a common fungal pathogen of aphids, scales, whiteflies and several other insect and non-insect hosts (Hall 1981, Evans and Samson 1986; McCoy et al. 1988). Genetic analysis of several isolates of *L. lecanii* using RAPD suggests that the fungus is a highly diverse species (Mor et al. 1996). The majority of natural infections of *L. lecanii* in whiteflies are reported from more temperate climates, predominantly in *T. vaporariorum*. Natural epizootics of *L. lecanii* are common in greenhouse insects in Europe (Hussey 1958; Galani and Almasan

1984; Samson and Rombach 1985) including more northerly countries, such as Sweden (Ekbom 1979). With the spread of *B. tabaci*, especially into greenhouses, the incidence of *L. lecanii* reported from this pest has been on the increase (Table 3.1). An epizootic of *L. lecanii* in *B. tabaci* (B biotype) was recorded by Lourenção et al. (2001).

3.2.2.1 Screening of Isolates

Various isolates of *L. lecanii* from aphids, whiteflies and other insects show distinct differences in their ability to infect whiteflies. Strains collected from *T. vaporariorum* are usually more active against whiteflies than those isolated from other insects including aphids (Hall 1982; Kitazawa et al. 1984; Masuda and Kikuchi 1992; Chandler et al. 1993), and strains that originate from whiteflies have greater recycling potential in whitefly cadavers and populations (Hall 1982; Chandler et al. 1993). Screening of several isolates of *L. lecanii* and other *Lecanicillium* species against *B. tabaci* have been reported by Nier et al. (1991), Mor et al. (1996), Negasi et al. (1998), Steenberg and Humber (1999), and Gindin et al. (2000). The LT_{50} reported by Gindin et al. (2000) for the most virulent strains was 3.2–3.8 days. The activity of toxins isolated from *L. lecanii* was studied by Gindin et al. (1994).

3.2.2.2 Factors Affecting Insecticidal Activity

Extensive research has been conducted on this fungus for control of greenhouse whitefly, *T. vaporariorum*, and aphids (Hall 1981, 1982, 1985; Ekbom 1979). Efficacy studies against *T. vaporariorum* and research elucidating a variety of factors that influence insecticidal activity (temperature, media and other conditions for optimal growth) were summarized by Hall (1981) and updated by Lacey et al. (1996). Some of the most important factors that limit or enhance the entomopathogenic activity of *L. lecanii* for control of *T. vaporariorum* are elevated humidity, a fairly narrow range of optimal temperatures, stage of whitefly, and strain of fungus. High humidity is an absolute requirement for germination, establishment of infection, sporulation and consequently the capacity to produce epizootics (Ekbom 1979, 1981; Hall 1981; Riba and Entcheva 1984; Milner and Lutton 1986; Drummond et al. 1987; Chandler et al. 1994). The daily duration of humidity is also an important factor in the infection process and the continuity of control from a single application of the fungus. At least 12h of high humidity (85–95%) are required daily for elevated levels of infection and spread of the fungus (Hall 1981; Samson and Rombach 1985; Drummond et al. 1987). Early *L. lecanii* products were based on hyphal bodies and hyphae produced in liquid culture, and the infectivity of these products was low (Barlett and Jaronski 1988; Ravensberg et al. 1990b). Efficacy was largely dependent upon secondary growth and sporulation, processes that require prolonged periods of high humidity. Subsequent improvement in production methods resulted in the development of conidia-based formulations containing a 50-fold greater spore concentration that are more stable and less dependent on high humidity (Ravensberg et al. 1990b). The optimal temperatures for growth of whitefly-active isolates are 24–26°C (Ekbom 1979; Hall 1982; Li et al. 1991; Hall et al. 1994a).

The age and stage of the whitefly can also have a profound effect on the susceptibility to *L. lecanii*. Several researchers reported that second through fourth instar nymphs and adults of *T. vaporariorum* were infected by the fungus, whereas eggs and first instars were not (Hussey 1958; Ekbom 1979; Hall 1982; Masuda and Maeda 1989). Although *Bemisia* spp. are highly susceptible to *L. lecanii* and related species, only limited research on this fungus has been conducted on factors that influence its activity against *B. tabaci*. Hall et al. (1994b) reported that *B. tabaci* first instars were refractory to infection by *L. lecanii*. On the other hand, Meade and Byrne (1991) observed moderate activity of a commercial preparation of *L. lecanii* against first through third instar nymphs of *B. tabaci* and *T. vaporariorum*. There were no significant differences in susceptibility due to age or species of whitefly. Negasi et al. (1998) and Gindin et al. (2000) also observed high mortality among newly hatched *B. argentifolii* nymphs exposed to the fungus. Gindin et al. (2000) reported maximum mortality in third instar nymphs, pupae and adults of 83%, 73%, and 53%, respectively. Similar observations were made by Negasi et al. (1998). No actvity against eggs was reported by Gindin et al. (2000) and Negasi et al. (1998).

3.2.2.3 Production

Lecanicillium lecanii can be grown on a variety of mycological media (Galani 1979; Hall 1981; Li et al. 1991; Grajek 1994). At least some strains can be repeatedly subcultured without loss of virulence (Hall 1980; Hirte et al. 1989b). The production of blastospores has been reported by Ekbom (1979), Hall and Latgé (1980), and Hirte et al. (1989b). Aerial conidia of *L. lecanii* are fairly stable under refrigeration. When kept at 2–6°C, formulated conidia can be stored for up to 6 months (Butt et al. 1999).

3.2.3 *Aschersonia* spp.

Species in this genus are found only in whiteflies and coccid scales and have been reported to cause spectacular epizootics in certain host species (Fransen 1990b). *Aschersonia* spp. occur predominantly in humid tropical and subtropical regions worldwide. Twenty-five species of *Aschersonia have* been described from whiteflies by a number of authors, but relatively few cases of *Aschersonia* infections are reported for *Bemisia* (Table 3.1). The genetic diversity and phylogeny of 23 isolates collected from several locations around the world were reported by Obornik et al. (1999).

Of the *Aschersonia* species associated with whiteflies, *A. aleyrodis* and *A. placenta* have the broadest host range (Fransen, 1990b). In the USA (Florida, Alabama, Mississippi, Texas and Louisiana), *A. aleyrodis* has been observed infecting whiteflies other than *B. tabaci*. Fawcett (1944) reported outbreaks of *Aschersonia* species in *Dialeurodes citri* and *D. citrifolii* in citrus groves in Florida during the summer months. In Florida, high ambient RH promotes infection and sporulation on the host and frequent rains enable effective dispersal of the conidia. *Aschersonia* species have been successfully applied against whitefly populations in citrus groves in the Black Sea area and China (Fawcett 1944; Ponomarenko et al. 1975; Gao et al. 1985). The reported incidence of *Aschersonia* infections in whiteflies, especially in the humid tropics, is expected to increase as the more polyphagous *B. tabaci* B biotype becomes more widely distributed and is observed by more entomologists around the world. For example, an epizootic of *A. aleyrodis* in *B. tabaci* was described by Lourenção et al. (1999) in Brazil.

3.2.3.1 Screening of Isolates

The pathogenicities of different *Aschersonia* species and isolates have been tested on *T. vaporariorum* by Kogan and Serjapin (1978), Spassova et al. (1980), Fang et al. (1985), and Hirte et al. (1989a). Hirte et al. (1989a) observed high infection rates for several isolates of *Aschersonia* against this species. In their study, *Aschersonia placenta* provided the highest spore production. Mortalities of up to 73% and 93% in greenhouse whitefly on cucumbers and tomatoes, respectively, were obtained by Spassova et al. (1980) using the same species. More recently 31 isolates of several *Aschersonia* species were evaluated against *B. tabaci* and *T. vaporariorum* by Meekes et al. (2002). They reported similar levels of susceptibility in the two whitefly species to the various isolates. Those with the greatest insecticidal activity were *A. aleyrodis* from Colombia, *A. placenta* from India and two undetermined *Aschersonia* isolates from Thailand and Malaysia.

3.2.3.2 Factors Affecting Insecticidal Activity

Environmental factors that influence infection rates, spore survival and spread of *Aschersonia* are both abiotic and biotic. These include temperature, humidity, sunlight, rain, dew, irrigation, wind, white fly stage and age, host-plant characteristics, crop structure, intercropping, and microbial interactions. *In vivo* experiments showed high levels of infection for *A. aleyrodis* on nymphs of *T. vaporariorum* at 15°C, 20°C, 25°C, and 30°C. The rate of development of infection, however, is very slow at 15°C (Fransen et al. 1987). High humidity is an absolute requirement for germination of the spore and penetration of the host. In nature, humidity generally fluctuates within a 24 h period from high during the night and early morning to low at midday. Apparently, change from high to low humidity does not necessarily impair the infection process (Fransen et al. 1987). During low fluctuations, adequate humidity may be provided by the microclimate on the leaf surface; direct sunlight and/or low ambient humidity may induce more transpiration by the plant. Raising or maintaining high humidity by manipulating environmental conditions during the initial infection process has been used to increase infection and mortality rates. Experimental application of *A. aleyrodis* in greenhouses against *B. tabaci* and *T. vaporariorum* on poinsettia at an average daily temperature of 19°C and 60–75% relative humidity resulted in infection rates of only 25–30% for both whitefly species. However, covering the plants with plastic for 48 h after spraying resulted in mortality rates of 45% and 85% for *B. tabaci* biotype B and *T. vaporariorum*, respectively (J. Fransen unpublished data).

The occurrence of epizootics of *Aschersonia* species in natural habitats is correlated with the amount of precipitation. From the observations made by Petch (1921, 1925), Mains (1959) and more recently during exploration for whitefly natural enemies by staff of the USDA-EBCL and others, it can be concluded that *Aschersonia* species are usually found in areas with a mean annual precipitation of 100 cm or more. This is probably due to the requirement of both high humidity for germination and the predominant mechanism of spore dispersal being by rain or splashes of water. In Brazil, for example, the occurrence of *Aschersonia* spp. on whiteflies is very common in all areas where citrus is grown and coincides with the periods of greatest rainfall (Alves 1998; Dolinski and Lacey 2007).

The relationship between climate, host plant, and whitefly–fungus interaction is a complex one that warrants further investigation. Host plant characteristics such as leaf configuration, pubescence, size, distance from moist soil, presence and activity of stomata, etc. will influence the microclimate. In experiments with cucumber and poinsettia as host plants, lower numbers of *Aschersonia*-infected whiteflies were found on poinsettia compared to cucumber (J. Fransen unpublished data). Whitefly spatial distribution is strongly influenced by host plant age, architecture, location of vascular bundles, and oviposition behavior of the whitefly species (van Lenteren and Noldus 1990; Sütterlin et al. 1991; Cohen et al. 1996). Levels of fungal infection, in turn, will be related to the spatial distribution of whiteflies in different developmental stages.

Fransen et al. (1987) observed differential mortality for the various life stages of the greenhouse whitefly treated with *A. aleyrodis*. Eggs were not infected, but nymphs that hatched from treated eggs and settled on the treated leaf surface were infected to the same degree as first instar nymphs that were directly treated. High mortality rates were found in younger instars (1–3) treated with the fungus, but fourth instars, prepupal and pupal stages were considerably less susceptible and adults seldom showed symptoms of infection.

Survival of conidia during periods of unfavorable climatic conditions may take place in or on colonized hosts, on plant surfaces, in soil or other habitats. No detailed reports of survival of *A. aleyrodis* in natural habitats are available. As reported above for *P. fumosoroseus*, the detrimental effect of ultraviolet light is seen as a main factor reducing the viability (survival) of fungal spores on plant surfaces (Fargues et al. 1997b). However, spores on abaxial leaf surfaces can survive longer. *Aschersonia* spores are produced in pycnidia on the colonized insect, and slimy masses of conidia may survive better in a mass of pigmented spores covered by a mucilaginous substance. Additionally, whiteflies are predominantly present on the underside of leaves which are less exposed to direct sunlight. Conidia of *A. aleyrodis* survive up to 22 days on cucumber leaves at 20°C (Fransen 1995). Conidia present on the leaf surface can infect whitefly nymphs hatching from eggs laid long after the conidia were applied. Fransen (1995) noted that only about 10% of the conidia germinate on the cucumber leaf surface, whereas germination on the whitefly cuticle was considerably higher. Meekes et al. (2000) studied persistence of *A. aleyrodis* on cucumber, gerbera and poinsettia; germination capacity was highest on cucumber and lowest on gerbera. Conidia on cucumber remained viable and were able to infect 90% of greenhouse whitefly nymphs 31 days after application.

3.2.3.3 Production

Aschersonia species can be easily grown on artificial media such as potato dextrose agar, Sabouraud dextrose agar, and whole grains (millet, rice or corn). Spassova (1980) describes the production of fungal material using beer wort, 10% sugar, 0.1% manganese, 0.01% iron or magnesium at 24–26°C under fluorescent light. Research on different additives and pH ranges have been carried out by Oho (1967), Hirte et al. (1989a), and Ibrahim et al. (1993). *Aschersonia* species do not sporulate in liquid culture, which has been a drawback for mass production of the fungus. However, interest in two-phase and solid phase fermentation technology has increased in recent years. *Aschersonia* species sporulate well on autoclaved millet and other natural substrates when light and oxygen are provided (Fransen 1995).

Hirte et al. (1989a) observed a decline in virulence of *A. placenta* after growing on artificial media for 1½ years; mortality of whitefly nymphs decreased from 75–95% to 20–40%. In addition to duration of culture on artificial media, the number of subcultures may play an important role in attenuation of virulence. Fransen et al. (1987) found no decrease of virulence of *A. aleyrodis* after 12 serial passages on millet, but observed an increase of about 2 days in the LT_{50} after 19 serial passages on semi-artificial medium. Kogan and Serjapin (1978) advise reisolating *A. aleyrodis* once a year after three to four host passages. Subsequent subculture may gradually reduce the spore production on semi-artificial media, and host passaging may not only restore the virulence but also improve production of spores.

The activity of *Aschersonia aleyrodis* against *T. vaporariorum* under greenhouse conditions has been investigated in Europe by Landa (1984), Fransen et al. (1987) and Meekes et al. (2002). Because of its specificity for certain homopterous insects, it is ideal for whitefly control in situations, such as greenhouses, in which beneficial insects are used for biological control and pollination. Meeks et al. (2002) also investigated several species and isolates of *Aschersonia* spp. for potential as microbial control agents of *Bemisia*. Several of the isolates tested originated from EBCL and other USDA foreign exploration and are stored in the ARSEF collection (Humber 2002).

3.2.4 *Beauveria bassiana*

This fungus is one of the most common pathogens reported from insects around the world and has been developed as a microbial biocontrol agent for a wide variety of insects (McCoy et al. 1988; Feng et al. 1994; Wraight and Carruthers 1999). It possesses relatively low epizootic potential against whiteflies (Wraight et al. 1998), and has only rarely been reported from *Bemisia* species under natural conditions (Table 3.1). Nevertheless, *B. bassiana* is highly virulent against the nymphal stages of whiteflies under laboratory conditions (Eyal et al. 1994; Wraight et al. 1998; Zaki 1998; Negasi et al. 1998; Ramos et al. 2000; Vicentini et al. 2001; Brownbridge et al. 2001), and spray applications of this pathogen have provided moderate to good control of whiteflies infesting various field and greenhouse crops (Table 3.2). In both laboratory bioassays and small-scale field tests, *B. bassiana* has exhibited a capacity to infect *B. tabaci* nymphs at rates comparable to *P. fumosoroseus*, a fungus generally considered one of the most important natural enemies of *Bemisia* (Lacey et al. 1996; Wraight et al. 1998, 2000).

In laboratory bioassays, highly virulent isolates of *B. bassiana* have exhibited mean LC_{50} of about 100 conidia/mm^2 against third-instar nymphs, and greater than 250 conidia/mm^2 against fourth instars (Wraight et al. 1998, unpublished observations in Wraight 1997). These findings are in accord with other reports of lower susceptibility of fourth-instar and pupal whiteflies compared to earlier instars treated with *B. bassiana* and other fungal pathogens (Fransen et al. 1987; Osborne and Landa 1992; Vicentini et al. 2001, unpublished observations in Faria and Wraight 2001). The greater susceptibility of early instars may be related to the molting process. During molt, the nymphs exit the old cuticle, expand, and then settle back down onto the leaf substrate, usually in a new location on the leaf and potentially onto fungal spores contaminating the leaf surface. This increases the dose to which the insects are exposed, and spores sequestered at the interface of the insect cuticle and phylloplane likely experience moisture conditions highly favorable for germination and infection. High susceptibility of nymphs treated as third instars has been noted by some researchers (Fransen et al. 1987; Negasi et al. 1998); this may be associated with the great expansion of the fourth-instar nymph after emergence from the third-instar cuticle (Wraight unpublished observations).

Virulence of *Beauveria* against adult *B. tabaci* has not been determined in comparable terms (rates applied directly against adults and expressed as conidia per unit area of treated surface). Garza Gonzalez and Arredondo Bernal (1993) observed 74–87% mortality among *B. tabaci* adults exposed to concentrations of 10^8 *B. bassiana* conidia/ml. Negasi et al. (1998) reported an adult LC_{50} of 1.3×10^6 conidia/ml for an isolate of *B. bassiana* isolated from thrips. This value was more than 100 times greater than the LC_{50} against third-instar nymphs. Zaki (1998) treated *B. tabaci* nymphs with commercially formulated *B. bassiana* at a concentration of approximately 10^7 conidia/ml and subsequently noted 68% mortality among emerged adults. The treated group of adults produced 82% fewer eggs than the controls. Impacts of *B. bassiana* applications on adult populations in the field have been highly variable (see Section 3.3). In most studies it was not determined if reductions in adult populations were the result of direct infection and mortality of adults or due to low emergence rates of adults from treated nymphal populations. During numerous field tests of *B. bassiana* against *B. tabaci* biotype B, Wraight et al. (2000) observed few fungus-killed adults, even in plots treated twice weekly with high rates of *B. bassiana* conidia. Much of the body of an adult whitefly is covered with waxy scales and shielded by the wings. These characteristics, plus the fact that adult whiteflies are not confined to the shaded, humid abaxial phyllosphere were hypothesized to account for the lower susceptibility of adults compared to nymphs under field conditions. The eggs of *B. tabaci* biotype B are not susceptible to *Beauveria* infection (Negasi et al. 1998; Ramos et al. 2000).

There is a great volume of literature on the pathobiology of *B. bassiana*. However, this fungus remains little studied with regard to its activity specifically as a pathogen of whiteflies. Studies of host-plant effects are reviewed above in the section on *P. fumosoroseus*. Most strains of this pathogen are severely limited by temperatures greater than approximately 32°C (Fargues et al. 1997a), and inhibition of insecticidal efficacy has been noted in some hosts at temperatures as low as 30°C (Long et al. 2000) or by brief exposures to high temperatures (Inglis et al. 1996). This sensitivity to temperatures commonly encountered in field and greenhouse environments bears serious consideration with regard to development of *B. bassisana* for control of *Bemisia* spp., as these insects are important pests of crops grown under hot, arid or semiarid conditions (Faria and Wraight 2001). On the other hand (and as described above), the undoubtedly humid microniche between the sessile nymphs and the leaf surface apparently makes it possible for fungal spores to germinate and infect nymphal whiteflies under dry conditions. Wraight et al. (2000) noted that *B. bassiana* and *P. fumosoroseus* were capable of infecting whitefly nymphs on leaf surfaces exposed directly to atmospheres with 25–30% RH. Furthermore, addition of a 24 h period of 100% relative humidity immediately posttreatment did not significantly increase fungal infection in low-humidity assays.

Selected isolates of *B. bassiana*, including many that are highly virulent against whiteflies, rank among the most efficiently mass produced of the entomopathogenic fungi. This advantage resulted in rapid development and registration of *B. bassiana*-based products for whitefly control within only a few years after the first devastating outbreaks of *B. tabaci* biotype B were experienced in the USA and elsewhere. These products are produced by many predominantly small enterprises worldwide. Conidia comprise the active ingredient in virtually all of the commercial *B. bassiana* products. These spores are mass produced on solid substrates based on various whole or processed grains or grain components, such as barley, rice, or wheat bran, which may be supplemented with nutrients. In most production systems, the solid substrates are inoculated with hyphal bodies produced in liquid fermentations (Feng et al. 1994; Wraight et al. 2001). Recent advances in mass production and formulation technologies for *B. bassiana* have led to production of both wettable powder and emulsifiable oil formulations of notably high quality supporting label rates for field applications as high as 2.5×10^{13} conidia/ha and label rates for greenhouse applications exceeding 4×10^{13} conidia/380 liters/0.05–0.1 ha. Some of these products exhibit room-temperature shelf lives of nearly 1 year. Much of the existing body of literature related to microbial biocontrol of *Bemisia* whiteflies in field and greenhouse crops is based on these products (see Section 3.3).

3.2.5 *Metarhizium anisopliae*

Like the other Hypocreales covered in this chapter, the fungus *Metarhizium anisopliae* also has a broad insect host range and has been developed or studied as a microbial control agent of several insect pests (McCoy et al. 1988; Zimmermann 1993). However, relatively little research has been conducted on this fungus for control of *Bemisia tabaci*. Herrera at al. (1999) bioassayed isolates of this and other fungi against whitefly nymphs and observed the highest mortalities, up to 97%, for five *M. anisopliae* isolates. Davidson et al. (1996a, b) studied the effects of destruxins (cyclodepsipeptide toxins isolated from filtrates of *M. anisopliae* cultures) on adults of *B. tabaci* biotype B. In their studies, the activity of the destruxins was equal to that of imidacloprid and ivermectin.

Malsam et al. (2002) evaluated the effect of sublethal concentrations of oils for increasing the efficacy of *M. anisopliae* against *B. tabaci* and *T. vaporariorum*. Formulated sunflower oil provided the highest synergistic effect. Batta (2003) produced and tested an invert emulsion (water in oil formulation) of *M. anisopliae* against *B. tabaci*. He reported a shelf life of up to 30.8 months at 20°C (50% loss of viability after 4.6 months) and 67–100% and 30–92% mortality in whitefly nymphs in the laboratory and field, respectively.

3.3 Implementation of Entomopathogenic Fungi for Control of *Bemisia tabaci* Biotype B

Microbial biocontrol of *B. tabaci* based on inundative spray applications of entomopathogenic fungi has been under investigation for only about 10 years. Applications of various fungi against *B. tabaci* infesting a broad variety of field and greenhouse crops have yielded highly variable results, and applications employing conventional spray equipment and economically acceptable spray schedules have provided, with few exceptions, no more than moderate levels of control of pest populations (Table 3.2). These inconsistent results have had a substantial negative impact on commercial development efforts, yet they are not surprising when one considers the numerous variables associated with the practical, economic application of these biological control agents and the problems associated with assessing their efficacy. Nearly all efficacy tests of fungal control agents have used formulated spore preparations. Spores (both conidia and blastospores) have been mixed with a variety of wetting agents or emulsifiable oils, which themselves possess insecticidal properties highly dependent upon environmental and other factors. Experimental designs have too often failed to attribute the observed population reductions to the various insecticidal components of these mycoinsecticide formulations. Relatively short shelf life of fungal formulations can also influence the outcome of tests conducted with separate batches of fungi or the same batch conducted at different times. Fungal spores or germ elements must contact the whitefly cuticle to initiate infection, and the vast majority of studies also have not quantified actual application rates in terms of infectious propagules per unit surface area of whitefly-infested foliage. This has been largely overlooked, despite enormous differences in targeting efficiencies of insecticide spray application equipment and technologies. In many cases, descriptions of the application equipment have been poor or lacking. Another difficult problem relates to the lack of any standard post-treatment sampling protocol to assess mortality of immature whitefly stages treated with slow-acting microbial control agents. The great majority of field and greenhouse tests of fungal pathogens against *Bemisia* spp. have been conducted with *P. fumosoroseus* and *B. bassiana* in cucurbits, tomatoes and cotton; however, the total number of comprehensive tests against even these crops is too low to permit more than the drawing of a few broad generalizations.

3.3.1 Field Crops

Some of the best and most consistent results in the field have been obtained against nymphal whitefly populations infesting cucurbits. This suggests that the physical and chemical characteristics of the cucurbit phylloplane and phyllosphere are highly favorable for fungal infection of these sessile, scale-like insects. Cucurbit plants typically produce large leaves, often with lower-surface concavities that limit exposure to solar radiation and trap humid air under calm conditions. The leaf undersides are also densely covered with stout trichomes that contribute to establishment and maintenance of a boundary layer. Unless staked, most cucurbits have a recumbent growth form, producing a simple (essentially mono-layered) canopy, which allows for intense focusing of spray applications, especially during the early part of the season before the vines fill the beds (Wraight and Carruthers 1999; Vandenberg et al. 2007). Foliage developing in close proximity to the soil on a recumbent plant is generally exposed to a cooler and more humid environment than foliage produced in the upper canopy of an erect plant.

Considering the reports summarized in Table 3.2, tests in cucurbits, including cucumber, melons and squash treated with *P. fumosoroseus* and *B. bassiana* have typically resulted in 50–70% reductions in nymphal *B. tabaci* populations. Higher levels of control were achieved by Wraight et al. (2000) in tests in which high rates of fungal conidia were applied with a portable, hand-targeted air-blast sprayer and coverage of leaf undersides was verified. High levels of control were also reported by Orozco-Santos et al. (2000), who employed a similar sprayer. Impacts of fungal treatments on populations of adult whiteflies have been less consistent; e.g., Wraight et al. (2000), while achieving excellent control of nymphs with *B. bassiana*, noted minimal impacts on adult populations until the end of the crop cycle. Liu et al. (1999) (see also Knowles and Jaronski 1997), in contrast, reported approximately 70% control of adults within 14 days following initial application of *B. bassiana*, and the data of Orozco-Santos et al. (2000) also reveal substantial reductions in adult populations during and following a *B. bassiana* spray program.

Tests of *B. bassiana* and *P. fumosoroseus* against *B. tabaci* infesting tomatoes have produced results generally similar to those reported from cucurbits, with virtually all studies achieving moderate or at least significant levels of population reduction. Liu et al. (1999) concluded that *B. bassiana* was ineffective in controlling *Bemisia* on field tomatoes; yet at the same time reported signs of *B. bassiana* infection (red pigmentation) in 40–50% of the nymphal population. Bolckmans et al. (1995) reported lower efficacy of *P. fumosoroseus* against *T. vaporariorum* infesting tomato versus cucumber in greenhouse cultures; however, the tomato trial results were confounded by applications of several pesticides, including fungicides for *Botrytis* control.

We are not aware of any published observations of fungal efficacy in tomatoes comparable to the >90% control levels reported from some cucurbit tests, and results have been published that indicate potentially negative interactions between entomopathogenic fungi and solanaceous plants including tomato. As related in the section on *P. fumosoroseus*, glycoalkaloids found in the foliage of tomato inhibit germination and growth of *B. bassiana* and *P. fumosoroseus* conidia in vitro (Costa and Gaugler 1989; Lacey and Mercadier 1998; Poprawski et al. 2000) and have been hypothesized to reduce efficacy of these fungal pathogens applied against greenhouse whiteflies infesting this plant (Poprawski et al. 2000). Vidal et al. (1998b), however, reported equivalent efficacy of *P. fumosoroseus* on tomato, cabbage, and cucumber against *B. tabaci* biotype B in a humid greenhouse (relative humidity >70%). These contradictory results might be explained by differences in protocols. The two studies involved different whitefly species, and in the study by Poprawski et al. (2000), the plants were enclosed in plastic bags for 24 h post-treatment to generate high-humidity conditions. The possibility that this procedure produced elevated concentrations of volatile compounds inhibitory to the fungus was not investigated.

In terms of microbial control potential, differences in leaf morphology/physiology and growth habits of the two plants also warrants consideration. Tomato is a tall, erect plant with a relatively sparse canopy and deeply incised leaves. It is likely that the abaxial phyllosphere of a tomato leaf is significantly less humid than that of a cucurbit leaf. The largely unknown and poorly defined variables of phyllosphere humidity and temperature certainly deserve greater attention in studies of fungus–insect–host plant tritrophic interactions. Precise characterization of microclimatic conditions associated with plants and specific plant parts is difficult, but such studies are being pursued. Although the entomopathogenic fungi are capable of initiating infections in early-mid instar whitefly nymphs under dry conditions (Wraight et al. 2000), whitefly pest populations are comprised of multiple, overlapping generations, and infection of fourth-instar nymphs, pupae and adults is probably more dependent upon high-humidity conditions. Clearly little can be concluded on the basis of so few published studies. Much additional work will be required before we achieve an understanding of the full range of factors determining efficacy of entomopathogenic fungi on different host plants.

Applications of entomopathogenic fungi (*B. bassiana* and *P. fumosoroseus*) against *B. tabaci* in cotton have produced highly variable results. Moderate to high levels of control (50–87%) of nymphal *B. tabaci* have been achieved in some tests (Jaronski and Lord 1996; Akey and Henneberry 1998), but little or no efficacy has been noted in other tests (Liu et al. 1999; Wraight et al. 1996). Results from a laboratory study by Poprawski and Jones (2001) suggest that *B. bassiana* may be inhibited by allelochemicals associated with the cotton leaf and possibly sequestered by the whitefly. However, additional studies of these plant–fungus interactions under field or simulated field conditions are needed, and allelochemistry is unlikely to be the sole explanation for the highly variable results. Most tests that have produced significant control were conducted in intensively irrigated crops (Jaronski and Lord 1996; Jaronski et al. 1996; Akey and Henneberry 1998). Wraight et al. (1996) observed no control in multiple field tests of *B. bassiana* and *P. fumosoroseus* against *B. tabaci* biotype B infesting infrequently irrigated cotton in southern Texas (irrigated at 2–3 week intervals) even though high rates of conidial deposition were verified on leaf undersides. Liu et al. (1999) also reported poor control from applications of *B. bassiana* against whiteflies on cotton in southern Texas. Certainly in the case of applications made during the summer months, and especially in dryland or infrequently irrigated crops, phylloplane temperature and humidity (and temperature–humidity interactions) must be considered potentially critical factors in fungal efficacy. While it is commonly believed that leaf surfaces are significantly cooler than ambient, this is not the case under all conditions. In situations where water is limited and plants become drought stressed, phylloplane temperatures on insolated leaves may exceed ambient by several degrees, and elevated leaf surface temperatures will profoundly influence boundary layer humidity. Under these conditions, the development and physiological defense reactions of infected whiteflies will be optimized while development and aggressiveness of invading fungal elements is minimized.

3.3.2 Greenhouse Crops

Few published studies have rigorously examined the potential of entomopathogenic fungi for control of *B. tabaci* under greenhouse conditions. There exists, however, a long history of commercial development of fungal pathogens (initially *L. lecanii* and more recently *P. fumosoroseus* and *B. bassiana*) for control of the greenhouse whitefly, *T. vaporariorum* (see Hajek et al. 2001). Methodology for evaluating fungi against whiteflies in greenhouses is presented by Burges (2007).

Economical mass production of *P. fumosoroseus* conidia has proven difficult, and the commercial products developed in the USA and Europe are based on blastospores produced in liquid fermentations. Greater than 90% control has been achieved in trials of these products against greenhouse whitefly infesting cucumber (Bolckmans et al. 1995; Sterk et al. 1996). The fungi developed for *T. vaporariorum* control are also highly pathogenic against *Bemisia* spp. (Lacey et al. 1996), and there is no reason to believe that the successes achieved against *T. vaporariorum* will not be repeatable with *B. tabaci*. Osborne and Landa (1994), testing blastospore-based preparations of *P. fumosoroseus* against *B. tabaci* on hibiscus and poinsettia, reported levels of control comparable to chemical insecticides. *Beauveria bassiana* also possesses demonstrated capacity to substantially reduce *B. tabaci* populations on greenhouse ornamentals, though it may not provide adequate control as a stand-alone product (Olson and Oetting 1999). There are numerous reasons to consider greenhouse crops as excellent targets for mycoinsecticide applications: (1) insect pathogenic fungi require specific environmental conditions for maximum efficacy, and both temperature and moisture conditions can be regulated to a significant degree in enclosed cropping systems; (2) many greenhouse glazings block UV light, which is lethal to fungal spores applied to crop foliage; (3) residues of toxic pesticides are more persistent in protected environments than in the field, and many floral crops must be harvested daily; workers are thus exposed to elevated risks, creating a strong demand for safer pesticides with short reentry periods and preharvest intervals; (4) intensive use of chemical insecticides on floral and nursery crops contributes to rapid development of resistance in pest populations, and alternative control agents are needed for resistance management; and (5) floral and nursery crops are generally of higher value than field crops, which creates potential to maximize mycoinsecticide efficacy through application of higher rates and use of highly efficient (e.g., hand targeted) spray application equipment. There is potential also for making efficient applications via sophisticated greenhouse and nursery crop irrigation systems. Shipp et al. (2003) demonstrated effective suppression of *T. vaporariorum* and *Frankliniella occidentalis* with whole greenhouse sprays of *B. bassiana* when high humidity was maintained.

On the other hand, there are many potential constraints to use of fungal pathogens. Economic thresholds for insect pests are exceedingly low for many ornamental crops, and such levels of control are difficult to achieve with biological control agents. Without costly environmental controls, the greenhouse environment can be quite harsh. Temperatures may be high during peak production seasons, and the struggle waged constantly by greenhouse growers against plant pathogenic fungi and bacteria has culminated in adoption of protocols recommended by plant pathologists to minimize formation or accumulation of free water on plant foliage. Protocols are also followed to prevent prolonged periods of high relative humidity. These conditions have effectively precluded exploitation of the great natural epizootic potential of such pathogens as *L. lecanii, P. fumosoroseus*, and *A. aleyrodis*, and sales of commercial mycoinsecticides based on these pathogens have stagnated after great initial interest.

Nevertheless, fungi remain an important option for control of greenhouse pest insects including whiteflies (Burges, 2007). Future research on the use of fungi for whitefly control in this environment should address the short term needs required for acceptable insecticidal activity of insect pathogens while minimizing conditions engendering plant pathogens.

3.4 Integrated Pest Management

Sustainable agriculture in the 21st century will rely increasingly on alternative interventions to chemical pesticides for pest management that are environmentally friendly and reduce the amount of human contact with pesticides. The integrated pest management (IPM) strategy, in which natural enemies (parasitoids, predators and pathogens) of pest arthropods and other alternative measures play significant roles in crop protection, is one aspect of sustainable agriculture that attempts to minimize negative environmental impact and other deleterious effects due to insecticide usage. In order to be most effective, the separate components of IPM should have minimal or no antagonistic interactions with one another.

3.4.1 Interaction of Fungi with Beneficial Insects Including Other Natural Enemies of Whiteflies

Many natural enemies, including parasitoids, predators and entomopathogens are responsible for the suppression of whitefly populations. They may occur concomitantly or separately, depending, for instance, on host density, climate, host plant and the availability of alternative prey or hosts. Studying the complex interaction between the parasitoid *Encarsia formosa*, the pathogen *A. aleyrodis*, and the whitefly *T. vaporariorum*, Fransen and van Lenteren (1993, 1994) observed that both natural enemies can act complementarily. When the fungus has penetrated the whitefly host, *Enc. formosa* females can detect the infection in the hemolymph with their ovipositors and refrain from laying eggs. When parasitized whitefly hosts are treated with *A. aleyrodis* spores 4 days after parasitization takes place, fungal infection will not establish, and healthy *Enc. formosa* females emerge. Similar observations have been made by several researchers regarding the interactions of entomopathogenic fungi, other insect pests and their parasitoids (Brooks 1993; Lacey and Mesquita 2002).

In greenhouses, *Enc. formosa* and *A. aleyrodis* have been used simultaneously resulting in successful suppression of whitefly populations (Hristova 1971; Ramakers and Samson 1984; Landa 1984). *Aschersonia* species are selective pathogens of whitefly and coccids and are unable to infect other insects in the environment. In addition, infected and sporulating whitefly nymphs can act as a food source for predatory mites (*Amblyseius* and *Tarsonemus* species) released for biological control in greenhouse crops (J. Fransen unpublished observations) and *Acalvolia* species in citrus orchards (Osborne and Landa 1992). Acting as an alternative food source to predators of other pests, *Aschersonia*-infected whiteflies may have a positive influence on the establishment of these predators. In dry seasons more parasitism and predation may occur compared with fungal infection, whereas in wet seasons epizootics of entomopathogens may contribute to the suppression of whitefly populations. Steinberg and Prag (1994) reported on the successful control of *B. tabaci* in greenhouse cucumber using a combined treatment of *A. aleyrodis* and the coccinellid predator *Delphastus catalinae* (reported as *D. pusillus*). No antagonistic interaction between the two agents was detected and a single application provided control for approximately 2 months.

Ekbom (1979) observed mortality in *Enc. formosa* 6–8 days after treatment with *L. lecanii* and subsequent continuous incubation in high humidity. However, Hall (1981) reported that *Enc. formosa* was not attacked by *L. lecanii* when the fungus occurred naturally in crops where the parasitoid was present. Because of the apparent synergistic activity of *Enc. formosa* and *L. lecanii*, Hall (1982, 1985) recommended their combined use to obtain the best possible control. Landa (1984) evaluated various combinations of several interventions, including *L. lecanii* and *A. aleyrodis*, pesticides, predatory insects and mites and parasitic insects and sticky traps for control of *T. vaporariorum* on cucumbers in greenhouses.

Published information on the interaction of whitefly isolates of *P. fumosoroseus* and non-target organisms is somewhat limited relative to other fungi. Sterk et al. (1995a, b, 1996) observed little or no effect of *P. fumosoroseus* on *Enc. formosa* and several predators used for control of greenhouse whitefly. Research conducted at EBCL and the USDA ARS Biological Control of Pests Research Unit (Weslaco, Texas) indicates that species of *Encarsia* and *Eretmocerus* are susceptible to the fungus under initially high humidity that permits infection (Lacey et al. 1997; T. Poprawski and M. Ciomperlik unpublished data). However, in the lag time between application of conidia and death of the parasitoids, some oviposition by female parasitoids still takes place (Jones and Poprawski 1996). Jones and Poprawski (1996) also observed a lack of infection of *B. tabaci* biotype B that were parasitized by *Eretmocerus* and subsequently treated with *B. bassiana*. Poprawski et al. (1998) reported on the effect of *B. bassiana* and *P. fumosoroseus* on *Serangium parcesetosum*, a coccinellid predator of whiteflies. Although there was some delay in the development of the beetle, survivorship was not significantly affected by *P. fumosoroseus*. However, *S. parcesetosum* was highly susceptible to *B. bassiana* resulting in 50% reduction in predation by beetle larvae. Jaronski et al. (1998) reported on the effect of *B. bassiana* on *Eretmocerus* sp. and the complex of predators attacking *B. tabaci* under field conditions. Their results indicated a negative impact on populations of the predatory bugs, *Orius* sp. and *Nabis* sp. They concluded that several abiotic and biotic factors may help to protect certain other non-target insects from the fungus, including species of reduviid, chrysopid, and *Georcoris*. Shipp et al. (2003) advise caution when using parasitoids and certain predators in greenhouses sprayed with *B. bassiana* under high humidity conditions.

3.4.2 Interaction of Entomopathogenic Fungi and Pesticides

The integration of fungi into comprehensive IPM programs will not only depend on compatibility with other biological control agents, but also with chemical methods of control, and the prevailing environmental conditions in a given cropping system and their effects on infectivity and persistence of spores. The concurrent use of pesticides in IPM, particularly fungicides, can have deleterious effects on entomopathogenic fungi. Samson and Rombach (1985) reported that epizootic development of *L. lecanii* has often been interrupted by the application of fungicides. In laboratory studies on fungicides and *L. lecanii*, Hall (1981) concluded that several fungicides, insecticides and acaricides were sufficiently innocuous to *L. lecanii* to warrant simultaneous application. Several compounds, however, were deleterious to the fungus and he recommended either avoiding their use in conjunction with *L. lecanii* or using separate spraying schedules for some of the less toxic pesticides. Osborne et al. (1990b) screened several pesticides against *P. fumosoroseus* and concluded that relatively few inhibited fungal growth. The use of *Aschersonia* species can be integrated with the use of fungicides if safety periods of 3–7 days are taken into account between microbial insecticide and fungicide applications (Fransen 1993). Studies in field crops have also indicated that asynchronous application can mitigate negative interactions between *B. bassiana* and fungicides (Jaros-Su et al. 1999; Costa, 1998). The deleterious effects of pesticides (including dichlorvos, fenobucarb and methidathion) on hyphal growth of *B. bassiana* are reported by Kurogi et al. (1993). James and Elzen (2001) and James (2003) investigated the effects of combining entomopathogenic fungi with imidacloprid and azadirachtin (neem) for control of *B. tabaci* biotype B. Adding imidacloprid to *B. bassiana* increased mortality of whitefly nymphs, but the augmentation of mortality was less than additive. James and Elzen (2001) hypothesized that the fungus caused a behavioral response that reduced insect feeding and thus uptake of imidacloprid. Similarly, James (2003) observed an increase in mortality when azadirachtin and *P. fumosoroseus* were combined, but the effects were less than additive. The azadirachtin moderately inhibited growth and germination of the fungus.

Experience gained from integrated control of the greenhouse whitefly and citrus whiteflies demonstrated the effectiveness of fungal pathogens as a major component in the integrated control of these pests (Ramakers and Samson 1984; Hall 1982, 1985; Landa 1984; Samson and Rombach 1985; Fransen 1990a, b, 1993; Ravensberg et al. 1990a, b; Van der Schaaf et al. 1991). Their minimal impact on the environment, safety to humans, and compatibility with other natural enemies makes them ideal components for integrated pest management.

3.5 Conclusions and Recommendations

Entomopathogenic fungi can be effective control agents and serve as alternatives to broad spectrum chemical insecticides under certain conditions. Yet, despite the encouraging results in a number of tests (especially cucurbit tests) by multiple research groups, fungal spray programs have rarely succeeded in reducing whitefly populations (nymphs or adults) to sub economic-threshold levels. Population sampling following fungal applications typically reveals impressive initial declines in whitefly numbers (especially nymphal populations), but these are followed by gradual increases, with populations building to very damaging levels by the end of the crop cycle (Olson and Oetting 1999; Wraight and Carruthers 1999; Liu and Meister 2001). This problem is well known to microbial control researchers pursuing development of slow-acting, moderately efficacious control agents like fungal pathogens. Levels of control as high 80–90% may not be adequate to prevent populations of highly fecund pests such as *B. tabaci* from increasing to damaging levels by the end of a crop cycle. This is especially problematical in long-cycle crops, such as melons, tomatoes, and cotton, which may support development of many whitefly generations; the problem is greatly exacerbated by hot weather conditions that support rapid development of pest populations. Moreover, under conditions of rapid plant growth and heavy pest pressure, achieving even these moderate levels of control requires, at minimum, weekly applications of any control agent with a strictly contact mode of action, and such intense spray programs based on mycoinsecticides have proven economically unsustainable. Consideration of the extraordinarily low pest thresholds tolerable in plant virus transmission systems raises the challenge to an even more daunting level.

Despite the numerous real and potential challenges outlined above, fungal entomopathogens may ultimately find greater success in greenhouse crops than in field crops. Such a prediction is based on the fact that greenhouses provide extensive opportunity for use of fungal pathogens as components of integrated pest management systems having a great variety of augmentative biological control agents. Populations of predators, parasitoids, and pathogens can be established and manipulated to a degree that is simply not possible under open field conditions. There is obviously great need for research on integration of fungi with other control agents; studies are also needed to develop application strategies that take into account crop and pest population phenology in order to optimize the efficacy of fungal microbial biocontrol agents (Faria and Wraight 2001).

The increased utilization of entomopathogenic fungi will require improvements in the pathogens and their formulation, careful selection of application windows and niches, and a better understanding of how they fit into integrated systems and their interaction with the environment and other components of integrated pest management. It will also require an increased awareness of their attributes by growers and the public (Lacey et al. 2001).

The regulatory restrictions placed on the use of so-called exotic strains in the USA are obstacles to implementation of entomopathogenic fungi. The utility of a diversified fungal germplasm will be severely encumbered until these issues are resolved. Consideration should be given to the fact that even the less selective species, such as *B. bassiana*, are much less harmful to nontarget organisms and the environment than are broad spectrum insecticides. Diversified germplasm repositories, such as the USDA-ARSEF collection, ultimately will provide a rich source of genetic material for future genetic manipulation.

References

Akey, D.H. and T.J. Henneberry. 1994. Sweetpotato whitefly control by Naturalis-L, the fungus *Beauveria bassiana*, in furrow and sub-drip irrigated upland cotton. Proceedings Beltwide Cotton Conferences, Memphis, TN, January 5–8, 1994, San Diego, CA, pp. 1089–1091, National Cotton Council of America, Memphis, TN.

Akey, D.H. and T.J. Henneberry. 1998. Control of silverleaf whitefly with the entomopathogenic fungi, *Paecilomyces fumosoroseus* and *Beauveria bassiana* in upland cotton in Arizona. Proceedings Beltwide Cotton Conferences, San Diego, CA, 5–9 January 1998, Vol. 2, pp. 1073–1077. National Cotton Council of America.

Alves, S.B. 1998. Fungos entomopatogênicos. In S.B. Alves, ed., Controle microbiano de insetos. pp. 289–381. Piracicaba, Brasil, Fundação de Estudos Agrários Luiz de Queiroz, 1163p.

Alves, S.B., C.A. Silveira, R.B. Lopes, M.A. Tamai, E.Q. Ramos, and S. de Salvo. 2001. Eficácia de *Beauveria bassiana*, imidacloprid e thiacloprid no controle de *Bemisia tabaci* e na incidência do BGMV. Manejo Integrado de Plagas y Agroecologia 61: 31–36.

Azevedo, A.C.S., M.C. Furlaneto, D.R. Sosa-Gomez, and M.H.P. Fungaro. 2000a. Molecular characterization of *Paecilomyces fumosoroseus* (Deuteromycotina: Hyphomycetes) isolates. Scientia Agricola 57: 729–732.

Azevedo, A.C.S., D.R. Sosa-Gomez, M.R. Faria, and M.H.P. Fungaro. 2000b. Effects of double-stranded RNA on virulence of *Paecilomyces fumosoroseus* (Deuteromycotina: Hyphomycetes) against the silverleaf whitefly, *Bemisia tabaci* strain B (Homoptera: Aleyrodidae). Genetics and Molecular Biology 23: 61–63.

Balakrishnan, S. and Y.L. Nene 1980. A note on the mode of penetration of the fungus *Paecilomyces farinosus* into the whitefly, *Bemisia tabaci*. Science and Culture 46: 231–232.

Bartlett, M.C. and S.T. Jaronski. 1988. Mass production of entomogenous fungi for biological control of insects. In R.N. Burge, ed., Fungi in Biological Control Systems, pp. 61–85. Manchester University Press, New York.

Bateman, R.P., M. Carey, and C. Prior. 1993. The enhanced infectivity of *Metarhizium flavoviride* in oil formulations to desert locusts at low humidities. Annals of Applied Biology 122: 145–152.

Batta, Y.A. 2003. Production and testing of novel formulations of the entomopathogenic fungus *Metarhizium anisopliae* (Metschinkoff) Sorokin (Deuteromycotina: Hyphomycetes). Crop Protection 22: 415–422.

Ben-Ze'ev, I.S. 1993. Check list of fungi pathogenic to insects and mites in Israel, updated through 1992. Phytoparasitica 21: 213–237.

Ben-Ze'ev, I.S., Y. Zelig, S. Bitton, and R.G. Kenneth. 1988. The Entomophthorales of Israel and their arthropod hosts: 1980–1988. Phytoparasitica 16: 247–257.

Ben-Ze'ev, I.S., G. Gindin, I. Barash, and B. Raccah. 1994. Entomopathogenic fungi attacking *Bemisia tabaci* in Israel. *Bemisia* Newsletter (special issue) 8: 36.

Berger, E.W. 1921. Natural enemies of scale insects and whiteflies in Florida. Quarterly Bulletin Florida Station Plant Board 5: 141–154.

Bolckmans, K., G. Sterk, J. Eyal, B. Sels, and W. Stepman. 1995. PreFeRal (*Paecilomyces fumosoroseus* strain Apopka 97), a new microbial insecticide for the biological control of whiteflies in greenhouses. Mededelingen – Faculteit Landbouwkundige en Toegepaste Biologische Wetenschappen, Universiteit Gent, 60: 707–711.

Boulard, T., M. Mermier, J. Fargues, N. Smits, M. Rougier, and J.C. Roy. 2002. Tomato leaf boundary layer climate: implications for microbiological whitefly control in greenhouses. Agricultural and Forest Meteorology 110: 159–176.

Brooks, W.M. 1993. Host-parasitoid-pathogen interactions. In N.E. Beckage, S.N. Thompson, and B.A. Federici, eds., Parasites and Pathogens of Insects. Vol. 2, Pathogens, pp. 231–272. Academic Press, San Diego.

Brownbridge, M, S. Costa, and S.T. Jaronski. 2001. Effects of in vitro passage of *Beauveria bassiana* on virulence to *Bemisia argentifolii*. Journal of Invertebrate Pathology 77: 280–283.

Burges, D., ed., 1998. Formulation of Microbial Biopesticides. Kluwer Academic, Dordrecht, The Netherlands.

Burges, D. 2007. Techniques for testing microbials for control of arthropod pests in greenhouses. In L.A. Lacey and H.K. Kaya, eds., Field Manual of Techniques in Invertebrate Pathology: Application and evaluation of pathogens for control of insects and other invertebrate pests, pp. 463–479. Springer, Dordrecht, The Netherlands.

Butt, T.M., J.G. Harris, and K.A. Powell. 1999. Microbial biopesticides: the European Scene. In F.R. Hall and J.J. Menn, eds., Methods in Biotechnology, Vol. 5, Biopesticides: Use and Delivery, pp. 23–44. Humana Press, Totowa, NJ.

Cantone, F.A. and J.D. Vandenberg. 1998. Intraspecific diversity in *Paecilomyces fumosoroseus*. Mycological Research 102: 209–215.

Cantone, F.A. and J.D. Vandenberg. 1999a. Genetic transformation and mutagenesis of the entomopathogenic fungus *Paecilomyces fumosoroseus*. Journal of Invertebrate Pathology 74: 281–288.

Cantone, F.A. and J.D. Vandenberg. 1999b. Use of the green fluorescent protein for investigations of *Paecilomyces fumosoroseus* in insect hosts. Journal of Invertebrate Pathology 74: 193–197.

Carruthers, R.I., S.P. Wraight, and W.A. Jones. 1993. An overview of biological control of the sweetpotato whitefly, *Bemisia tabaci*. Proceedings of the Beltwide Cotton Conference, January 10–14, 1993, New Orleans, LA, 2: 680–685. National Cotton Council, Memphis, TN.

Castineiras, A. 1995. Natural enemies of *Bemisia tabaci* (Homoptera: Aleyrodidae) in Cuba. Florida Entomologist 78: 538–540.

Chandler, D., J.B. Heale, and A.T. Gillespie. 1993. Competitive interaction between strains of *Verticillium lecanii* on two insects hosts. Annals of Applied Biology 122: 435–440.

Chandler, D., J.B. Heale, and A.T. Gillespie. 1994. Effect of osmotic potential on the germination of conidia and colony growth of *Verticillium lecanii*. Mycological Research 98: 84–388.

Chen, W.W. and M.G. Feng. 1999. Current status in basic and applied research on the entomopathogenic fungus, *Paecilomyces fumosoroseus*. Natural Enemies of Insects 21: 140–144.

Cliquet, S. and M.A. Jackson. 1997. Comparison of air-drying methods for evaluating the desiccation tolerance of liquid culture-produced blastospores of *Paecilomyces fumosoroseus*. World Journal of Microbiology and Biotechnology 13: 299–303.

Cohen, A.C., T.J. Henneberry, and C.C. Chu. 1996. Geometric relationships between whitefly feeding behavior and vascular bundle arrangements. Entomologia Experimentalis et Applicata 78: 135–142.

Costa, H. 1998. Interaction of *Beauveria bassiana* with various fungicides under exposed conditions, 1997. Arthropod Management Tests 23: 349.

Costa, S.D. and R.R. Gaugler. 1989. Sensitivity of *Beauveria bassiana* to solanine and tomatine: Plant defensive chemicals inhibit an insect pathogen. Journal of Chemical Ecology 15: 97–706.

Davidson, E.W., R.B.R. Patron, L.A. Lacey, R. Frutos, A. Vey, and D.L. Hendrix. 1996a. Activity of natural toxins against the silverleaf whitefly, *Bemisia argentifolii*, using a novel feeding bioassay system. Experimentalis et Applicata 79: 25–32.

Davidson, E.W., R.B.R. Patron, A. Vey, R. Frutos, R. St. Leger, L.A. Lacey, and D.L. Hendrix. 1996b. Effect of destruxins from *Metarhizium anisopliae*, ivermectin and *Bacillus thuringiensis* deltaendotoxins on adult silverleaf whitefly, *Bemisia argentifolii*. In Proceedings Beltwide Cotton Conference, Nashville, TN, January 9–12, 1996, 2: 739–740, National Cotton Council, Memphis, TN.

De la Torre, M. and H.M. Cardenas-Cota. 1996. Production of *Paecilomyces fumosoroseus* conidia in submerged culture. Entomophaga 41: 443–453.

Dolinski, C. and L.A. Lacey. 2007. Microbial control of arthropod pests of tropical tree fruits. Neotropical Entomology 36: 161–179.

Drummond, J., J.B. Heale, and A.T. Gillespie. 1987. Germination and effect of reduced humidity on expression of pathogenicity in *Verticillium lecanii* against the glasshouse whitefly *Trialeurodes*. Annals of Applied Biology 111: 193–201.

Ekbom, B.S. 1979. Investigations on the potential of a parasitic fungus (*Verticillium lecanii*) for biological control of the greenhouse whitefly (*Trialeurodes*). Swedish Journal of Agricultural Research 9: 129–138.

Ekbom, B.S. 1981. Humidity requirements and storage of the entomopathogenic fungus *Verticillium lecanii* for use in greenhouses. Annales Entomologici Fennici 47: 61–62.

Evans, H.C. and R.A. Samson. 1986. The genus *Verticillium*: Taxonomic problems in species with invertebrate hosts. Proceedings of the IV International Colloquium of Invertebrate Pathology, pp. 186–189, August 18–22, 1986, Veldhoven, The Netherlands, Society for Invertebrate Pathology, Knoxville, TN.

Eyal, J., M.D.A. Mabud, K.L. Fischbein, J.F. Walter, L.S. Osborne, and Z. Landa. 1994. Assessment of *Beauveria bassiana* Nov. EO-1 strain, which produces a red pigment, for microbial control. Applied Biochemistry Biotechnology 44: 65–80.

Fang, Q.-X., Y.-X. Gong, Y.-K. Zhou, Y.-M. Hu, S.F. Yang, and Y.-Y. Zhou. 1985. Selecting strains of the fungus *Aschersonia*. Acta Microbiologia Sinica 25: 208–213.

Fargues, J., N.K. Maniania, J.C. Delmas, and N. Smits. 1992. Influence de la temperature sur la croissance in vitro d'hyphomycètes entomopathogènes. Agronomie 12: 557–564.

Fargues, J., M.S. Goettel, N. Smits, A. Ouedraogo, C. Vidal, L.A. Lacey, C.J. Lomer, and M. Rougier. 1996. Variability in susceptibility to simulated sunlight of conidia among isolates of entomopathogenic Hyphomycetes. Mycopathologia 135: 171–181.

Fargues, J., M.S. Goettel, N. Smits, A. Ouedraogo, and M. Rougier. 1997a. Effect of temperature on vegetative growth of *Beauveria bassiana* isolates from different origins. Mycologia 89: 383–392.

Fargues, J., M. Rougier, R. Goujet, N. Smits, C. Coustere, and B. Itier. 1997b. Inactivation of conidia of *Paecilomyces fumosoroseus* by near-ultraviolet (UVB and UVA) and visible radiation. Journal of Invertebrate Pathology 69: 70–78.

Fargues, J., M.C. Bon, S. Manguin, and Y. Couteaudier. 2002. Genetic variability among *Paecilomyces fumosoroseus* isolates from various geographical and host insect origins based on the rDNA-ITS regions. Mycology Research 106: 1066–1074.

Faria, M. and S. Wraight. 2001. Biological control of *Bemisia tabaci* with fungi. Crop Protection 20: 767–778.

Fawcett, H.S. 1944. Fungus and bacterial diseases of insects as factors in biological control. Botanical Review 10: 327–348.

Feng, M.G., T.J. Poprawski, and G.G. Khachatourians. 1994. Production, formulation and application of the entomopathogenic fungus *Beauveria bassiana* for insect control: Current status. Biocontrol Science and Technology 4: 3–34.

Fransen, J.J. 1990a. Fungi on aphids, thrips and whitefly in the greenhouse environment. Proceedings of the Vth International Colloquium on Invertebrate Pathology and Microbial Control, pp. 376–379, August 20–24, 1990, Adelaide, Australia, Society for Invertebrate Pathology, Knoxville, TN.

Fransen, J.J. 1990b. Natural enemies of whiteflies: Fungi. In D. Gerling, ed., Whiteflies: Their Bionomics, Pest Status and Management, pp. 187–210. Intercept, Andover, UK.

Fransen, J.J. 1993. Integrated pest management in glasshouse ornamentals in the Netherlands: a step by step policy. International Organization of Biological Control Western Palearctic Regional Section Bulletin. 16: 35–38.

Fransen, J.J. 1995. Survival of spores of the entomopathogenic fungus *Aschersonia aleyrodis* (Deuteromycotina: Coelomycetes) on leaf surfaces. Journal of Invertebrate Pathology 65: 73–75.

Fransen, J.J and J.C. van Lenteren. 1993. Host selection and survival of the parasitoid *Encarsia formosa* on greenhouse whitefly, *Trialeurodes vaporariorum*, in the presence of hosts infected with the fungus *Aschersonia aleyrodis*. Entomologia Experimentalis et Applicata 69: 39–249.

Fransen, J.J. and J.C. van Lenteren. 1994. Survival of the parasitoid *Encarsia formosa* after treatment of parasitized greenhouse whitefly larvae with fungal spores of *Aschersonia aleyrodis*. Entomologia Experimentalis et Applicata 71: 235–243.

Fransen, J.J., K. Winkelman, and J.C. van Lenteren. 1987. The differential mortality at various life stages of the greenhouse whitefly, *Trialeurodes* (Homoptera: Aleyrodidae), by infection with the fungus *Aschersonia aleyrodis* (Deuteromycotina: Coelomycetes). Journal of Invertebrate Pathology 50: 158–165.

Galani, G. 1979. Studies on the variation of the pathogenicity of *Verticillium lecanii* (Zimm.) Viegas to the larvae of *Trialeurodes vaporariorum* Westw. Analele Instituteului de Cercetari de Protectia Plantelor 25: 243–245.

Galani, G. and L. Almasan. 1984. Depistarea unei epizootii naturale micotice la musculita alba, *Trialeurodes vaporariorum* Westw. Buletinul de Protectia Plantelor 1: 33–36.

Gao, R., Z. Ouyang, Z. Gao, and J. Zheng. 1985. A preliminary report on the application of *Aschersonia aleyrodis* for the control of citrus whitefly. Chinese Journal of Biological Control 1: 45–46.

Garza Gonzalez, E. 1993. Enemigos naturales de accion microbial de mosquita blanca. In L.A. Rodríguez del Bosque, ed., II Taller sobre control biologico de mosquita blanca, pp. 54–66. SARH. Sinaloa, Mexico.

Garza Gonzalez, E. and H.C. Arredondo Bernal. 1993. Sensibilidad in vitro de adultos de mosquito blanca de las hortalizas *Bemisia tabaci* (Genn.) (Homoptera: Aleyrodidae) a diferentes aislamientos de *Beauveria bassiana* (Bals.) *Paecilomyces* spp. y *Verticillium lecanii* (Vuill.). Proceedings of the XVI Congreso Nacional de Control Biologico Sociedad Mexicana de Control Biologico, October 7–8, 1993, Monterrey, N.L., Mexico. Sociedad Mexicana de Control Biologico, Mexico City.

Gindin, G., I. Barash, N. Harari, and B. Raccah. 1994. Effect of endotoxic compounds isolated from *Verticillium lecanii* on the sweetpotato whitefly, *Bemisia tabaci*. Phytoparasitica 22: 189–196.

Gindin, G., N.U. Geschtovt, B. Raccah, and I. Barash. 2000. Pathogenicity of *Verticillium lecanii* to different developmental stages of the silverleaf whitefly, *Bemisia argentifolii*. Phytoparasitica 28: 229–239.

Gindin, G. and I.S. Ben-Ze'ev. 1994a. Virulence and persistence of *Conidiobolus coronatus* and *Conidiobolus* sp. in glasshouse populations of *Bemisia tabaci*. International Organization of Biological Control Western Palearctic Regional Section Bulletin 17: 185–188.

Gindin, G. and I.S. Ben-Ze'ev. 1994b. Natural occurrence of and inoculation experiments with *Conidiobolus coronatus* and *Conidiobolus* sp. in glasshouse populations of *Bemisia tabaci*. Phytoparasitica 22: 197–208.

Goettel, M.S. and G.D. Inglis. 1997. Fungi: Hyphomycetes. In L. A. Lacey, ed., Manual of Techniques in Insect Pathology, pp. 213–249. Academic Press, London.

Goettel, M.S., T.J. Poprawski, J.D. Vandenberg, Z. Li, and D.W. Roberts. 1990. Safety to nontarget invertebrates of fungal biocontrol agents. In M. Laird, L.A. Lacey and E.W. Davidson, eds., Safety of Microbial Insecticides, pp. 209–231. CRC Press, Boca Raton.

Grajek, W. 1994. Sporogenesis of the entomopathogenic fungus *Verticillium lecanii* in solid-state cultures. Folia Microbiologica 39: 29–32.

Hall, R.A. 1980. Effect of repeated subculturing on agar and passaging through an insect host on pathogenicity, morphology, and growth rate of *Verticillium lecanii*. Journal of Invertebrate Pathology 36: 216–222.

Hall, R.A. 1981. The fungus *Verticillium lecanii* as a microbial insecticide against aphids and scales. In H.D. Burges, ed., Microbial Control of Pests and Plant Diseases 1970–1980, pp. 483–498. Academic Press, London.

Hall, R.A. 1982. Control of whitefly, *Trialeurodes* and cotton aphid, *Aphis gossypii* in glasshouses by two isolates of the fungus, *Verticillium lecanii*. Annals of Applied Biology 10: 1–11.

Hall, R.A. 1985. Whitefly control by fungi. In N.W. Hussey and N.E.A. Scopes, eds., Biological Pest Control – The Glasshouse Experience, pp. 116–118. Cornell University Press, Ithaca, NY.

Hall, R.A. and J.P. Latgé. 1980. Etude de quelques facteurs stimulant la formation in vitro des blastospores de *Verticillium lecanii* (Zimm.) Viegas. Comptes Rendus l'Académie de Séances, Paris 291: 75–78.

Hall, R.A., D. Peterkin, and B. Ali. 1994a. Fungal control of whitefly, *Thrips palmi* and sugarcane froghopper in Trinidad and Tobago. In Proceedings of the VIth International Colloquium of Invertebrate Pathology and Microbial Control, pp. 277–282, August 28–September 2, 1994, Montpellier, France, Society for Invertebrate Pathology, Knoxville, TN.

Hall, R.A., D. Peterkin, and G.V. Pollard. 1994b. Influence of culture age on rate of fungal spore germination in four entomogenous fungi. Mycology Research 98: 763–768.

Hajek, A.E., S.P. Wraight, and J.D. Vandenberg. 2001. Control of arthropods using pathogenic fungi. In S.B. Pointing and K.D. Hyde, eds., Bio-Exploitation of Filamentous Fungi, pp. 309–347. Fungal Diversity Press, Hong Kong.

Herrera, F., M. Carballo, and P. Shannon. 1999. Eficacia de cepas nativas de hongos entomopatogenos sobre *Bemisia tabaci*, en el laboratorio. Manejo Integrado de Plagas. 54: 37–43.

Hirte, W.F., I. Glathe, and H. Adam. 1989a. Produktion und Applikation eines mikrobiellen Praeparates auf der Basis von *Aschersonia* -Sporen. Zentralblatt fur Mikrobiologie 144: 151–162.

Hirte, W.F., C. Walter, M. Grünberg, H. Sermann, and H. Adam. 1989b. Selektion von Pathotypen von *Verticillium lecanii* für verschiedene tierisch Schaderreger in Gewächshauskulturen und Aspekte der biotechnologischen Massensporenproduktion. Zentralblatt fur Mikrobiologie 144: 405–420.

Hristova, E. 1971. Possibilities of using *Encarsia formosa* Gahan and *Aschersonia aleyrodis* Webber in the control of *Trialeurodes vaporariorum* Westwood. Acta Horticulturae 17: 396–399.

Humber, R.A. 1997a. Fungi: Identification. In L.A. Lacey, ed., Manual of Techniques in Insect Pathology, pp. 153–185. Academic Press, London.

Humber, R.A. 1997b. Fungi: Preservation of cultures. In L. A. Lacey, ed., Manual of Techniques in Insect Pathology, pp. 269–279. Academic Press, London.

Humber, R.A. 2002. Collection of entomopathogenic fungal cultures. Catalogue of isolates. United States Department of Agriculture, Agriculture Research Service Publication, ARS-110. ARSEF webpage: http://arsef.fpsnl.cornell.edu

Hussey, N.W. 1958. Notes on a fungus parasitic on greenhouse whitefly. Plant Pathology 7: 71–72.

Ibrahim, Y.B., T.K. Lim, M.K. Tang, and H.M. Teng. 1993. Influence of temperature, pH and selected growth media on germination, growth and sporulation of *Aschersonia placenta* and *Hypocrella raciborskii*. Biocontrol Science and Technology 3: 55–61.

Inglis, G.D., D.L. Johnson, and M.S. Goettel. 1996. Effects of temperature and thermo regulation on mycosis by *Beauveria bassiana* in grasshoppers. Biological Control 7: 131–139.

Jackson, M.A., M.R. McGuire, L.A. Lacey, and S.P. Wraight. 1997. Liquid culture production of desiccation tolerant blastospores of the bioinsecticidal fungus *Paecilomyces fumosoroseus*. Mycology Research 101: 35–41.

James, R.R. 2001. Effects of exogenous nutrients on conidial germination and virulence against the silverleaf whitefly for two Hyphomycetes. Journal of Invertebrate Pathology 77: 99–107.

James, R.R. 2003. Combining azadirachtin and *Paecilomyces fumosoroseus* (Deuteromycotina: Hyphomycetes) to control *Bemisia argentifolii* (Homoptera: Aleyrodidae). Journal of Economic Entomology 96: 25–30.

James, R.R. and G.W. Elzen. 2001. Antagonism between *Beauveria bassiana* and imidacloprid when combined for *Bemisia argentifolii* (Homoptera: Aleyrodidae) control. Journal of Economic Entomology 94: 357–361.

Jaronski, S.T. and J.C. Lord. 1996. Evaluation of *Beauveria bassiana* (Mycotrol WP) for control of whitefly in spring cantaloupes, 1995. Arthropod Management Tests 21: 103.

Jaronski, S.T., J.C. Lord, and R. Paden. 1996. Evaluation of *Beauveria bassiana* (Mycotrol® WP) with pyrethroids for control of whitefly in spring cantaloupes. Arthropod Management Tests 21: 102–103.

Jaronski, S.T., J. Lord, J. Rosinska, C. Bradley, K. Hoelmer, G. Simmons, R. Osterlind, C. Brown, R. Staten, and L. Antilla. 1998. Effect of a *Beauveria bassiana*-based mycoinsecticide on beneficial insects under field conditions. Brighton Crop Protection Conference: Pests & Diseases – 1998,

Vol. 2. Proceedings of an International Conference, Brighton, UK, 16–19 November 1998, pp. 651–656. British Crop Protection Council, Thornton Heath, UK.

Jaros-Su, J., E. Groden, and J. Zhang. 1999. Effects of selected fungicides and the timing of fungicide application of *Beauveria bassiana*-induced mortality of the Colorado potato beetle (Coleoptera: Chrysomelidae). Biological Control 15: 259–269.

Jones, W.A. and T.J. Poprawski. 1996. *Bemisia argentifolii* parasitized by *Eretmocerus* sp. is immune to infection by *Beauveria bassiana*. In T.J. Henneberry, N.C. Toscano, R.M. Faust, and J.R. Coppedge, eds., Silverleaf Whitefly (Formerly Sweetpotato Whitefly, Strain B): 1996 Supplement to the 5-Year National Research and Action Plan, p. 118. United States Department of Agriculture, Agricultural Research Service, 1996–01.

Kaya, H.K. and L.A. Lacey. 2007. Introduction to microbial control. In L.A. Lacey and H.K. Kaya, eds., Field Manual of Techniques in Invertebrate Pathology: Application and Evaluation of Pathogens for Control of Insects and Other Invertebrate Pests, pp. 3–7. Springer, Dordrecht, The Netherlands.

Kirk, A.A. and L.A. Lacey. 1996. A systematic approach to foreign exploration for natural enemies of *Bemisia*. In D. Gerling and R. Mayer, eds., *Bemisia* 1995: Taxonomy, Biology, Damage, Control and Management, pp. 531–533. Intercept, Andover, UK.

Kirk, A.A., L.A. Lacey, and J. Goolsby. 2001. Classical biological control of *Bemisia* and successful integration of management strategies in the United States. Pages 309–329 in: Virus-Insect-Plant Interactions. K.F. Harris, O.P. Smith, and J.E. Duffus, eds. Academic Press. San Diego, CA.

Kirk, A.A., L.A. Lacey, N. Roditakis, and J.K. Brown. 1993. The status of *Bemisia tabaci* (Hom.: Aleyrodidae), *Trialeurodes vaporariorum* (Hom.: Aleyrodidae) and their natural enemies in Crete. Entomophaga 38: 405–410.

Kitazawa, K., I. Fujisawa, and S. Imabayashi. 1984. Isolation of *Verticillium lecanii* (Zimm.) Viegas affecting aphids and greenhouse whitefly in Japan. Annals of the Phytopathological Society of Japan 50: 574–581.

Knowles, T.C. and S.T. Jaronski. 1997. Control of silverleaf whitefly in fall cantaloupe with *Beauveria bassiana*, 1995. Arthropod Management Tests 22: 115.

Kogan, V.S. and A.A. Serjapin. 1978. Against the greenhouse whitefly: *Aschersonia*. Zashchita Rastenii 1: 29–31.

Kurogi, S., F. Kuroki, Y. Kawasaki, and K. Nonaka. 1993. Studies on a fungus, *Beauveria bassiana*, isolated from *Thrips parmi* [*T. palmi*] Karny. 1. Pathogenicity to *Thrips parmi* and *Bemisia tabaci* and effect of pesticides (including Dichlorvos fenobucarb and Methidathion) on hyphal growth. Proceedings of the Association for Plant Protection, Kyushu 39: 111–113.

Labanowski, G.S. and G. Soika. 1999. Effectiveness of microbial and botanical insecticides in the control of *Bemisia tabaci* and *Frankliniella occidentalis* on ornamental plants. Bulletin OEPP 29(1–2): 77–80.

Lacey, L. A. and W. M. Brooks. 1997. Initial handling and diagnosis of diseased insects. In L.A. Lacey, ed., Manual of Techniques in Insect Pathology, pp. 1–15. Academic Press, London.

Lacey, L.A. and G. Mercadier. 1998. The effect of selected allelochemicals on germination of conidia and blastospores and mycelial growth of the entomopathogenic fungus, *Paecilomyces fumosoroseus* (Deuteromycotina: Hyphomycetes). Mycopathologia 142: 17–25.

Lacey, L.A. and A.L. Mesquita. 2002. Interaction of entomopathogenic fungi, insect parasitoids and their hosts. Proceedings VIIIth International Colloquium on Invertebrate Pathology and Microbial Control, August 18–23, 2002, Foz do Iguaçu, Brazil, pp. 31–35, Society for Invertebrate Pathology, Knoxville, TN.

Lacey, L.A., A.A. Kirk, and R.D. Hennessey. 1993. Foreign exploration for natural enemies of *Bemisia tabaci* and implementation in integrated control programs in the United States. Proceedings ANPP International Conference on Pests in Agriculture. 1: 351–360. December 7–9, 1993, Montpellier, France, Association Nacionale de Protection des Plantes, Paris, France.

Lacey, L.A., J.J. Fransen, and R. Carruthers. 1996. Global distribution of naturally occurring fungi of *Bemisia*, their biologies and use as biological control agents. In D. Gerling and R. Mayer, eds., *Bemisia* 1995: Taxonomy, Biology, Damage, and Management, pp. 401–433. Intercept, Andover, UK.

Lacey, L.A., A.L.M. Mesquita, G. Mercadier, R. Debire, D.J. Kazmer, and F. Leclant. 1997. Acute and sublethal activity of the entomopathogenic fungus, *Paecilomyces fumosoroseus* (Deuteromycotina: Hyphomycetes) on adult *Aphelinus asychis* (Hymenoptera: Aphelinidae). Environmental Entomology 26: 1452–1460.

Lacey, L.A., A.A. Kirk, L. Millar, G. Mercadier, and C. Vidal. 1999. Ovicidal and larvicidal activity of conidia and blastospores of *Paecilomyces fumosoroseus* (Deuteromycotina: Hyphomycetes) against *Bemisia argentifolii* (Homoptera: Aleyrodidae): Description of a bioassay system allowing prolonged survival of control insects. Biocontrol Science and Technology 9: 9–18.

Lacey, L.A., R. Frutos, H.K. Kaya, and P. Vail. 2001. Insect pathogens as biological control agents: Do they have a future? Biological Control 21: 230–248.

Landa, Z. 1984. Ochrana proti molici sklenikove (*Trialeurodes vaporarium* Westw.) V Programech integrovane ochrany sklenikovych okurek. Sbornik UVTIZ, Zahradnictvi 11: 215–228.

Latgé, J.P. and B. Papierok. 1988. Aphid pathogens. In A.K. Minks and P. Harrewijn, eds., Aphids Their Biology, Natural Enemies and Control, Vol. B, pp. 323–335. Elsevier Science Publishers B. V., Amsterdam.

Li, G., Y. Yan, and L. Wang. 1991. Influence of temperature and nutrition on the growth of an entomopathogenic fungus, *Verticillium lecanii* (Beijing strain). Chinese Journal of Biological Control 7: 115–119.

Liu, T.-X. and C.W. Meister. 2001. Managing *Bemisia argentifolii* on spring melons with insect growth regulators, entomopathogens and imidacloprid in South Texas. Subtropical Plant Science: Journal of the Rio Grande Valley Horticultural Society 53: 44–48.

Liu, T.-X, P.A. Stansly, A.N. Sparks, Jr., T.C. Knowles, and C.C. Chu. 1999. Application of Mycotrol and Naturalis-L (*Beauveria bassiana)* for management of *Bemisia argentifolii* (Homoptera, Aleyrodidae) on vegetables, cotton and ornamentals in the southern United States. Subtropical Plant Science: Journal of the Rio Grande Valley Horticultural Society 51: 41–48.

Long, D.W., E. Groden, and F.A. Drummond. 2000. Horizontal transmission of *Beauveria bassiana* (Bals.) Vuill. Agricultural Forest and Entomology 2: 11–17.

Lopez-Avila, A., C. Cardona Mejia, J. Garcia Gonzalez, F. Rendon, and P. Hernandez. 2001. Reconocimiento e identificacion de enemigos naturales de moscas blancas (Homoptera: Aleyrodidae) en Colombia y Ecuador. Revista Colombiana de Entomologia 27: 137–141.

Lourenção, A.L., V.A. Yuki, and S.B. Alves. 1999. Epizootia de *Aschersonia* cf. *goldiana* em *Bemisia tabaci* (Homoptera: Aleyrodidae) Biotipo B no Estado de São Paulo. Anais da Sociedade Entomologica do Brasil 28: 343–345.

Lourenção, A.L., M.A.C. Miranda, and S.B. Alves. 2001. Ocorrência epizootica de *Verticillium lecanii* em *Bemisia tabaci* biotipo B (Hemiptera: Aleyrodidae) no Estado do Maranhão. Neotropical Entomology 30: 183–185.

Mains, E.B. 1959. North American species of *Aschersonia* parasitic on Aleyrodidae. Journal of Invertebrate Pathology 1: 43–47.

Malsam, O., M. Kilian, E.-C. Oerke, and H.-W. Dehne. 2002. Oils for increased efficacy of *Metarhizium anisopliae* to control whiteflies. Biocontrol Science and Technology 12: 337–348.

Masuda, T. and O. Kikuchi. 1992. Pathogenicity of *Verticillium lecanii* isolates to whitefly and aphids. Japanese Journal of Applied Entomology and Zoology 36: 239–245.

Masuda, T. and M. Maeda. 1989. Pathogenicity of a fungus *Verticillium lecanii* to the whitefly, *Trialeurodes vaporariorum*, in various stages. Japanese Journal of Applied Entomology and Zoology 33: 101–104.

McCoy, C.W., R.A. Samson, and D.G. Boucias. 1988. Entomogenous fungi. In C.M. Ignoffo and N.B. Mandava, eds., Handbook of Natural Pesticides, Vol. V, Microbial Insecticides, Part A: Entomogenous Protozoa and Fungi, pp. 151–236. CRC Press. Boca Raton, FL.

Meade, D.L. and D.N. Byrne. 1991. The use of *Verticillium lecanii* against subimaginal instars of *Bemisia tabaci.* Journal of Invertebrate Pathology 57: 296–298.

Meekes, E.T.M., S. Van Voorst, N.N. Joosten, J.J. Fransen, and J.C. van Lenteren. 2000. Persistence of the fungal whitefly pathogen, *Aschersonia aleyrodis*, on three different plant species. Mycology Research 104: 234–1240.

Meekes, E.T.M.., J.J. Fransen, and J.C. van Lenteren. 2002. Pathogenicity of *Aschersonia* spp. against whiteflies *Bemisia argentifolii* and *Trialeurodes vaporariorum*. Journal of Invertebrate Pathology 81: 1–11.

Mellín Rosas, M.A. and E. Garza Gonzalez. 1993. Sensibilidad in vitro de estados ninfales de mosquita blanca *Bemisia tabaci* (Homoptera: Aleyrodidae) a 12 aislamientos de *Paecilomyces* spp. Proceedings of the XVI Congreso Nacional de Control Biologico Sociedad Mexicana de Control Biologico, Monterrey, NL, Mexico. October 7–8, 1993. Sociedad Mexicana de Control Biologico, Mexico City.

Mesquita, A.L.M., L.A. Lacey, C.S. Ceianu, and R. Debire. 1999. Predatory and parasitic activity of *Aphelinus asychis* (Hymenoptera: Aphelinidae) following exposure to the entomopathogenic fungus *Paecilomyces fumosoroseus* (Deuteromycotina: Hyphomycetes) under different humidity regimes. Anais da Sociedade Entomologica do Brasil 28: 661–673.

Milner, R.J. and G.G. Lutton. 1986. Dependence of *Verticillium lecanii* (Fungi: Hyphomycetes) on high humidities for infection and sporulation using *Myzus persicae* (Homoptera: Aphididae) as host. Environmental Entomology 15: 380–382.

Mohamed, M.I. and R.V. Varma. 1992. Fungal pathogens of *Eligma narcissus indica* Roth (Lepidoptera: Noctuidae), the *Ailanthus* defoliator, with special reference to *Paecilomyces farinosus* and *P. fumosoroseus*. Phytophaga 4: 101–109.

Mor, H., G. Gindin, I.S. Ben-Ze'ev, B. Raccah, N.U. Geschtovt, N. Ajtkhozhina, and I. Barash. 1996. Diversity among isolates of *Verticillium lecanii* as expressed by DNA polymorphism and virulence towards *Bemisia tabaci*. Phytoparasitica 24: 111–118.

Negasi, A., B.L. Parker, and M. Brownbridge. 1998. Screening and bioassay of entomopathogenic fungi for the control of silverleaf whitefly, *Bemisia argentifolii*. Insect Science and its Application 18: 37–44.

Nene, Y.L. 1973. Note on a fungus parasite of *Bemisia tabaci* Genn., a vector of several plant viruses. Indian Journal of Agricultural Science 43: 514–516.

Nier, T., F. Rivera, J.C. Bermudez, Y. Dominguez, C. Benavides, and M. Ulloa. 1991. Primer reporte en Mexico del aislamento de *Verticillium lecanii* a partir de la mosquito blanca e pruebas de patogenicidad in vitro sobre este insecto. Revista Mexicana de Micologia 7: 149–156.

Obornik, M., R. Stouthamer, E. Meekes, and M. Schilthuittzen. 1999. Molecular characterization and phylogeny of the entomopathogenic fungus *Aschersonia* spp. Plant Protection Science 35: 1–9.

Oho, N. 1967. Possible utilization of *Aschersonia aleyrodis* for control of citrus whitefly, *Dialeurodes citri* Ashmead. Proceedings Joint US-Japan Seminar on Microbial Control of Insect Pests, Fukuoka, April 21–23, pp. 119–128.

Ohta, M., A. Ozawa, and H. Kobayashi. 1998. The efficacy of a *Paecilomyces fumosoroseus* preparation against whitefly on tomato in the greenhouse [in Japan]. Proceedings of the Kanto-Tosan Plant Protection Society 45: 181–184.

Olson, D.L. and R.D. Oetting. 1999. The efficacy of mycoinsecticides of *Beauveria bassiana* against silverleaf whitefly (Homoptera: Aleyrodidae) on poinsettia. Journal of Agricultural and Urban Entomology 16: 179–185.

Orozco-Santos, M., J. Farias-Larios, J. Lopez-Perez, and N.R. Ramirez-Vazquez. 2000. Uso de *Beauveria bassiana* para el control de *Bemisia argentifolii* en melon. Manejo Integrado de Plagas, 56: 45–51.

Osborne, L.S. and Z. Landa. 1992. Biological control of whiteflies with entomopathogenic fungi. Florida Entomologist 75: 456–471.

Osborne, L.S. and Z. Landa. 1994. Utilization of the entomogenous fungus *Paecilomyces fumosoroseus* against sweetpotato whitefly, *Bemisia tabaci*. International Organization of Biological Control Western Palearctic Regional Section Bulletin 17: 201–206.

Osborne L.S., K. Hoelmer, and D. Gerling. 1990a. Prospects for biological control of *Bemisia tabaci*. International Organization of Biological Control Western Palearctic Regional Section Bulletin 13: 153–160.

Osborne, L.S., G.K. Storey, C.W. McCoy, and J.F. Walter. 1990b. Potential for controlling the sweetpotato whitefly, *Bemisia tabaci*, with the fungus, *Paecilomyces fumosoroseus*. Proceedings of the Vth International Colloquium on Invertebrate Pathology and Microbial Control. pp. 386–390. August 20–24, 1990, Adelaide, Australia, Society for Invertebrate Pathology, Knoxville, TN.

Ota, M., A. Ozawa, and H. Kobayashi. 1999. Efficacy of *Beauveria bassiana* preparation against whitefly on tomato. Annual Report of the Kanto-Tosan Plant Protection Society, 46: 109–112.

Petch, T. 1921. Studies on entomogenous fungi, II: The genera *Hypocrella* and *Aschersonia*. Annals of the Royal Botanical Garden of Peradeniya 7: 167–278.

Petch, T. 1925. Entomogenous fungi: Additions and corrections. Transactions of the British Mycological Society 10: 190–201.

Peterkin, D. and R.A. Hall. 1994. The effect of accelerated spore germination on virulence of *Paecilomyces fumosoroseus*. Proceedings of the VIth International Colloquium on Invertebrate Pathology and Microbial Control. 2: 128. August 28–September 2, 1994, Montpellier, France, Society for Invertebrate Pathology, Knoxville, TN.

Ponomarenko, N.G., N.A. Prilepskaja, M. Murvanidze, and L.A. Stoljarova. 1975. *Aschersonia* against whitefly. Zashchita Rastenii 6: 44–45.

Poprawski, T.J. 1999. Control of *Bemisia argentifolii* on collards using different formulations and rates of *Beauveria bassiana*, 1997. Arthropod Management Tests 24: 122–123.

Poprawski, T.J. and L.A. Lacey. 2000. European Parasite Laboratory - European Biological Control Laboratory, Sèvres, Béhoust, and Montpellier, France. In J.R. Coulson, P.V. Vail, M.E. Dix, D.A. Nordlund and W.C. Kauffman, eds., 110 Years of Biological Control Research and Development in the United States Department of Agriculture: 1883–1993, pp. 329–331. United States Department of Agriculture, Agricultural Research Service.

Poprawski, T.J. and W.J. Jones. 2001. Host plant effects on activity of the mitosporic fungi *Beauveria bassiana* and *Paecilomyces fumosoroseus* against two populations of *Bemisia* whiteflies (Homoptera: Aleyrodidae). Mycopathologia 151: 11–20.

Poprawski, T.J., P.H. Robert, I. Majchrowicz, and G. Boivin. 1985. Susceptibility of *Delia antiqua* (Diptera: Anthomyiidae) to eleven isolates of entomopathogenic Hyphomycetes. Environmental Entomology 14: 557–561.

Poprawski, T.J., J.C. Legaspi, and P.E. Parker. 1998. Influence of entomopathogenic fungi on *Serangium parcesetosum* (Coleoptera: Coccinellidae), an important predator of whiteflies (Homoptera: Aleyrodidae). Environmental Entomology 27: 785–795.

Poprawski, T.J., S.M. Greenberg, and M.A. Ciomperlik. 2000. Effect of host plant on *Beauveria bassiana*- and *Paecilomyces fumosoroseus*-induced mortality of *Trialeurodes vaporariorum* (Homoptera: Aleyrodidae). Environmental Entomology 29: 1048–1053.

Ramakers, P.M.J. and R.A. Samson. 1984. *Aschersonia aleyrodis*, a fungal pathogen of whitefly. II. Application as a biological insecticide in glasshouses. Zeitschrift fuer Angewandte Entomologie 97: 1–8.

Ramos, E.Q., S.B. Alves, M.R. Tanzini, and R.B. Lopes. 2000. Susceptibilidade de *Bemisia tabaci* a *Beauveria bassiana* en condiciones de laboratorio. Manejo Integrado de Plagas 56: 65–69.

Rao, N.V. and A.S. Reddy. 1992. The natural enemies of cotton whitefly, *Bemisia tabaci* Gennadius - a review. Agriculture Review 13: 12–20.

Rao, N.V., A.S. Reddy, and K.T. Rao. 1989. Natural enemies of cotton whitefly, *Bemisia tabaci* Gennadius in relation to host population and weather factors. Journal of Biological Control 3: 10–12.

Ravensberg, W.J., M. Malais, and D.A. van der Schaaf. 1990a. Applications of *Verticillium lecanii* in tomatoes and cucumbers to control whitefly and thrips. International Organization of Biological Control Western Palearctic Regional Section Bulletin/Bulletin OILB SROP 13: 173–178.

Ravensberg, W.J., M. Malais, and D.A. van der Schaaf. 1990b. *Verticillium lecanii* as a microbial insecticide against glasshouse whitefly. Proceedings of the Brighton Crop Protection Conference 1: 265–268. November 19–22, 1990, Brighton, UK. British Crop Protection Council, Thornton Heath, UK.

Riba, G. and L. Entcheva. 1984. Influence de l'hygrométrie ambiante sur l'aggressivité comparée de plusieurs Hyphomycetes entomopathogénes a l'égard de l'aleurode des serres, *Trialeurodes vaporariorum* (Westw.). Comptes Rendus Séances l'Académie d'Agriculture, France 70: 521–526.

Ruiz-Vega, J. and T. Aquino-Bolaños. 1999. Manejo de *Bemisia tabaci* mediante barreras vivas y *Paecilomyces* en Oaxaca, Mexico. Manejo Integrado de Plagas 52: 68–73.

Ruiz V.J. and Z.J. Medina. 2001. Avances en el manejo integrado de *Bemisia tabaci* en tomate y chile en Oaxaca, Mexico. Manejo Integrado de Plagas 59: 34–40.

Rushtapakornchai, W. and P. Petchwichit. 1996. Efficiency of some insecticides for controlling tobacco whitefly *Bemisia tabaci* and leaf miner *Liriomyza trifolii* on tomato. Kaen Kaset Khon Kaen Agriculture Journal 24: 184–189.

Ryckewaert, P. and C. Alauzet. 2002. The natural enemies of *Bemisia argentifolii* in Martinique. BioControl 47: 115–126.

Saito, T. 1992. Control of *Thrips palmi* and *Bemisia tabaci* by a mycoinsecticidal preparation of *Verticillium lecanii*. Proceedings Kanto-Tosan Plant Protection Society 39: 209–210.

Saito, T. 1993. Control of two whitefly species, *Bemisia tabaci* and *Trialeurodes vaporariorum*, by a preparation of *Verticillium lecanii*. Proceedings of the Kanto Tosan Plant Protection Society 40: 221–222.

Sakamoto M., Y. Inove, and J. Aoki. 1985. Effect of light on the conidiation of *Paecilomyces fumosoroseus*. Transactions of the Mycological Society of Japan 26: 499–509.

Samson, R.A. 1974. *Paecilomyces* and some allied Hyphomycetes. Studies in Mycology 6: 31–41.

Samson, R.A. and M.C. Rombach. 1985. Biology of the fungi *Verticillium* and *Aschersonia*. In N.W. Hussey and N.E.A. Scopes, eds., Biological Pest Control – The Glasshouse Experience, pp. 34–42. Cornell University Press. Ithaca, NY.

Sandoval-Coronado, C.F., H.A. Luna-Olvera, K. Arevalo-Nino, M.A. Jackson, T.J. Poprawski, and L.J. Galan-Wong. 2001. Drying and formulation of blastospores of *Paecilomyces fumosoroseus* (Hyphomycetes) produced in two different liquid media. World Journal of Microbiology and Biotechnology 17: 423–428.

Shipp, J.L., Y. Zhang, D.W.A. Hunt, and G. Ferguson. 2003. Influence of humidity and greenhouse microclimate on the efficacy of *Beauveria bassiana* (Balsamo) for control of greenhouse arthropod pests. Enviromental Entomology 32: 1154–1163.

Silvie, P. and B. Papierok. 1991. Les ennemis naturels d'insectes du cotonnier au Tchad: premières données sur les champignons de l'ordre des Entomophthorales. Coton et Fibres Tropicales 46: 293–308.

Smith, P. 1993. Control of *Bemisia tabaci* and the potential of *Paecilomyces fumosoroseus* as a biopesticide. Biocontrol News and Information 14: 71N–78N.

Smith, P.A. 1997. Oil and emulsion formulations of a microbial control agent increase the potency against a wider range of pest life stages. Phytoparasitica 25: 93S–100S.

Smits, N., M. Rougier, J. Fargues, R. Goujet, and R. Bonhomme. 1996. Inactivation of *Paecilomyces fumosoroseus* conidia by diffuse and total solar radiation. FEMS Microbiology Ecology 21: 167–173.

Smits, N., J. Fargues, and M. Rougier. 1997. Modelling the persistence of quiescent conidia of the entomopathogenic hyphomycete *Paecilomyces fumosoroseus* exposed to solar radiation. Biocontrol Science and Technology 7: 365–375.

Sosa-Gomez, D.R., F. Moscardi, and M. Santos. 1997. *Bemisia* spp. na cultura da soja: ocorrência, controle quimico e incidência de fungo entomopatogênico *Paecilomyces* spp. 16th Congresso Brasileiro de Entomologia, Resumos, p. 144. SEB/Embrapa-CNPMF, Salvador, Brazil. March, 2 to 7 of 1997, Salvador, Brazil.

Spassova, P. 1980. Investigation on the entomopathogenic fungus *Aschersonia* sp. Horticultural and Viticultural Science 17: 77–84.

Spassova, P., E. Hristova, and E. S. Elenkov. 1980. Pathogenicity of various species of fungi of the genus *Aschersonia* to the larvae of the greenhouse whitefly (*Trialeurodes vaporariorum* Westw.) on tomatoes and cucumbers. Horticulture Viticulture Science 17: 70–75.

Steenberg, T. and R.A. Humber. 1999. Entomopathogenic potential of *Verticillium* and *Acremonium* species (Deuteromycotina: Hyphomycetes). Journal of Invertebrate Pathology 73: 309–314.

Steinberg, S. and H. Prag. 1994. Efficacy of the fungus *Aschersonia aleyrodis* and the coccinellid predator *Delphastus pusillus*, used to control *Bemisia tabaci* in greenhouse cucumber. *Bemisia* Newsletter 8: 33.

Sterk, G., K. Bolckmans, R. de Jonghe, L. de Wael, and J. Vermeulen. 1995a. Side-effects of the microbial insecticide PreFeRal WG (*Paecilomyces fumosoroseus*, strain Apopka 97) on *Bombus terrestris*. Mededelingen - Faculteit Landbouwkundige en Toegepaste Biologische Wetenschappen, Universiteit Gent 60: 713–717.

Sterk, G., K. Bolckmans, M. van de Veire, B. Sels, and W. Stepman. 1995b. Side-effects of the microbial insecticide PreFeRal (*Paecilomyces fumosoroseus*, strain Apopka 97) on different species of beneficial arthropods. Mededelingen – Faculteit Landbouwkundige en Toegepaste Biologische Wetenschappen, Universiteit Gent 60: 719–724.

Sterk, G., K. Bolckmans, and J. Eyal. 1996. A new microbial insecticide, *Paecilomyces fumosoroseus* strain Apopka 97, for the control of the greenhouse whitefly. Brighton Crop Protection Conference: Pests & Diseases, Vol. 2. Proceedings of an International Conference, Brighton, UK, November 18–21, 1996, pp. 461–466. British Crop Protection Council, Thornton Heath, UK.

Sütterlin, S., G.J. Van der Mey, and J.C. Van Lenteren. 1991. Distribution in space and time of *Trialeurodes vaporariorum* (Westwood) on gerbera: Does host plant architecture influence the dispersal and distribution of the whitefly? Proceedings of the Experimental and Applied Entomology Section, Netherlands Entomological Society 2: 115–120.

Tigano-Milani, M.S., M.R. de Faria, I. Martins, and R.E. Lecuona. 1992. Ocorrência natural de *Beauveria bassiana* (Bals.) Vuill., *Metarhizium anisopliae* (Metch.) Sorok. e *Paecilomyces* sp. em solos de diferentes regiões do Brasil. Anais da Sociedade Entomologica do Brasil 22: 391–393.

Tigano-Milani, M.S., R.J. Honeycutt, L.A. Lacey, R. Assis, M. McClelland, and B.W.S. Sobral. 1995. Genetic variability of *Paecilomyces fumosoroseus* strains revealed by molecular markers. Journal of Invertebrate Pathology 65: 274–282.

Vandenberg, J.D., A.M. Shelton, and S.P. Wraight. 2007. Application and evaluation of entomopathogens in crucifers and cucurbits. In L.A. Lacey and H.K. Kaya, eds., Field Manual of Techniques in Invertebrate Pathology: Application and Evaluation of Pathogens for Control of Insects and Other Invertebrate Pests, pp. 361–374. Springer, Dordrecht, The Netherlands.

van der Schaaf, D.A., W.J. Ravensberg, and M. Malais. 1991. *Verticillium lecanii* as a microbial insecticide against whitefly. International Organization of Biological Control Western Palearctic Regional Section Bulletin 14: 120–123.

van Lenteren, J.C. and L. P.J.J. Noldus. 1990. Whitefly-plant relationships: behavioural and ecological aspects. In D. Gerling, ed., Whiteflies: Their Bionomics, Pest Status and Management, pp. 47–89. Intercept, Andover, UK.

Vega, F.E., P.F. Dowd, M.R. McGuire, M.A. Jackson, and T.C. Nelsen. 1997. *In vitro* effects of secondary plant compounds on germination of blastospores of the entomopathogenic fungus *Paecilomyces fumosoroseus* (Deuteromycotina: Hyphomycetes). Journal of Invertebrate Pathology 70: 209–213.

Vestergaard, S., A. Cherry, S. Keller, and M. Goettel. 2003. Safety of Hyphomycete fungi as microbial control agents. In H.M.T. Hokkanen and A.E. Hajek, eds., Environmental Impacts of Microbial Insecticides: Need and Methods for Risk Assessment. pp. 35–62. Kluwer Academic, Dordrecht, The Netherlands.

Vicentini, S., M. Faria, and M.R.V. de Oliveira. 2001. Screening of *Beauveria bassiana* (Deuteromycotina: Hyphomycetes) isolates against nymphs of *Bemisia tabaci* (Genn.) Biotype B (Hemiptera: Aleyrodidae) with description of a new bioassay method. Neotropical Entomology 30: 97–103.

Vidal, C., J. Fargues, and L.A. Lacey. 1997a. Intraspecific variability of *Paecilomyces fumosoroseus*: effect of temperature on vegetative growth. Journal of Invertebrate Pathology 70: 18–26.

Vidal, C., L.A. Lacey, and J. Fargues. 1997b. Pathogenicity of *Paecilomyces fumosoroseus* (Deuteromycotina: Hyphomycetes) against *Bemisia argentifolii* (Homoptera: Aleyrodidae) with a description of a bioassay method. Journal of Economic Entomology 90: 765–772.

Vidal, C., J. Fargues, L.A. Lacey, and M.A. Jackson. 1998a. Effect of various liquid culture media on morphology, growth, propagule production, and pathogenic activity to *Bemisia argentifolii* of the entomopathogenic *Paecilomyces fumosoroseus*. Mycopathologia 143: 33–46.

Vidal, C., L.S. Osborne, L.A. Lacey, and J. Fargues. 1998b. Effect of host plant on the potential of *Paecilomyces fumosoroseus* (Deuteromycotina: Hyphomycetes) for controlling the silverleaf whitefly, *Bemisia argentifolii* (Homoptera: Aleyrodidae) in greenhouses. Biological Control 12: 191–199.

Wraight, S.P., 1997. Evaluation of entomopathogenic fungi in field crops: methods for research and development. In Hacia un Futuro Agricola Sostenible, Proceedings of the XX Congreso Nacional de Control Biológico, pp. 241–24. November 13–14, 1997, Guadalajara, Mexico. Sociedad Mexicana de Control Biologico, Mexico City.

Wraight, S.P. and R.I. Carruthers. 1999. Production, delivery, and use of mycoinsecticides for control of insect pests of field crops. In F.R. Hall and J. Menn, eds., Methods in Biotechnology, Vol. 5, Biopesticides: Use and Delivery, pp. 233–269. Humana Press, Totowa, NJ.

Wraight, S.P., R.I. Carruthers, and C.A. Bradley. 1996. Development of entomopathogenic fungi for microbial control of whiteflies of the *Bemisia tabaci* complex. Proceedings V Siconbiol Simpósio de Controle Biológico, Foz. de Iguacu, Brazil, June 9–14,1996, pp. 28–34. EMBRAPA/CNPSo, Foz do Iguaçú, Brazil. Empresa Brasileira de Pesquisa Agropecuaria – Centro Nacional de Pesquisa de Soja, Londrina, PR, Brazil.

Wraight, S.P., R.I. Carruthers, C.A. Bradley, S.T. Jaronski, L.A. Lacey, P. Wood, and S. Galaini-Wraight. 1998. Pathogenicity of the entomopathogenic fungi *Paecilomyces* spp. and *Beauveria bassiana* against the silverleaf whitefly, *Bemisia argentifolii*. Journal of Invertebrate Pathology 71: 217–226.

Wraight, S.P., R.I. Carruthers, S.T. Jaronski, C.A. Bradley, C.J. Garza, and S. Galaini-Wraight. 2000. Evaluation of the entompathogenic fungi *Beauveria bassiana* and *Paecilomyces fumosoroseus* for microbial control of the silverleaf whitefly, *Bemisia argentifolii*. Biological Control 17: 203–217.

Wraight, S.P., M.A. Jackson, and S.L. De Kock. 2001. Production, stabilization, and formulation of fungal biocontrol agents. In T. Butt, C. Jackson, and N. Magan, eds., Fungi as Biocontrol Agents: Progress, Problems and Potential, pp. 253–287. CAB International, Wallingford, UK.

Wraight, S.P., G.D. Inglis, and M.S. Goettel. 2007. Fungi. In L.A. Lacey and H. K. Kaya, eds., Field Manual of Techniques in Invertebrate Pathology: Application and Evaluation of Pathogens for Control of Insects and Other Invertebrate Pests, pp. 223–248. Springer, Dordrecht, The Netherlands.

Yen, D.F. and Y.T. Tsai. 1969. Entomogenous fungi of Homoptera in Taiwan. Plant Protection Bulletin, Taiwan 11: 1–10.

Zaki, F.N. 1998. Efficiency of the entomopathogenic fungus, *Beauveria bassiana* (Bals), against *Aphis craccivora* Koch and *Bemisia tabaci* Gennadius. Journal of Applied Entomology 122: 397–399.

Zare, R. and W. Gams. 2001. A revision of *Verticillium* section *Prostrata*. IV. The genera *Lecanicillium* and *Simplicillium* gen. nov. Nova Hedwigia 73: 1–50.

Zimmermann, G. 1993. The entomopathogenic fungus *Metarhizium anisopliae* and its potential as a biocontrol agent. Pesticide Science 37: 375–379.

Chapter 4
Systematics and Biology of *Encarsia*

John M. Heraty[1], Andrew Polaszek[2], and Michael E. Schauff[3]

Abstract The genus *Encarsia* includes 343 described species and numerous undescribed species. Immatures are parasitoids of various whiteflies, armored scales, aphids, Lepidoptera or even the opposite sex of the same species. Several species are known to attack *Bemisia*, but so far, none have proved effective in the control of the pest species in this genus. The taxonomy and classification of *Encarsia* species is undergoing rapid changes using both morphological and molecular techniques. *Wolbachia* and newly discovered bacteria are associated with sex ratio distortion in species of *Encarsia*. Whiteflies appear to be the basal host associated with members of this genus, with only a few species potentially host specific for *Bemisia* whiteflies.

4.1 Introduction

Species of *Encarsia* Förster (Hymenoptera: Aphelinidae) are minute, solitary, endoparasitic wasps found worldwide. *Encarsia* is the largest genus within Aphelinidae, with a total of 343 described species (Noyes 2001; Heraty et al. 2007). However, a large number of species are undescribed and often even the named species of *Encarsia* are difficult to identify. Adults are known to attack the sessile stages of whiteflies, armored scale insects, aphids and lepidopteran eggs (Viggiani 1984; Polaszek 1991; Williams and Polaszek 1996). Most species are autoparasitic with females developing as primary endoparasitoids and males as hyperparasitic endoparasitoids of the same or other species (Williams and Polaszek 1996; Hunter and Woolley 2001). Males of only two species, *Enc. inaron* (Walker) and *Enc. longicornis* Mercet, have been shown to develop as primary parasitoids of their

[1] Department of Entomology, University of California, Riverside, CA, 92521

[2] Department of Entomology, The Natural History Museum, London, SW7 5BD UK

[3] USDA-ARS, Plant Sciences Institute, BARC West, Building 003, 10300 Baltimore Ave., Beltsville, MD 20705

J. Gould et al. (eds.), *Classical Biological Control of* Bemisia tabaci *in the United States.*

whitefly host (Mazzone 1983; Viggiani 1988). In some species, males are rare or absent, with these aberrant sex ratios associated with either a specialized group of bacteria or with *Wolbachia* (Zchori-Fein et al. 2001). *Encarsia* is one of the most important parasitic groups exploited for biological control (Noyes and Hayat 1994). Several species have demonstrated their importance in the control of citrus blackfly (*Enc. clypealis* (Silvestri)), *Enc. perplexa* Huang and Polaszek [as *Enc. opulenta* (Silvestri)] and *Enc. smithi* (Silvestri)) (Clausen 1978), spiny blackfly (*Enc. smithi*) (Kuwana 1934), San Jose scale (*Enc. perniciosi* (Tower)) (Clausen 1978), greenhouse whitefly (*Enc. formosa* Gahan) (Clausen 1978; van Lenteren and Woets 1988; Hoddle et al. 1998), and ash whitefly (*Enc. inaron* (Walker)) (Bellows et al. 1992).

Many whitefly species in the genus *Bemisia* (Aleyrodidae) are severe pests of agricultural crops in North America (McAuslane et al. 1993; Toscano et al. 1998) and elsewhere. *Encarsia* species may be important native parasitoids of *Bemisia*, but the species imported as biological control agents have yet to demonstrate their ability to establish and suppress these whiteflies (Goolsby et al. 1998). Even with an extensive worldwide search for parasitoids of *Bemisia*, relatively few of the 41 described species that are known to attack *Bemisia* have been used in these biological control programs. The parasitoids that were successfully recovered and released may simply be the most common species in agricultural situations, but given that many species are cryptic and either difficult or impossible to identify without refined behavioral or genetic data, many possibilities for new control agents exist. In this chapter we review the changes in our knowledge of *Encarsia* from a taxonomic and phylogenetic perspective, and how this may ultimately affect our ability to use these species as biological control agents of whiteflies or armored scales.

4.2 Taxonomy

The majority of species now placed in *Encarsia* were previously assigned to one of three genera: *Encarsia* Förster 1878, *Prospaltella* Ashmead 1904, and *Aspidiotiphagus* Howard 1894. Other generic names applied to these species include *Doloresia* Mercet 1912, *Mimatomus* Cockerell 1911, *Paraspidiotiphagus* Alam 1956, *Prospaltoides* Brèthes 1914 and *Trichaporus* Mercet 1930, but all have been synonymized. Until recently, biological characteristics were used to separate genera with *Encarsia* species parasitic on whiteflies and *Aspidiotiphagus* and *Prospaltella* attacking armored scales. Viggiani and Mazzone (1979) synonymized all these under *Encarsia*. DeBach and Rose (1981) argued that a set of morphologically distinct species with a narrowed fore wing having an asetose patch that were parasitic on scales should remain as the distinct genus *Aspidiotiphagus*, and they erected *Aleurodiphilus* to contain species having a similar fore wing and parasitizing whiteflies. Hayat (1983) treated all these genera as *Encarsia* and we follow that convention here.

The number of described species of *Encarsia* is increasing at a rapid rate. Just since 1995, 127 species (37.0% of the total number of 343) have been described (Evans et al. 1995; Jasnosh 1995; Chou et al. 1996; Krishnan and Vasantharaj David 1996; Evans and Angulo 1996; Evans 1997; Evans and Polaszek 1997; Evans and Castillo 1998; Hayat 1998; Evans and Polaszek 1998; Huang and Polaszek 1998; Polaszek et al. 1999; Gomez and Garcia 2000; Heraty and Polaszek 2000; Myartseva 2001; Schmidt et al. 2001; Manzari et al. 2002; Pedata and Polaszek 2003; Hernández-Suárez et al. 2003; Polaszek et al. 2004; Schmidt and Polaszek 2007a, b). However, these may represent only a small proportion of the total number of species that are in existence today. Most *Encarsia* species are described from material that is reared, which generally means a focus on species of agricultural importance. However, it is interesting that given the intense focus on species of *Bemisia* over the past several years, only nine of these 127 new species are known to attack this host. As we move from agricultural to natural ecosystems, we can expect the number of species to increase dramatically. For example, in one canopy fogging sample in Sulawesi, Indonesia, more than 156 species of *Encarsia* were recognized, which is more than half of the known species (Noyes 1989). Current descriptions are based on morphological characters that are relatively easy to recognize, but these do not take into account differences in behavior and reproductive incompatibility that distinguish cryptic or sibling species (Heraty and Polaszek 2000).

Molecular sequence data are being used to help establish the identity of species. Differences in the 28S-D2 rDNA transcript gene regions were used to differentiate two closely related species, *Encarsia formosa* Gahan and *Enc. luteola* Howard (Babcock and Heraty 2000). The nature of these species has been debated, and these data provide evidence to support the use of very minor morphological characters to recognize these species (antennal sensillum, number of cells across the axilla and degree of scutellar sculpture) (Polaszek et al. 1992; Schauff et al. 1996). A similar approach was taken using the same gene region to establish the identity of *Encarsia estrellae* Manzari and Polaszek, *Enc. dichroa* Mercet and *Enc. inaron* (Walker) and also for species within the *meritoria*-complex (Manzari et al. 2002; Polaszek et al. 2004). More species are likely to be discovered as they are analyzed at the molecular level. Closely related species are more readily distinguished by their ITS2, COI or COII sequence divergence than by their morphological differences (Stouthamer et al. 1999; Giorgini and Monti 2003; R. Stouthamer, personal communication). Within *Encarsia*, morphologically similar but genetically distinct and geographically isolated populations of *Enc. smithi* (Babcock et al. 2001) would suggest that they are different species. On the other hand, some species exhibit considerable behavioral divergence that is not demonstrated by a corresponding genetic divergence. *Encarsia sophia* (Girault) has varying levels of reproductive isolation and host choice that are not reflected in either their morphological or genetic differences for the 28S gene region (Heraty and Polaszek 2000; Babcock et al. 2001; Hernández-Suárez et al. 2003), although they may be reflected in COI (Giorgini and Monti 2003). Mating and host choice differences in populations of *Enc. formosa* attacking *Bemisia* on *Poinsettia* are neither reflected in 28S-D2 or ITS2 sequences, nor in a more extensive survey of AFLP (amplified fragment-length polymorphism) differences (Nemec and Stary 1984; Y. Gai and R. Stouthamer, personal communication). Clearly, we are only just beginning to understand the true diversity of the genus using molecular parameters.

Unfortunately, much of the current descriptive effort is focused on the redescription and illustration of species already described. Even recently described species such as *Enc. protransvena* Viggiani have been subsequently redescribed and illustrated as many as six different times. *Encarsia sophia* (Girault and Dodd) (=*Enc. transvena* Timberlake) has been redescribed nine times, and *Enc. formosa* Gahan at least 10 times. Often these redescriptions are produced as part of regional treatments, and as such are necessary, because unfortunately most *Encarsia* are recognized by an overall combination of characters, and not a set of unique characters. Thus any diagnosis requires a fairly complete treatment of the overall character set pertaining to each species. As identification keys and species group placement are better developed, perhaps this redundant aspect of *Encarsia* taxonomy can be overcome. Morphometrics has been an aid in delimiting species boundaries in closely related species (Heraty and Polaszek 2000; Manzari et al. 2002; Polaszek et al. 2004). High resolution digital photography is a significant breakthrough that may simplify future descriptions and allow for better recognition of described species. The digital illustrations of the body, antenna and wings of *Encarsia* species by Manzari et al. (2002), Pedata and Polaszek (2003) and Polaszek et al. (2004) are superb examples of how imaging technology can enhance our means of describing features of a species.

4.3 Identification Keys to Species

Progress is being made toward providing reliable keys to the species of *Encarsia*. Hayat (1989) provided the first reliable key to the species of India. While regionally limited in scope, the key includes many of the species found elsewhere. Comprehensive regional keys in the traditional couplet format have been developed for species in China (Huang 1994; Huang and Polaszek 1998), India (Hayat 1989, 1998), Egypt (Polaszek et al. 1999), and Russia (Jasnosh 1989). More user-friendly pictorial-format keys were developed for *Encarsia* parasitic on whiteflies in North America (Schauff et al. 1996), and Australia and the Polynesian islands (Schmidt et al. 2001). These regional keys are useful, but they always fall short of a satisfactory identification tool because they are not comprehensive on a worldwide level; and with the consistent importation, natural spread and discovery of species, it is difficult to name species with confidence unless representative material is available. Several recent studies have focused instead on worldwide reviews of species within a species group, which include the *cubensis* group (Evans and Polaszek 1998), the *flavoscutellum* group (Evans et al. 1995), part of the *strenua* group (Heraty and Polaszek 2000), the *longifasciata* group (Pedata and Polaszek 2003) and part of the *luteola* group (Polaszek et al. 2004). Polaszek et al. (1992) focused on a comprehensive review of the *Encarsia* attacking *Bemisia*, but this addressed only 19 of the 41 species now known to attack *Bemisia*, although it did address the species most

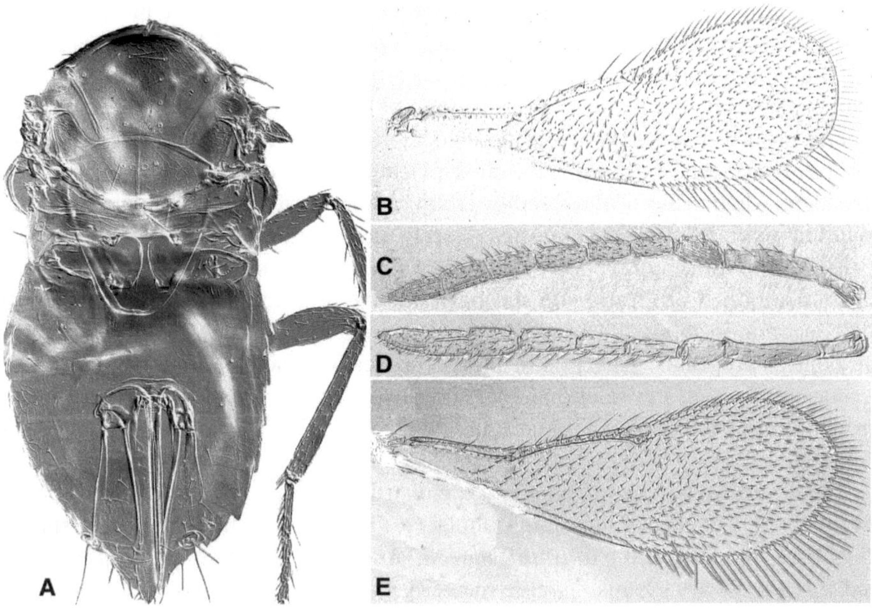

Fig. 4.1 Digital images of *Encarsia*. (**A–C**) *Encarsia bimaculata*; (**D–E**) *Encarsia protransvena.*

commonly encountered. Nobody has yet tried to produce a digital identification key using some of the standard packages now available (i.e., Lucid; www.lucidcentral.com). With the availability of digital imaging technology (Fig. 4.1), this may be the next logical step.

4.4 Species Relationships

A goal of systematics is to group species into evolutionary units that are presumed to share a common ancestor. Many species of *Encarsia* are undescribed, but we still must be able to accurately recognize species with the greatest potential for biological control. A common assumption is that closely related species may share similar habits and host preferences to known species, and are therefore desirable candidates for biological control. Species of the *citrina* group attack armored scale insects and species in the *flavoscutellum* group attack Hormaphididae. Because of their shared evolutionary history, closely related species are expected to have similar insect or plant host relationships, courtship patterns, environmental preferences or other behavioral attributes. If we can readily characterize these groups, and they have an evolutionary basis, then hopefully we can make accurate predictions of their host associations and other behavioral characteristics of interest. These relationships are most commonly determined by the presence of shared derived morphological characters (synapomorphies). Unfortunately, species groups of *Encarsia*, which are our first approximation of related species, are often defined by combinations of characters, many of which are characteristic of one or more species placed in other species groups. Even obvious group characteristics are found in clearly unrelated groups of species; for example, the close placement of scutellar sensillae, which are considered diagnostic of the *strenua* group, are now known to be convergent and found in unrelated groups of species (Heraty and Polaszek 2000).

Delimiting the natural species groups of *Encarsia* is important. Currently, species are grouped arbitrarily on the basis of overall similarity. This can lead to misconceptions about behavior and host associations that are crucial for biological control programs. Analysis of morphological characters alone has led to differing opinions on the relationships, composition and placement of species into groups within *Encarsia* (Viggiani and Mazzone 1979; Hayat 1998; Huang and Polaszek 1998).

The described species of *Encarsia* are distributed among 25 recognized species groups, with 60 species remaining unplaced (Table 4.1). The majority of species (52%) are included in the *aurantii, inaron, lahorensis, luteola, opulenta, parvella* and *strenua* species groups. Twelve species groups were first suggested by Viggiani and Mazzone (1979), and 32 species group names have been proposed by various authors. Seventeen species groups were recognized by Hayat (1998), who chose not to recognize four groups (*elegans, inquirenda, luteola* and *perflava*), which have been recognized by subsequent authors (Huang and Polaszek 1998; Polaszek et al. 1999; Babcock et al. 2001), and other groups were either proposed but not addressed (*scapeata, singularis, tremblayi*) or proposed after Hayat's review (*citrella, cubensis, divergens*). Over the last 20 years there has been an effort to accurately define and place species into these groups (Evans et al. 1995; Hayat 1998; Huang and Polaszek 1998; Evans and Polaszek 1998; Heraty and Polaszek 2000; Babcock et al. 2001; Manzari et al. 2002; Pedata and Polaszek 2003; Polaszek et al. 2004), but there is a need for a more comprehensive review of groups beyond that of Hayat (1998).

Many species can be allocated into morphologically and behaviorally distinct groups. For instance, species in the *citrina* group (previously *Aspidiotiphagus*) are all armored scale parasites and can be readily distinguished by a narrowed fore wing with a concave posterior margin, asetose patch on the fore wing, and the propodeum with reticulate sculpture (DeBach and Rose 1981). The *strenua* group can be recognized by having one or more setae on the margin of the wing at the apex of the costal cell, a bare area just anterior to the stigmal vein, and closely spaced placoid sensillae on the scutellum (Heraty and Polaszek 2000). Not all of these characters are unique. A distinct asetose wing spot also is found in members of the *cubensis* and *parvella* (previously *Aleurodiphilus*) groups, but a vague bare spot is also found in some species in the *perflava* group. Certain characters may or may not indicate relationships. For example, *Enc. quercicola* has close sensilla on the scutellum, but not the wing characteristics of the *strenua* group. In an opposite pattern, some characteristics, perhaps mistakenly identified, may artificially group taxa.

Table 4.1 Attributes of species groups of *Encarsia*. (Adapted from an earlier version of Heraty et al. 2007).

Attacked group	No. spp.	Scale	White-Fly	*Bemisia*	*Trial.*	OtWh	OtHo	No. of host genera 1	2	3	4	5+
Unplaced	60	10	8	1	1	5	1a?	16	1	1	–	–
aurantii	43	19	11	2	0	11	2a?, 2b	18	5	3	–	–4
citrella	4	0	4	3	0	4	0	1	1	2	–	–
citrina	9	7	0	–	–	–	0	4	–	1	–	2
cubensis	7	0	5	3	1	5	0	2	1	1	–	1
divergens	2	1	0	–	–	–	0	1	–	–	–	–
duorunga	3	0	2	2	0	0	0	2	–	–	–	–
elegans	4	0	2	0	0	2	0	1	–	1	–	–
flavoscutellum	4	0	0	–	–	–	3c	2	1	–	–	–
inaron	19	0	14	7	2	13	1b, 1d?, 1e	7	2	2	1	2
inquirenda	4	3	0	–	–	–	0	2	1	–	–	–
lahorensis	13	0	10	2	2	9	1a?, 1b, 1e	6	–	2	1	1
longifasciata	5	0	3	1	0	2	0	1	1	–	–	1
lutea	9	0	9	2	1	9	1a?, 1b	6	1	–	–	2
luteola	11	0	10	6	5	7	0	6	–	–	1	3
merceti	3	0	4	0	0	4	0	3	1	–	–	–
opulenta	16	0	15	0	1	14	1a?, 1b	11	3	1	–	–
parvella	14	0	13	4	2	13	1b	7	2	3	–	1
perflava	6	0	4	1	1	4	0	1	–	1	–	2
scapeata	2	0	1	0	0	1	0	1	–	–	–	–
septentrionalis	1	0	1	0	0	1	1a?	–	–	1	–	–
singularis	3	3	0	–	–	–	0	3	–	–	–	–
strenua	27	1	24	8	6	21	1a?, 1f, 1g?	16	3	3	–	3
tremblayi	1	0	1	0	0	1	0	1	–	–	–	–
tricolor	6	0	4	0	1	4	1b	2	2	–	–	–
tristis	1	0	1	0	0	1	0	–	1	–	–	–
Total	277	44	146	42	23	133	25	120	26	22	3	22

Abbreviations: *Trial.* = *Trialeurodes* spp., OtWh = other whiteflies, OtHo = other hosts (a = Coccidae or Pseudococcidae, b = Hymenoptera, c = Hormaphididae, d = Thysanoptera, e = Lepidoptera, f = Psyllidae, g = Aphididae). Numbers of host genera attacked do not include questionable records.

For example, 26 of the species in the *strenua* group attack whiteflies, but *Enc. titilliata* (Girault), which possesses all three of the *strenua*-group characters, attacks armored scale (Heraty and Polaszek 2000). Other characteristics such as sculpture of the thorax, shape of the scutellar sensilla, and coloration, suggest that *Enc. titilliata* may actually belong elsewhere, but without a larger scale analysis this cannot be verified. Because *Encarsia* are small, many of their characteristics appear to be simple reductions and possibly not valuable for assessing phylogenetic relationships. The *luteola* group all have a 4-segmented mid tarsus, but this also occurs in the *cubensis* group. Because this character represents a simple reduction from 5 to 4 segmented tarsi, Hayat (1998) presumed that this would not be a good character for assessing group relationships and placed the included species into other species groups.

To properly evaluate which features can accurately assess the relationships of species, morphological characters need to be identified and assessed in a phylogenetic analysis. In an analysis of relationships between *Encarsia* and two closely related sister taxa, Polaszek and Hayat (1992) were able to find only 24 characters to assess relationships between genera, of which only eight were variable within *Encarsia*. Babcock et al. (2001) used a morphological matrix of 14 characters to assess the relationships of 24 species in 10 species groups. The results were not very satisfactory, with resolution of only the *luteola* and *strenua* species groups, and little resolution of relationships within these species groups (Fig. 4.2). Identifying and evaluating phylogenetically significant morphological characters for more than 200 species is likely impossible.

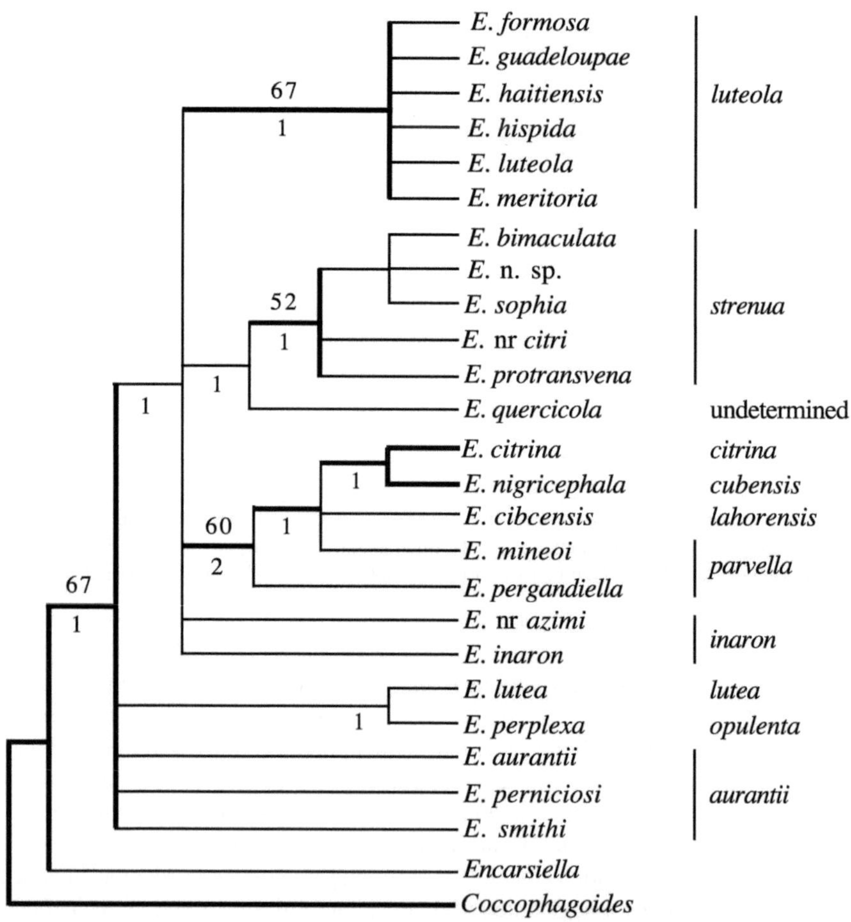

Fig. 4.2 Strict consensus of four trees from a morphological analysis of 14 characters after Successive Approximations Character Weighting of 50 most parsimonious (MP) trees (Adapted from Babcock et al. 2001). Thin lines collapse in the consensus of 50 MP trees.

The analysis of nucleotide sequence data provides an opportunity to assess relationships and test the validity of morphological features currently used for placing species of *Encarsia* into groups (Heraty 2003). To evaluate the species of *Encarsia* on a large scale, only sequence data will be useful. Protein or RAPD (randomly amplified polymorphic DNA) comparisons may be useful for identification of populations or limited to comparisons of a few closely related species (Kirk et al. 2000), but they cannot address large numbers of taxa, and there are inherent problems associated with phylogenetic comparisons using these techniques (Blackeljau et al. 1995). Babcock and Heraty (2000) used sequences of the D2 expansion region of 28S rDNA to evaluate the relationships of four species in the *luteola* group. These not only indicated the expected sister group relationship between *Enc. luteola* and *Enc. formosa*, but also contained conserved marker sites for two restriction enzyme sites that could distinguish these two species. Expanding on this analysis, Babcock et al. (2001) used the same sequences to evaluate the relationships of the same 26 species discussed above. The results were similar for an analysis of 70 populations or a subset of 26 representative populations (species) and two outgroup taxa (*Coccophagoides* and *Encarsiella*). The resulting hypothesis of relationships was almost completely resolved and supported monophyly of the *luteola*, *inaron* and *strenua* groups, with strong resolution of the species included within these groups. Importantly, the results established that 4-segmented tarsi, male antennal complexes, and closely spaced scutellar sensillae were homologous, phylogenetically informative characters. Manzari et al. (2002) built upon this data set by adding four different species (*Enc. dichroa*, *Enc. estrellae*, *Enc. tricolor* and *Enc.* nr *azimi*). The results were similar and again supported the monophyly of the expanded *inaron* group.

We reanalyzed a new data set for *Encarsia* that combined the species of three papers (Babcock et al. 2001; Manzari et al. 2002 [but without *Eretmocerus*], Pedata and Polaszek 2003). Together with *Dirphys*, *Encarsiella* was proposed as the monophyletic sister group to *Encarsia* (Polaszek and Hayat 1992). The sequence alignment was used from Babcock et al. (2001), and only one additional insertion event was needed to accommodate an extra base found in all populations of *Enc. estrellae*. A strict consensus of the three resulting trees is presented in Fig. 4.3, which conflicted only in the relationships of species within the *inaron* group. This tree is identical to a similar analysis from Heraty (2003), but with the addition of *Enc. arabica* Pedata and Polaszek. Other than the *parvella* group, species in groups represented by more than one species are all supported as monophyletic. The results of five analyses (Babcock et al. 2001; Manzari et al. 2002; Heraty 2003; Pedata and Polaszek 2003; and current analysis) are generally the same, but with different placement of *Enc. nigricephala*, *Enc. aurantii* and in some of the between group relationships. In all of the results, *Encarsiella* falls within *Encarsia*. To constrain *Encarsia* as monophyletic, with *Encarsiella* as a sister taxon, an additional 32 steps are required to explain the difference for both of the new analyses (34 and 82 taxa). As reported in Babcock et al. (2001), the relationships within species groups do not change under the constraint hypothesis.

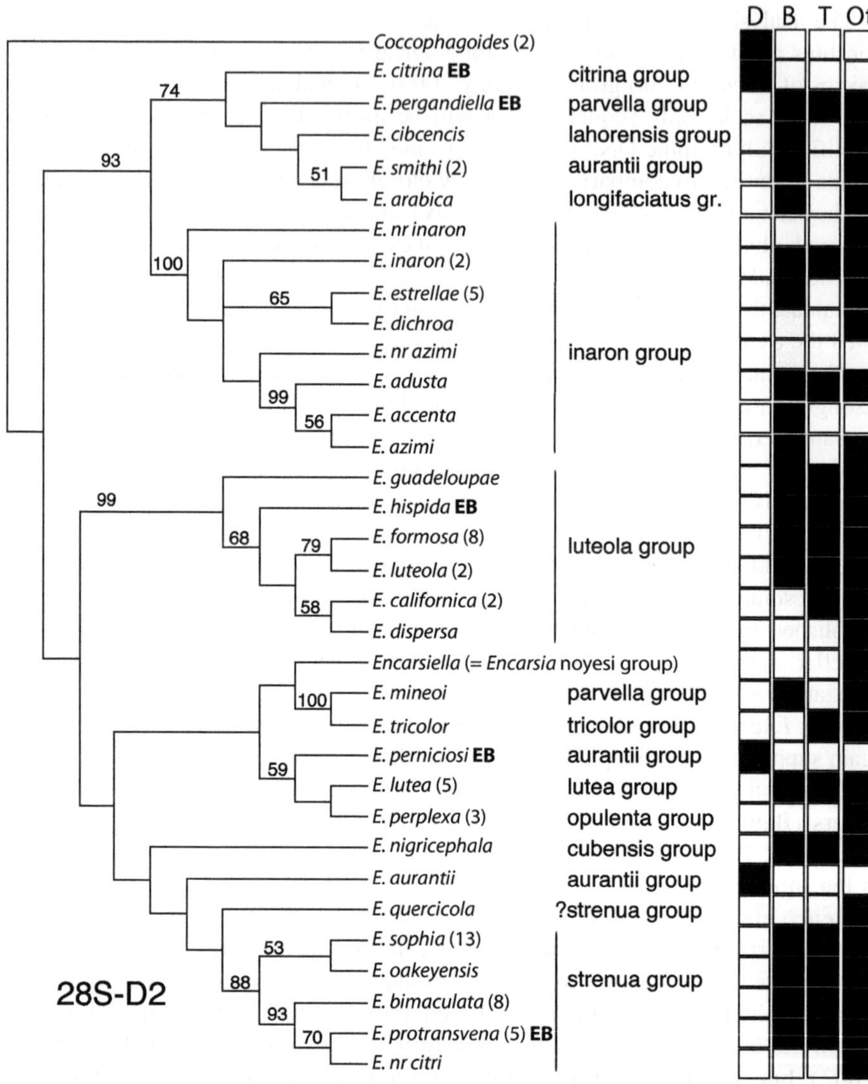

Fig. 4.3 Strict consensus of three trees of 856 steps (c.i. 0.41, r.i. 0.62) recovered from a parsimony analysis of 28S-D2 rDNA from 32 species of *Encarsia* and two closely related genera (*Coccophagoides* and *Encarsiella*). Data were analyzed using PAUP 4.0*b9 (Swofford 2002) using 100 random addition sequences and TBR branch swapping. Only the 28S-D2 transcript region without the first 134 bases (highly conserved region not sequenced for the Manzari et al. 2002 species) was used. Bootstrap proportions greater than 50% are shown above branches. The same results were obtained when additional populations, identified by the numbers in parentheses, were added for a total of 82 terminal taxa (sequences from Babcock et al. 2001; and Manzari et al. 2002; some names corrected from Schmidt et al. 2001), but with 17 trees of 905 steps. These were identical to the 3 trees after pruning out the extra populations. Behavioral attributes indicated: D = Diaspididae host; B = *Bemisia* host; T = *Trialeurodes vaporariorum* host; O = other whitefly host. Species with bacterial associate marked with EB.

At this point it is difficult to interpret these results. Either *Encarsiella* is not a valid genus but simply an aberrant group within *Encarsia*, *Encarsia* are not a monophyletic group (potentially many genera), or we have simply not sampled enough species or gene regions to develop a satisfactory result. Based on similar information, *Encarsiella* was recently transferred and placed as the *Encarsia noyesi* species group (Schmidt and Polaszek 2007b). As only one gene region has currently been sampled in 32 of 343 species, the latter is probably true for now. However, the amount of genetic divergence within *Encarsia* may suggest an old divergence. For both the D2 and D3 expansion regions of 28S rDNA, the level of divergence found between species groups is equivalent to the variation between subfamilies or families of other Chalcidoidea (Heraty 2003). Adding more species of *Encarsia*, as well as other genera of Coccophaginae, will be necessary to resolve the phylogeny of the group, but at least an initial framework has been developed.

4.5 Biological Attributes of *Encarsia*

It is difficult to assess biological changes within a phylogenetic perspective when most of the specialized aspects of behavior are known for only a few species or appear to be unique characteristics within *Encarsia*. Behavioral traits associated with mating and reproduction are very clearly elucidated in Hunter and Woolley (2001) and we do not plan to review these biologies here.

4.5.1 Sex Ratio Distortion in Encarsia

Wolbachia is recognized as a sex-ratio determining Proteobacteria, but within *Encarsia* has been found only in *Enc. formosa*, which is parthenogenetic (Zchori-Fein et al. 2001). The recognized sister group, *Enc. luteola*, is bisexual, and of the other species of *Encarsia* assayed none has *Wolbachia* (Fig. 4.3; marked by EB). A new bacterial associate belonging to the Cytophaga-Flexibacter-Bacteriod group of bacteria (EB) was, however, identified in six species of *Encarsia* that are dispersed across the genus (Fig. 4.3; Zchori-Fein et al. 2001). In all species, but excluding one population of *Enc. pergandiella*, the bacterium was associated with parthenogenesis. A phylogeny of the EB bacteria using 16S rDNA placed the bacteria within *Encarsia* as monophyletic with a divergence between the sequence of EB in *Enc. hispida* versus that of *Enc. berlesei*, *Enc. citrina* and *Enc. pergandiella* that had bootstrap support of 88% (Zchori-Fein et al. 2001). This divergence may correspond to the phylogenetic divergence between *Enc. hispida* and both *Enc. citrina* and *Enc. pergandiella* in Fig. 4.3. As more sequences of the EB bacteria and associated *Encarsia* become known, it will be interesting to note if the two phylogenies are concordant.

4.5.2 Host Relationships of *Encarsia*

Encarsia species are endoparasitoids, with one potential case of ectoparasitism known for both sexes of *Enc. ectophaga* (Silvestri) on armored scale (Hunter and Woolley 2001). Most species of *Encarsia* are autoparasitoids, with female eggs deposited on a primary host and male eggs deposited as parasitoids of the same or other species of *Encarsia* (Walter 1983a, b; Viggiani 1984; Polaszek 1991; Williams and Polaszek 1996; Hunter and Woolley 2001); however, males and females of *Enc. inaron* and *Enc. longicornis* Mercet are both primary parasitoids of whiteflies and in some species males develop as primary parasitoids of lepidopteran eggs (Hunter and Woolley 2001).

Whiteflies are the recorded host for 146 species of *Encarsia* (Table 4.1). The *aurantii* group, as defined by Heraty et al. (2007), includes 11 species attacking whiteflies and 19 species parasitic on armored scale. Three species groups comprised of 15 species are parasitic only on armored scale, and another 10 unplaced species of *Encarsia* have been reared from scales (Table 4.1). Only the *flavoscutellum* group is exclusively parasitic on another group of Hemiptera, the Hormaphididae (Evans et al. 1995). Hosts in the Coccidae, Pseudococcidae, Psyllidae and Thysanoptera are all considered as doubtful records (Polaszek 1991; Williams and Polaszek 1996). Lepidopteran eggs are parasitized by two species, in different species groups. *Encarsia porteri* (Mercet) is heterotrophic with females developing in whiteflies and males only in lepidopteran eggs (Polaszek 1991), and an undescribed species closely related to *Enc. inaron* has both males and females developing only in eggs of Lepidoptera (Williams and Polaszek 1996).

The outgroup used in the phylogenetic analysis, *Coccophagoides*, is a parasite of armored scale, whereas the two proposed sister taxa of *Encarsia, Encarsiella* and *Dirphys*, are whitefly parasitoids (Noyes 2001). A shift to scale parasitism is not a unique event within *Encarsia* (Babcock et al. 2001, Fig. 4.3). Given the distribution of armored scale parasitism for the taxa in Fig. 4.3, it is more parsimonious to assume that parasitism of whiteflies is ancestral, and that armored scale parasitism occurred independently at least three times.

A total of 42 species distributed across 12 species groups (and unplaced species) have been reared from *Bemisia* (Table 4.1). The *inaron, luteola, parvella* and *strenua* groups have the largest number of species known to attack *Bemisia*. These groups also have among the largest numbers of *Encarsia* species, although the *aurantii* group (43 species) has only two *Bemisia* parasitoids and the *opulenta* group (16 species) has no species attacking *Bemisia*. Only eight species of *Encarsia* have been reared exclusively from *Bemisia* (*Enc. accenta* Schmidt, *Enc. desantisi* Viggiani, *Enc. duorunga* Hayat, *Enc. mohyuddini* Shafee and Rizvi, *Enc. polaszeki* Evans, *Enc. reticulata* Rivnay and *Enc. silvestrii* Viggiani and Mazzone) (Heraty et al. 2007). In some cases, this apparent host specificity at the generic level may be simply due to not having encountered enough specimens to record them from different hosts. For example, *Enc. polaszeki* has been encountered and reared only once (Evans 1997). Most species of *Encarsia* that attack *Bemisia* also parasitize at least one other genus of whiteflies, with six of the most common species (*Enc. lutea, Enc. formosa, Enc. nigricephala, Enc. pergandiella, Enc. protransvena* and *Enc. sophia*) attacking more than five host genera (Noyes 2001; Heraty et al. 2007). *Encarsia bimaculata* was found exclusively on *Bemisia* as part of an extensive survey of whiteflies in Florida by Fred Bennett where it was introduced (Heraty and Polaszek 1999); however, Schmidt et al. (2001) reared this species from *Bemisia* and the invasive greenhouse whitefly, *Trialeurodes vaporariorum* (Westwood). *Encarsia bimaculata* may be host specific under natural conditions or may attack different genera within its native range (Southeast Asia). There does not seem to be a correlation between the phylogeny of the *Encarsia* groups and the number and/or the type of host utilized (Fig. 4.3).

Of the species of *Encarsia* with known host associations, 119 have been reared from a single host and 73 from multiple hosts, and of these, 22 species have been reared from more than five host genera (Table 4.1). Because of problems of correct parasitoid or host identification, a general focus on rearing records from agricultural systems, and the chance of encountering both host and parasitoid in the field, the numbers of host genera attacked in Table 4.1 may not be overly representative of most species. Although the number of species that attack 1–3 host genera may be debatable, there is little doubt that 11 of the species groups include species that are extreme generalists.

4.6 Biological Control of *Bemisia*

From 1991 to 1998, the Mission Biological Control Laboratory processed 18 foreign shipments of seven species of *Encarsia*, which included *Enc. bimaculata* (as *Enc.* nr *strenua*) (India, Thailand), *Enc. formosa* (Greece, Egypt, Thailand), *Enc.* near *hispida* (Brazil), *Enc. lutea* (Cyprus, Israel, Spain), *Enc.* near *pergandiella* (Brazil), *Enc. sophia* (as *Enc. transvena*) (Malaysia, Philippines, Spain, Taiwan, Thailand) and *Encarsia* sp. (*parvella* group) (Dominican Republic) (Goolsby et al. 1998). Only three of these species were released into the field in the Lower Rio Grande Valley of Texas (*Enc.* nr *hispida* [ex Brazil, 2,400 specimens], *Enc. lutea* [ex Cyprus, 5,600 specimens] and *Enc. sophia* [ex Spain, 60,000]). All species were initially recovered in Texas, but in very low numbers. *Encarsia bimaculata*, which had the second highest laboratory evaluation but a poor field cage evaluation, was not released in Texas. *Encarsia bimaculata* is widespread in Southeast Asia (Huang and Polaszek 1998; Schmidt et al. 2001). This species was released in Florida (climatically similar to Southeast Asia) and was recovered from field collections of *Bemisia* on *Euphorbia, Sesamum, Chamaesyce*, and *Magnolia* in 1992 and 1993 (Heraty and Polaszek 1999). However, recent recoveries have not been made in Florida (Evans, personal communication). Geographic populations of six of the aforementioned *Encarsia* species (excepting *Enc.* sp. nr. *parvella*) were released in desert valleys of Arizona and California, but only *Enc. sophia* (*ex* Pakistan) eventually became established (see Chapter 13). This is noteworthy because large numbers ("hundreds of thousands") of *Enc. sophia* from Pakistan (obtained earlier through another project) had been released into California's Imperial Valley from 1991 to

1992 against earlier outbreaks of *B. tabaci* biotype "A", but without any recoveries (Hoelmer 1995, see Chapter 13). Most of the common species attacking *Bemisia* (*Enc. formosa*, *Enc. lutea*, *Enc. pergandiella*, *Enc. protransvena* and *Enc. sophia*) are essentially cosmopolitan (Polaszek et al. 1992; Huang and Polaszek 1998; Schmidt et al. 2001). However, different populations often exhibit very different behavioral and ecological responses (Goolsby et al. 1998). As an example, the Nile and Netherlands strains of *Enc. formosa* are successful in attacking *Bemisia* on *Poinsettia*, whereas other strains of *Enc. formosa* are not (Heinz 1995). The number of host species attacked and the success of the parasitoid is likely to be determined by ecological factors such as plant characteristics and habitat, as well as historical associations with particular host groups (Hoelmer 1995).

We would contend that insufficient effort has been focused on the evaluation of species and populations in the *inaron* and *strenua* groups, both of which have the highest proportion of *Bemisia* parasitoids. Notably, in the control of the citrus blackfly, *Aleurocanthus woglumi* Ashby, more than 25,000 individuals of five species of *Encarsia* were imported for the control program in Mexico, and more than 4 million parasitoids captured and re-released in Mexico (Clausen 1978). This was one of the first programs to demonstrate the importance of different species or populations in different habitats and the differential success of various species at different stages of the control program. Control by species of *Encarsia* has ranged from single species introductions followed by immediate success (Bellows et al. 1992), augmentative releases for economic control (van Lenteren and Woets 1988; Hoddle et al. 1998), to multiple releases of locally important species (Clausen 1978). Along with the large numbers of species that remain to be discovered, *Encarsia* will continue to have significant impact on the control of whitefly and scale pests.

Acknowledgements We would like to thank Eric Grissell (USDA-SEL, Washington, DC) and Kim Hoelmer (USDA, Montpellier, France) for their comments on this manuscript. Figure 4.2 was reprinted from Babcock et al. (2001) with permission from Elsevier Science. This chapter was supported by a National Science Foundation Grant BSR-9978150 to JMH.

References

Babcock, C.S. and J.M. Heraty. 2000. Molecular markers distinguishing *Encarsia formosa* Gahan and *Encarsia luteola* Howard (Hymenoptera: Aphelinidae). Annals of the Entomological Society of America 93: 738–744.

Babcock, C.S., J.M. Heraty, P.J. De Barro, F. Driver, and S. Schmidt. 2001. Preliminary phylogeny of *Encarsia* Förster (Hymenoptera: Aphelinidae) based on morphology and 28S rDNA. Molecular Phylogenetics and Evolution 18: 306–323.

Bellows, T.S., T.D. Paine, J.R. Gould, L. Bezark, J.C. Ball, W. Bentley, R. Coviello, J. Downer, P. Elam, D. Flaherty, P. Gouveia, K. Koehler, R. Molinar, N. O'Connell, E. Perry, and G. Vogel. 1992. Biological control of ash whitefly: A success in progress. California Agriculture 46: 24–28.

Blackeljau, T., L. De Bruyn, H. De Wolf, K. Jordaens, S. Van Dongen, R. Verhagen, and B. Winnepenninckx. 1995. Random amplified polymorphic DNA (RAPD) and parsimony methods. Cladistics 11: 119–130.

Chou, K.-C., Y.-S. Su, L.-Y.Chou, and C.-C. Ko. 1996. New records of Aphelinidae (Hymenoptera) from Taiwan. Journal of Agricultural Research 45: 195–202.

Clausen, C.P. 1978. Introduced parasites and predators of arthropod pests and weeds: A world review. US Department of Agriculture, Agricultural Research Service Agricultural Handbook 480.

DeBach, P. and M. Rose. 1981. A new genus and species of Aphelinidae with some synonymies, a rediagnosis of *Aspidiotiphagus* and a key to pentamerous and heteromerous Prospaltellinae (Hymenoptera: Chalcidoidea: Aphelinidae). Proceedings of the Entomological Society of Washington 83: 658–679.

Evans, G.E. 1997. A new *Encarsia* (Hymenoptera: Aphelinidae) species reared from the *Bemisia tabaci* complex (Homoptera: Aleyrodidae). Florida Entomologist 80: 24–27.

Evans, G.E. and C.L. Angulo. 1996. A new species of *Encarsia* (Hymenoptera: Aphelinidae) from Costa Rica. Florida Entomologist 79: 582–586.

Evans, G.E. and J.A. Castillo. 1998. Parasites of *Aleurotrachelus socialis* (Homoptera: Aleyrodidae) from Colombia including descriptions of two new species (Hymenoptera: Aphelinidae Platygasteridae). Florida Entomologist 81: 171–178.

Evans, G.E., A. Polaszek, and F.D. Bennett. 1995. The taxonomy of the *Encarsia flavoscutellum* species-group (Hymenoptera: Aphelinidae), parasitoids of Hormaphididae (Homoptera: Aphidoidea). Oriental Insects 29: 33–45.

Evans, G.E. and A. Polaszek. 1997. Additions to the *Encarsia* parasitoids (Hym.: Aphelinidae) from Costa Rica. Florida Entomologist 79: 582–586.

Evans, G.E. and A. Polaszek. 1998. The *Encarsia cubensis* species-group (Hymenoptera: Aphelinidae). Proceedings of the Entomological Society of Washington 100: 222–233.

Giorgini, M. and M.M. Monti. 2003. Molecular differentiation of closely related species (Hymenoptera: Aphelinidae) based on mitochondrial COI gene. Abstracts from the XIII International Entomophagous Insects Workshop. Journal of Insect Science 3 (33): 9.

Gomez, J. and O. Garcia. 2000. A new species of *Encarsia* (Hymenoptera: Aphelinidae), a parasitoid of whitefly *Aleurodicus* sp. (Homoptera: Aleyrodidae) in Mexico. Pan-Pacific Entomologist 76: 49–51.

Goolsby, J.A., M.A. Ciomperlik, B.C. Legaspi Jr., J.C. Legaspi, and L.E. Wendel. 1998. Laboratory and field evaluation of exotic parasitoids of *Bemisia tabaci* (Gennadius) (Biotype "B") (Homoptera: Aleyrodidae) in the lower Rio grande Valley of Texas. Biological Control 12: 127–135.

Hayat, M. 1983. The genera of Aphelinidae (Hymenoptera) of the world. Systematic Entomology 8: 63–102.

Hayat, M. 1989. A revision of the species of *Encarsia* Foerster (Hymenoptera: Aphelinidae) from India and the adjacent countries. Oriental Insects 23: 1–131.

Hayat, M. 1998. Aphelinidae of India (Hymenoptera: Chalcidoidea): A taxonomic revision. Memoirs of Entomology, International, Vol. 13.

Heinz, K. 1995. Predators and parasitoids as biological control agents of *Bemisia* in Greenhouses. In D. Gerling and R.T. Mayer, eds., Bemisia: 1995: Taxonomy, Biology, Damage, Control and Management, pp. 435–449. Intercept Press, Andover.

Heraty, J.M. 2003. Molecular Systematics, Chalcidoidea and Biological Control. In L. Ehler, R. Sforza, and T. Mateille, eds., Genetics, Evolution and Biological Control, pp. 39–71. CAB International, Wallingford, Oxfordshire.

Heraty, J.M. and A. Polaszek. 2000. Morphometric analysis and descriptions of selected species in the *Encarsia strenua* group (Hymenoptera: Aphelinidae). Journal of Hymenoptera Research 9: 142–169.

Heraty, J.M., J.B. Woolley, and A.P. Polaszek. 2007. Catalog of the *Encarsia* of the World. http://cache.ucr.edu/~heraty/Encarsia.cat.pdf.

Hernández-Suárez, E., A. Carnero, A. Aguiar, G. Prinsloo, J. LaSalle, and A. Polaszek. 2003. Whitefly parasitoids (Hemiptera: Aleyrodidae; Hymenoptera: Aphelinidae, Euophidae, Platygastridae) from the Macaronesian archipelagos of the Canary Islands, Madiera and the Azores. Systematics and Biodiversity 1: 55–108.

Hoelmer, K.A. 1995. Whitefly parasitoids: Can they control field populations of *Bemisia*? In D. Gerling and R.T. Mayer, eds., *Bemisia*: 1995: Taxonomy, Biology, Damage, Control and Management, pp. 451–476. Intercept Press, Andover.

Hoddle, M.S., R.G. Van Dreische, and J.P. Sanderson. 1998. Biology and use of the whitefly parasitoid *Encarsia formosa*. Annual Review of Entomology 43: 645–669.

Huang, J. 1994. Systematic Studies on Aphelinidae of China, Chongqing Publishing House, Chongqing, Sichuan.

Huang, J. and A. Polaszek. 1998. A revision of the Chinese species of *Encarsia* Förster (Hymenoptera: Aphelinidae): Parasitoids of whiteflies, scale insects and aphids (Hemiptera: Aleyrodidae, Diaspididae, Aphidoidea). Journal of Natural History 32: 1825–1966.

Hunter, M.S. and J.B. Woolley. 2001. Evolution and behavioral ecology of heteronomous aphelinid parasitoids. Annual Review of Entomology 46: 251–290.

Jasnosh, V.A. 1989. Species of the genus *Encarsia* Foerster (Hymenoptera: Aphelinidae) – parasites of aleyrodids in the USSR. Proceedings of the Zoological Institute, Leningrad 191: 109–121.

Jasnosh, V.A. 1995. 48. Family Aphelinidae. In P.A. Ler, ed., Keys to Insects of the Far East of Russia, 4:506–551. Vladivostok: Dal'nauka.

Kirk, A.A., L.A. Lacey, J.K. Brown, M.A. Ciomperlik, J.A. Goolsby, D.C. Vacek, L.E. Wendel, and B. Napompeth. 2000. Variation in the *Bemisia tabaci s. l.* species complex (Hemiptera: Aleyrodidae) and its natural enemies leading to successful biological control of *Bemisia* biotype B in the USA. Bulletin of Entomological Research 90: 317–327.

Krishnan, B. and B.V. Vasantharaj David. 1996. Records and descriptions of some aphelinid parasitoids of Aleyrodidae (Homoptera: Insecta) from India. Jai Research Foundation, Valvada, Gujarat.

Kuwana, I. 1934. Notes on a newly imported parasite of the spiny whitefly attacking citrus in Japan. 5th Pacific Science Congress Proceedings (1933) 5: 3521–3523.

Manzari, S., A. Polaszek, R. Belshaw, and D.L.J. Quicke. 2002. Morphometric and molecular analysis of the *Encarsia inaron* species-group (Hymenoptera: Aphelinidae), parasitoids of whiteflies (Hemiptera: Aleyrodidae). Bulletin of Entomological Research 92: 165–175.

Mazzone, P. 1983. Contributo alla conocenza morfo-biologica degli Afenlinidi. 1. Osservazioni sugli stadi preimmaginali e sulla biologia de *Encarsia partenopea* Masi. ca delle specie del complesso *Encarsia* Foerster - *Prospaltella* Ashmead (Hym. Aphelinidae). Bollettino del Laboratario di Entomologia Agraria 'Filippo Silvestri' di Portici 40: 125–136.

McAuslane, H.J., F.A. Johnson, D.A. Knauft, and D.L. Colvin. 1993. Seasonal abundance and within-plant distribution of parasitoids of *Bemisia tabaci* (Homoptera: Aleyrodidae) in peanuts. Environmental Entomology 22: 1043–1050.

Myartseva, S.N. 2001. A new species of parasitoid wasp of the genus *Encarsia* (Hymenoptera: Aphelinidae) from Tamaulipas, Mexico. Acta Zoologica Mexicana (Nueva Serie) 82: 13–18.

Nemec, V. and P. Stary. 1984. Population diversity of *Encarsia formosa*, a biological control agent in glasshouses (Hymenoptera: Aphelinidae). Entomologica Generalis 9: 231–236.

Noyes, J.S. 1989. The diversity of Hymenoptera in the tropics with special reference to Parasitica in Sulawesi. Ecological Entomology 14: 197–207.

Noyes, J.S. 2001. Encyrtidae of Costa Rica (Hymenoptera: Chalcidoidea), 1. Memoirs of the American Entomological Institute 62: 1–355.

Noyes, J.S. and M. Hayat. 1994. Oriental Mealybug Parasitoids of the Anagyrini (Hymenoptera: Encyrtidae). CAB International, London.

Pedata, P. and A. Polaszek. 2003. A revision of the *Encarsia longifasciata* species group (Hymenoptera: Aphelinidae). Systematic Entomology 28: 361–374.

Polaszek, A. 1991. Egg parasitism in Aphelinidae (Hymenoptera: Chalcidoidea) with special reference to *Centrodora* and *Encarsia* species. Bulletin of Entomological Research 81: 97–106.

Polaszek, A., S. Abd-Rabou, and J. Huang. 1999. The Egyptian species of *Encarsia* (Hymenoptera: Aphelinidae): a preliminary review. Zoologische Medelingen Leiden 73: 131–163.

Polaszek, A.P., G.A. Evans, and F.D. Bennett. 1992. *Encarsia* parasitoids of *Bemisia tabaci* (Hymenoptera: Aphelinidae, Homoptera: Aleyrodidae): A preliminary guide to identification. Bulletin of Entomological Research 82: 375–392.

Polaszek, A. and M. Hayat. 1992. A revision of the genera *Dirphys* Howard and *Encarsiella* Hayat (Hymenoptera: Aphelinidae). Systematic Entomology 17: 181–197.

Polaszek, A., S. Manzari, and D.L.J. Quicke. 2004. Morphological and molecular taxonomic analysis of the *Encarsia meritoria* species-complex (Hymenoptera, Aphelinidae), parasitoids of whiteflies (Hemiptera, Aleyrodidae) of economic importance. Zoologica Scripta 33: 403–421.

Schauff, M.E., G.A. Evans, and J.M. Heraty. 1996. A pictorial guide to the species of *Encarsia* (Hym: Aphelinidae) parasitic on whiteflies (Homoptera: Aleyrodidae) in North America. Proceedings of the Entomological Society of Washington 98: 1–35.

Schmidt, S., I.D. Naumann, and P.J. De Barro. 2001. *Encarsia* species (Hymenoptera: Aphelinidae) of Australia and the Pacific Islands attacking *Bemisia tabaci* and *Trialeurodes vaporariorum* (Hemiptera: Aleyrodidae): A pictorial key and descriptions of four new species. Bulletin of Entomological Research 91: 369–387.

Schmidt, S. and A. Polaszek 2007a. The Australian species of *Encarsia* Foerster (Hymenoptera, Chalcidoidea: Aphelinidae), parasitoids of whiteflies (Hemiptera, Sternorrhyncha, Aleyrodidae) and armoured scale insects (Hemiptera, Coccoidea: Diaspididae). Journal of Natural History 41: 2099–2265.

Schmidt, S. and A. Polaszek 2007b. *Encarsia* or *Encarsiella*? - redefining generic limits based on morphological and molecular evidences (Hymenoptera, Aphelinidae). Systematic Entomology 32: 81–94.

Stouthamer, R., H. Jianguo, F. van Kan, G.R. Platner, and J.D. Pinto. 1999. The utility of internally transcribed spacer 2 DNA sequences of the nuclear ribosomal gene for distinguishing sibling species of *Trichogramma* (Hymenoptera: Trichogrammatidae). Biocontrol 43: 421–440.

Swofford, D.L. 2002. *PAUP**. Version 4.0*b9. Sinauer, Sunderland, MA.

Toscano, N.C., S.J. Castle, T.J. Henneberry, and N.P. Castle. 1998. Persistent silverleaf whitefly exploits desert crop systems. California Agriculture 52: 29–33.

van Lenteren, J.C. and J. Woets. 1988. Biological and integrated control in greenhouses. Annual Review of Entomology 33: 239–269.

Viggiani, G. 1984. Bionomics of Aphelinidae. Annual Review of Entomology 29: 257–276.

Viggiani, G. 1988 [1987]. Le specie italiane del genere *Encarsia* Foerster (Hymenoptera: Aphelinidae). Bollettino del Laboratario di Entomologia Agraria 'Filippo Silvestri' di Portici 44: 121–179.

Viggiani, G. and P. Mazzone. 1979. Contributi alla conoscenza morfo-biologica delle specie del complesso *Encarsia* Foerster - *Prospaltella* Ashmead (Hym. Aphelinidae). Bollettino del Laboratario di Entomologia Agraria 'Filippo Silvestri' di Portici 36: 42–50.

Walter, G.H. 1983a. Differences in host relationships between male and female heteronomous parasitoids (Aphelinidae: Chalcidoidea): A review of host location, oviposition and pre-imaginal physiology and morphology. Journal of the Entomological Society of South Africa 46: 261–282.

Walter, G.H. 1983b. Divergent male ontogenies in Aphelinidae (Hymenoptera: Chalcidoidea): A simplified classification and a suggested evolutionary sequence. Biological Journal of the Linnean Society 19: 63–82.

Williams, T. and A. Polaszek. 1996. A re-examination of host relations in the Aphelinidae (Hymenoptera: Chalcidoidea). Biological Journal of the Linnean Society 57: 35–45.

Zchori-Fein, E. Y. Gottlieb, J.K. Brown, J.M. Wilson, T.L. Karr, and M.S. Hunter. 2001. A newly discovered bacterium associated with parthenogenesis and a change in host selection behavior in parasitoid wasps. Proceedings of the National Academy of Sciences 98: 12555–12560.

Chapter 5
The Genus *Eretmocerus*

Gregory Zolnerowich[1] and Mike Rose[2]

Abstract The genus *Eretmocerus* includes 65 nominal species; undoubtedly there are innumerable undescribed species as well. All *Eretmocerus* species known are primary parasites of whitefly, and many nominal and undescribed species are known to attack *Bemisia*. This chapter discusses systematics and taxonomy, species introduced into the USA to attack *Bemisia* biotype B, naturally occurring species, utilization in biological control, possible competition for host resources between introduced and naturally occurring species, and the use of non-*Bemisia* whitefly to monitor for possible non-target effects.

5.1 Introduction

Introduced exotic *Eretmocerus* (Hymenoptera: Chalcidoidea: Aphelinidae) were released as natural enemies of *Bemisia* (*tabaci* group) (Homoptera: Aleyrodidae: Aleyrodinae) in Arizona, California, Florida, North Carolina, and Texas as part of the National Research and Action Plan for Development and Control Methodology for Sweetpotato Whitefly, *Bemisia tabaci* (Gennadius) (US Department of Agriculture 1992). Because of confusion regarding use of the names *Bemisia tabaci* (Gennadius) biotype A/B, and *B. argentifolii* Bellows and Perring (Bellows et al. 1994) on specimen labels and in the literature (Brown et al. 1995, 2000; Rosell et al. 1997), we refer to this group as *Bemisia* (*tabaci* group).

Before further discussion, the authors direct the readers to "Systematics, *Eretmocerus* and Biological Control" (Rose et al. 1996) for discussion and considerations of the early phases of *Eretmocerus* species importations into the USA for release against *Bemisia* (*tabaci* group), the need for and purposes of systematic and taxonomic studies, and the general morphological features used to identify species of *Eretmocerus*.

[1] Department of Entomology, Kansas Sate University, Manhattan, KS 66506

[2] (deceased) Department of Entomology, Montana State University, Bozeman, MT 59717

J. Gould et al. (eds.), *Classical Biological Control of* Bemisia tabaci *in the United States.*

Systematics and taxonomy are the basic means of organization that allow meaningful communication in natural history. With organization, a system can be catalogued, entities identified, and predictions made. Names of organisms provide information to researchers that enable the study of biology, behavior, ecology, and relationships between and among organisms, both plant and animal. Compere (1961) succinctly stated that, "… theories, plans and practices are no sounder than the systematics on which they are based." McKenzie (1956) emphasized the significant differences between systematics and taxonomic research. In essence, McKenzie said, "taxonomy must be based upon sound systematics" that "… develop systems of classification which will reveal significant relationships". Taxonomy, the science of naming things so they can be identified, relies on morphology. Again, from McKenzie, "For the purposes of identification we must seize upon structures that can be readily seen."

Obviously, then, identification of the species of *Eretmocerus* introduced, released, and recovered from *Bemisia* (*tabaci* group) and other whitefly species in the aforementioned states is absolutely essential to the evaluation of the importation and colonization aspects of the national biological control program. For example, specimens of *Eretmocerus* reared from *Bemisia* (*tabaci* group) in the USA frequently were misidentified as *Eretmocerus californicus* due to a lack of understanding and systematics research on the native fauna (Rose and Zolnerowich 1997a). Further, identification of naturally occurring *Eretmocerus* species that live in *Bemisia* (*tabaci* group) and other whitefly species in these areas are crucial to the evaluation of the effectiveness of species complexes regulating *Bemisia* (*tabaci* group), the development of sustainable pest management programs, and the evaluation of possible non-target host utilization and interspecies competition for host resources.

Thus, our approach to characterizing and describing species of naturally occurring and introduced *Eretmocerus* has been to develop means to recognize species based upon observable and/or measurable character sets. However, as we emphasized in Rose et al. (1996), biosystematic, behavioral, and molecular studies (see Hunter et al. 1996) are often essential research elements.

Taxonomic placement of natural enemies and their hosts are essential to effective classical biological control programs, understanding natural control, and evaluations of both (DeBach 1960; Rosen 1986; Wharton et al. 1990; Schauff and LaSalle 1998). Because of their importance to biological control, the systematics and taxonomy of parasitic Hymenoptera, and their host associations and geographic ranges have been the focus of treatment by a number of practitioners (see above references). Of the parasitic Hymenoptera that attack whitefly, members of the genus *Eretmocerus* are regarded as very effective, or potentially so (Rose et al. 1996). Historically, *Eretmocerus serius* Silvestri and *Eretmocerus debachi* Rose and Rosen were effective in major international biological control projects directed against the citrus blackfly, *Aleurocanthus woglumi* Ashby (Aleyrodidae: Aleyrodinae) (Clausen and Berry 1932), and the bayberry whitefly, *Parabemisia myricae* (Kuwana) (Aleyrodidae: Aleyrodinae) (Rose and DeBach 1991–1992), respectively.

Generally, *Eretmocerus* are solitary, internal parasites of whitefly that oviposit external to developing whitefly nymphs, complete larval development in the whitefly nymph, and pupate in the mummified fourth instar nymphs. Adult *Eretmocerus* emerge from the mummified whitefly nymphs through an exit hole cut in the dorsum of the host (see Rose and Zolnerowich 1997a, and Mound and Halsey 1978, for terminology).

Recently, M. Coombs (CSIRO, Queensland, Australia) discovered an undescribed gregarious *Eretmocerus* species attacking *Dumbletoniella* sp. (Aleyrodidae) in Queensland. *Dumbletoniella* is a large whitefly whose fourth stage larvae have been found to harbor as many as six *Eretmocerus* that emerge through multiple exit holes in the dorsum of the mummified host.

5.2 Techniques and Reference Collection

Specimens used for examination and comparison with reference specimens for identification, description, and illustration were mounted individually on microslides in Hoyer's mounting medium as described for species of *Aphytis* (Hymenoptera: Aphelinidae) (Rosen and DeBach 1979). Whenever possible, specimens are also mounted in balsam to serve as type specimens when species are described or redescribed. Type series specimens were compared to nominal species and descriptions following characterization of both male and female habitus and chaetotaxy, and linear measurements of 61 defined morphological features (see Rose 2000) of female type series specimens utilizing custom data acquisition software (Rosebud II) linked with a calibrated digitizing tablet, a Macintosh G4 computer, and a compound microscope equipped with Nomarski contrast enhancement.

The authors have developed a microslide reference collection of *Eretmocerus* from various whitefly and plant hosts from much of the world. Included in this collection are large representative series, including paratypes, of all the introduced exotic species of *Eretmocerus* released against *Bemisia* (*tabaci* group) in the USA, introduced exotic species that were not released, and species reared from *Bemisia* (*tabaci* group) collected in Arizona, California, Florida, Georgia, Mississippi, and Texas, and some more northern areas of the USA, prior to and after the release of exotic species. Extensive collections of *Eretmocerus* reared from *Bemisia* and many other whitefly genera by researchers from the USA, Australia, the Caribbean, Central and South America, China, Israel, Japan, Mexico, New Zealand, Thailand, and other areas of the world have been provided for examination and incorporation into the reference collection.

We currently are examining and comparing series of specimens reared from *Bemisia* (*tabaci* group), and other whitefly species collected in areas where exotic *Eretmocerus* were released, to reference specimens. This includes all available US and foreign type material from which photographs and drawings of key morphological features have been compiled.

5.3 The Genus *Eretmocerus*

Haldeman (1850) erected the genus *Eretmocerus* for the type species, *Eret. corni*, he reared from *Tetraleurodes corni* (Haldeman). The name *Eretmocerus* refers to the prominent "oar-shaped" antennal clubs of females. The type specimens of *Eret.*

corni were not available when Dozier (1932) designated a neotype from collections made in Pennsylvania in 1929. Haldeman's specimens are still unavailable, and Rose and Zolnerowich (1997a) accepted Dozier's neotype as representative of this species. Generic characters generally in use today are: female with 5-segmented antennae composed of radicle/scape, pedicel, two funicular segments and club; male with 3-segmented antennae composed of radicle/scape, pedicel and club; female antennal club undifferentiated, often large and oar-shaped (Fig. 5.1); male antennal club undifferentiated, elongate and exaggeratedly large; funicular segments reduced (Fig. 5.2); forewing with linea calva (or partial linea calva); stigmal vein very short (Fig. 5.3); all tarsi tetramerous; mandibles tridentate (Rose and Zolnerowich 1997a, b).

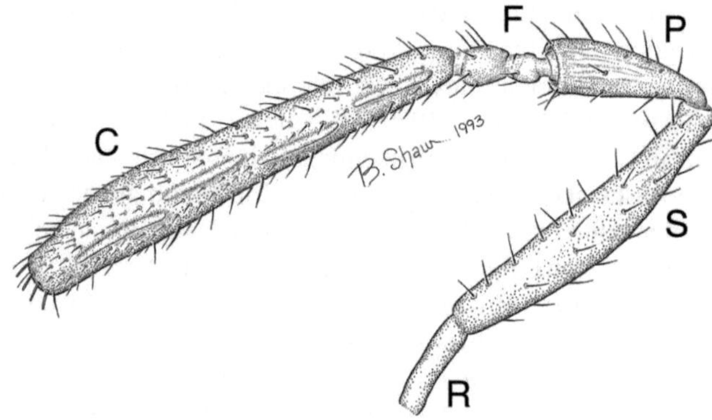

Fig. 5.1 *Eretmocerus* female antenna. C = club, F = funicle, P = pedicel, S = scape, R = radicle.

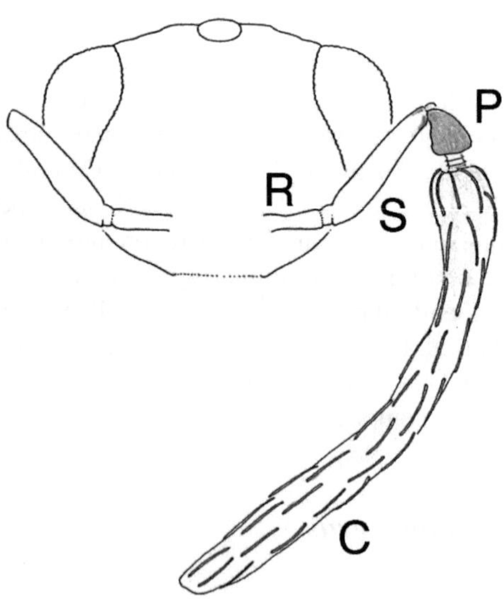

Fig. 5.2 *Eretmocerus* male antenna. C = club, P = pedicel, S = scape, R = radicle.

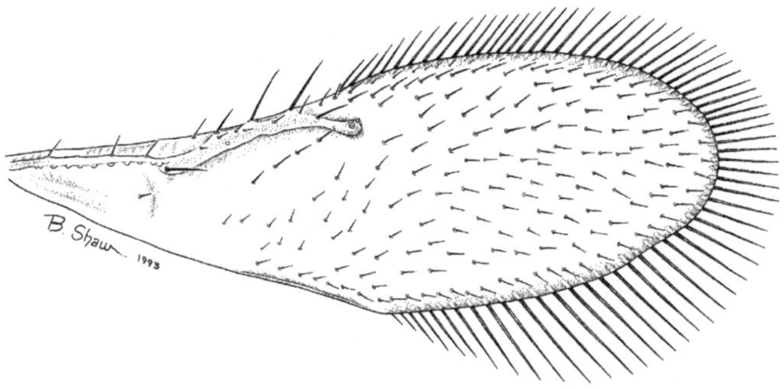

Fig. 5.3 *Eretmocerus haldemani* forewing.

Sixty-five species of *Eretmocerus* are now named worldwide (see ending note re: *Eretmocerus masii*). These are: *Eret. corni* Haldeman 1850; *Eret. californicus* Howard (1895); *Eret. paulistus* Hempel (1904); *Eret. haldemani* Howard (1908); *Eret. aleyrodesii* (Cameron) 1912 (see also Rosen and DeBach 1970); *Eret. diversiciliatus* Silvestri 1914; *Eret. australis* Girault 1921; *Eret. serius* Silvestri 1928; *Eret. nativus* Girault 1930; *Eret. mundus* Mercet 1931; *Eret. lativentris* Girault 1932; *Eret. illinoisensis*, *Eret. portoricensis* and *Eret. pallidus* Dozier 1932; *Eret. clauseni* and *Eret. longipes* Compere 1936; *Eret. aleurobi* Ishi 1938; *Eret. delhiensis* Mani 1941; *Eret. aleyrodiphaga* (Risbec) 1951; *Eret. orientalis* (= *Eret. serius* var. *orientalis*) and *Eret. silvestrii* Gerling 1969; *Eret. nairobii* Gerling 1970; *Eret. roseni* Gerling 1972; *Eret. gunturiensis* and *Eret. indicus* Hayat 1972; *Eret. nikolskajae* Myartseva 1973; *Eret. neobemisiae* Yashnosh 1974; *Eret. rajasthanicus* Hayat 1976; *Eret. cadabae* Viggiani 1982; *Eret. hydrabadensis* Husain and Agarwal 1982; *Eret. longicornis* and *Eret. siphonini* Viggiani and Battaglia 1983; *Eret. breviclavus* Subba Rao 1984; *Eretmocerus zippanguiphagus* Hulden 1986; *Eret. debachi* Rose and Rosen 1991–1992; *Eret. furuhashii* Rose and Zolnerowich 1994; *Eret. rosei* Evans and Bennett 1996; *Eret. adustiscutum*, *Eret. dialeurolonga* and *Eret. flavus* Krishnan and David 1996; *Eret. eremicus*, *Eret. joeballi*, *Eret. staufferi* and *Eret. tejanus* Rose and Zolnerowich 1997a; *Eret. emiratus*, *Eret. hayati* and *Eret. melanoscutus* Zolnerowich and Rose 1998; *Eret. bisetae, Eret. longiscapus*, *Eret. sculpturatus* and *Eret. trialeurodis* Hayat 1998; *Eret. queenslandensis* and *Eret. warrae* Naumann and Schmidt (in DeBarro et al. 2000); *Eret. ampliatus*, *Eret. aleurothrixus*, *Eret. comperei*, *Eret. desantisi*, *Eret. dozieri*, *Eret. exilis*, *Eret. gracilis*, *Eret. jimenezi* and *Eret. longiterebrus* Rose 2000; *Eret. picketti* Rose and Zolnerowich 2003; *Eret. perseae* Rose and Zolnerowich 2004; and *Eret. rui* Zolnerowich and Rose 2004. *Eretmocerus masii* was named by Silvestri in 1934, but no description was written, nor were types designated.

Of these 65 species, 16 of them are reported to have been reared from *Bemisia* (*tabaci* group). However, reports of rearing records in the literature often must be viewed with the understanding that misidentification of parasites and hosts is not uncommon. For example, although there are published reports of *Eret. californicus*, *Eret. corni*, and *Eret. haldemani* being reared from *Bemisia* (*tabaci* group), these records are without a doubt erroneous, and represent misidentifications. Such misidentifications can result from a lack of expertise in the host or parasite group, inadequate availability of current identification keys or literature, or a lack of understanding of the fauna that exists in a given location. A list of current species of *Eretmocerus* and reported hosts appears at the end of this chapter.

Species differentiation by various authors has relied heavily on several basic criteria: antennal club shape, antennal club size and configuration of antennal segments of females (i.e., Haldeman 1850; Howard 1895, 1908; Hempel 1904; Girault 1921, 1930, 1932; Silvestri 1928, 1934; Mercet 1931; Dozier 1932; Ishi 1938; DeSantis 1946, 1948, 1967; Gerling 1969, 1972; Hayat 1972, 1998; Myartseva 1973; Yashnosh 1974; Viggiani 1982; Husain and Agarwal 1982; Rose and Rosen 1991–1992; Rose and Zolnerowich 1994, 1997a, b, 2003; Zolnerowich and Rose 1998; Debarro et al. 2000, Rose 2000). Various authors have also utilized combinations of the preceding antennal characters with outstanding features such as: pigment, exaggerated or pronounced features (i.e., ovipositor length, tarsal terminae), forewing characters (i.e., chaetotaxy, relative shape and size, relative vein size), sculpture, chaetotaxy of mesosoma, chaetotaxy of gaster, relative proportions of body part lengths, characters of pronotum and characters of female genitalia (Howard 1895, 1908; Silvestri 1928; Girault 1921; Mercet 1931; Dozier 1932; Compere 1936; Ishi 1938; Gerling 1969, 1972; Hayat 1972, 1998; Myartseva 1973; Yashnosh 1974; Viggiani 1982; Husain and Agarwal 1982; Rose 2000; Rose and Zolnerowich 1994, 1997a, 2003; Zolnerowich and Rose 1998).

It can be readily seen that authors agreed that antennal characteristics – the outstanding feature of the genus – are also valuable species indicators. Antennal characters, together with certain other outstanding features, also were stressed in the keys to species by Mercet (1931), Compere (1936), Hayat (1972, 1998), Rose and Zolnerowich (1997a, b), Zolnerowich and Rose (1998), and Rose (2000).

5.4 Morphological Features

5.4.1 Antennae

The general habitus and configuration of female antennal segments, the radicle/scape, pedicel, funicle, and club are very consistent features in long series, although aberrant antennae are occasionally observed, particularly deformed funicular segments. However, most deformities are apparent and it is doubtful, even as individuals, that an aberrant specimen would be considered to be a unique species.

The position of specimens on microslides is critical to observations and measurements of all antennal segments. Antennae should be viewed in profile from the lateral aspect (Rose and Zolnerowich 1997a). Often, the head with antennae must be removed from the body to correctly position the antennae. When looking down on the dorsal aspect of the antennal segments (i.e., specimen mounted with antennae straight forward) nearly all clubs will appear cylindrical and important features of the club and club apex will be obscured. The same is true of the shape of funicular segments and pedicel. Antennal segments can be best examined and measured when the antennae are positioned at a 90° angle to the head and viewed laterally.

5.4.2 Head and Mouthparts

Head and mouthparts are very similar on all species examined. Mandibles are tridentate with an exaggerated "tooth," labial palpi with 1 segment, and maxillary palpi 2-segmented. Heraty and Schauff (1998) note the presence of the "tooth", but state that *Eretmocerus* do not have a true tooth as in Coccophaginae, rather they have an excision of the lower surface of the mandible.

5.4.3 Mesosoma

The female dorsum – pronotum, mesoscutum, parapsides, axillae, scutellum, metanotum, and propodeum – are very similar in all material examined. Differences between species are primarily discernable by the number of setae found on the mesoscutum and parapsides, the placement and size of setae on the scutellum, and relative proportions of mesosomal components. Sculpture and pigment of these features must also be considered.

Although female pigment has proved important in only a few species of *Eretmocerus*, the converse has proved true for male *Eretmocerus*, and pigment and pigment patterns can be important characters. Male pigment patterns have been used as a means to aid species separation from a number of whitefly hosts (Rose and Rosen 1991–1992; Rose and Zolnerowich 1994, 1997a, b; Zolnerowich and Rose 1998, Rose 2000).

The propodeum bears a distinct central lobe or projection, and the authors are examining the utility of this feature. Hayat (1998) noted the presence of an apically bifurcate seta near each propodeal spiracle, stated that this feature is a character unique to the genus, and placed *Eretmocerus* in the tribe Eretmocerini within the subfamily Aphelininae.

5.4.4 Wings and Wing Veins

Although forewings appear superficially similar, there are numerous differences between species. Wings range from very broad and round to narrow, even slightly inflexed. Length-to-width ratios, relative lengths of the anterior and posterior marginal alary fringe versus wing widths, density of setae in the disc, number of setae in the hyaline area of the wing base, number of long ("prominent" in Rosen and DeBach 1979) setae on the dorsum of veins, and general chaetotaxy can be utilized. The authors are also examining the number of setae in the "delta" area of the forewing (see Rosen and DeBach 1979).

There are also striking differences in configurations of forewing venation (Fig. 5.3). These are generally obvious and can be expressed as relative lengths of the submarginal, marginal, and stigmal veins. Two new species characterizations (Rose and Zolnerowich, in review/in press) include these features in diagnoses.

5.4.5 Legs

Other than as an occasional outstanding feature in a few species (e.g., *Eret. haldemani*), few leg characters have been utilized for species separation. Relative lengths of certain leg segments versus lengths of other body parts such as the ovipositor and antennal club are important to provide a relative index of size and proportion.

5.4.6 Gaster

The chaetotaxy of the dorsal abdomen can be variable within series reared from some whitefly species, but this feature has been used as a means to aid in the separation of species (Rose and Zolnerowich 1997a, b). Cuticular sculpture also must be considered, as portions of the tergites may have sculpture that ranges from substrigulate to imbricate to reticulate. This sculpture can be useful in distinguishing species. The ventral abdomen of all specimens examined bears a small group of setae located anterior to the base of the ovipositor. The number of setae in the group varies within series of specimens. Position of this group of setae also varies greatly depending on the position of specimens on microslides. In one case the number and position of these setae was fairly consistent and was utilized as part of the characterization of a species from *Aleurothrixus floccosus* (Maskell) (Homoptera: Aleyrodidae: Aleyrodinae) (Rose 2000).

The ovipositor in *Eretmocerus* ranges from tiny to greatly enlarged; this detail, generally presented as relative length to other body parts, has been utilized in species descriptions, and as an outstanding feature of a species from *A. floccosus* (Rose 2000).

5.5 Biosystematic Studies

The importance of biosystematic studies to biological control has long been recognized (Compere 1961; DeBach 1969; Rosen and Debach 1976; Rosen 1978, 1985). We refer to biosystematics as the study of the systematics of living organisms via experimentation. Examples include crossing tests, mating and ovipositional behavior, and host associations with animals and plants. The purpose of biosystematic studies is to clarify inherent biological and behavioral characteristics that distinguish taxa, particularly species and so-called "lower" taxonomic hierarchies between and among closely related species, cryptic species, and semispecies that demonstrate relative degrees of sexual isolation. This approach is, as Rosen and DeBach (1979) remarked, "a very useful concept stressing the often-forgotten fact that evolution is a continuous process."

Pinto (1998) pointed out in his monograph on species of *Trichogramma* that crossing data were "never used as apodictic evidence for species as some students of *Trichogramma* have done in the past". Pinto's reasons for this stance are based on two major concerns about defining species reproductively: first, that "reproduction is a purely relational trait", and second, concerns about understanding species limits under circumstances imposed by the logistical limitations of laboratory cultures. Rao and DeBach (1969a–c) utilized reciprocal crossing tests between three indistinguishable allopatric sibling forms of *Aphytis lingnanensis* Compere to demonstrate that species concepts are relative. This does not, however, reduce the importance of biosystematic studies to aid in separation of closely related species.

For example, Hunter et al. (1996) examined three populations of indigenous *Eretmocerus* reared from *Bemisia* (*tabaci* group) from Arizona, California and Texas. Previous examinations by the authors had shown that female morphology in all three populations was nearly identical, but that there were differences in male pigment patterns between the Texas population and the Arizona and California populations where male pigment appeared conspecific. Crossing tests showed that the Arizona and California populations were reproductively compatible but reproductively isolated from the Texas population. Rose and Zolnerowich (1997a) utilized the male pigment differences with reference to Hunter et al. (1996) to help characterize these three populations as *Eret. eremicus* (Arizona and California) and *Eret. tejanus* (Texas) (Fig. 5.4).

Host insect effects must also be considered. As an example, Woolley et al. (1995) showed that *Aphytis* sp. nr. *lingnanensis* from *Lepidosaphes gloveri* (Packard)

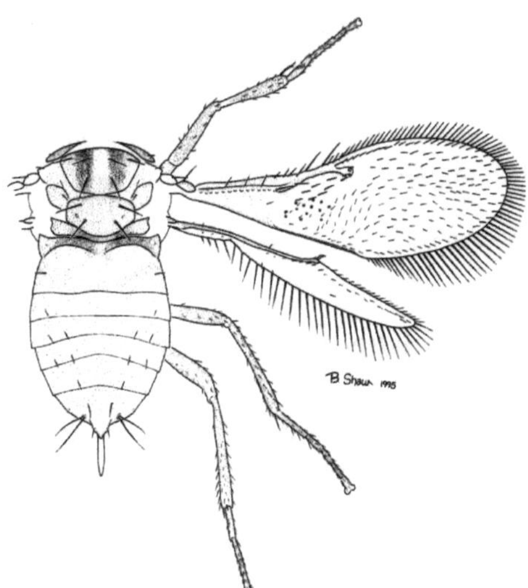

Fig. 5.4 *Eretmocerus tejanus* male pigment.

(Homoptera: Diaspididae) originally collected in Mazatlan, Mexico, was altered by the host. In this case, *A*. sp. nr. *lingnanensis* original material reared from *L. gloveri* collected in Mexico was morphometrically compared to the same species reared in the laboratory on an alternate armored scale insect, *Quadraspidiotus pernicious* Comstock. They showed how size differences due to *A*. sp. nr. *lingnanensis* living in a narrow, elongate host (*L. gloveri*) and a larger, round host (*Q. perniciousus*) caused the two populations to segregate under principal component analyses. Woolley et al. (1995) stated, "We do not know how one could see a host effect more clearly than this."

The authors are examining specimens that originated from a commercial *Eretmocerus* culture released and recovered from *Trialeurodes* sp. (Aleyrodidae) on commercial glasshouse tomatoes. The specimens meet the morphological criteria for *Eret. eremicus*, but are much larger specimens than those reared from *Bemisia* (*tabaci* group), which can be a smaller host than *Trialeurodes*. Interestingly, there are no males in this population, whereas all *Eret. eremicus* populations we have examined from *Bemisia* (*tabaci* group) have been biparental. In this case, biosystematic research to understand species limits would include host insect, microorganism, and molecular/biochemical studies.

The effect of microorganisms on parthenogenic reproduction by Aphelinidae (Werren 1997; Zchori-Fein et al. 1995, 2001) further expands some aspects of developing biosystematic criteria. It may be possible to use antibiotics to treat morphologically identical, or "nearly identical", populations that differ reproductively, with one being biparental and the other uniparental. If treatment with antibiotics produces no males, then the uniparental population is sexually isolated from the biparental population and distinguishing characters must be found for species status to be confirmed. If the treated uniparental population is "cured", reciprocal crossing tests can be conducted and male characters examined and compared.

Rose and Zolnerowich (2003) describe just such a situation for two populations of *Eretmocerus* that share the same whitefly host in the interior (biparental *Eretmocerus* population) and coastal regions (uniparental population) of southern California. The two populations are similar, and the biparental one was described as *Eret. picketti*. The uniparental population is unnamed, pending additional study.

5.6 Molecular and Biochemical Studies

Molecular and biochemical techniques are being emphasized in systematics as tools to discriminate populations and/or species. Such techniques are complementary to morphologically based systematics and biosystematics, and vice versa, and should be used together. In the cases presented herein, some background data have been developed for a few species of *Eretmocerus*.

Three *Eretmocerus* populations were examined by Hunter et al. (1996) using morphological characters, biosystematics, and allozyme patterns to help discriminate two new *Eretmocerus* species, *Eret. eremicus* and *Eret. tejanus*, reared from *Bemisia* (*tabaci* group) in Arizona, California, and Texas.

During importation and rearing of many exotic populations of *Eretmocerus* associated with *Bemisia* (*tabaci* group), the USDA-APHIS Mission Biological Control Laboratory in Mission, Texas, employed randomly amplified polymorphic DNA (RAPD) technology to discriminate between naturally occurring and exotic populations in culture and from field trials (Goolsby et al. 2000). However, it must be noted that on occasion the RAPD pattern for particular cultures changed, possibly as a result of contamination and mixing of cultures.

DNA "squash-blot" tests to discriminate between certain species of *Eretmocerus* reared from *Bemisia* (*tabaci* group) were under development by Heilmann (1997) and initial results with three exotic species were reported at the Fifth Annual Review of the national *Bemisia* program.

DeBarro et al. (2000) utilized both DNA sequence data and morphological characters to differentiate between three species of *Eretmocerus* that attack *Bemisia* (*tabaci* group) and *Trialeurodes vaporariorum* (Westwood) in Australia. They used their results to describe two new species of *Eretmocerus*, and examined inter- and intraspecific variation in several gene regions in samples that also included *Eret. mundus*, *Eret. hayati*, and undescribed populations from Ethiopia and Hong Kong.

5.7 Current Research

Systematics research on the genus *Eretmocerus* contributes to knowledge about parasitic Hymenoptera, and has direct and immediate relevance to biological control, sustainable pest management, species interactions, and non-target host utilization. All but one of the *Eretmocerus* species introduced and released against *Bemisia* (*tabaci* group) in the USA are new to science, are morphologically very similar, and are considered by us to be closely related to each other. The exotic species most intensively released were described by Zolnerowich and Rose (1998), who noted that populations provided for examination from cultures were often "mixed"; that is, samples could include specimens that did not meet specific criteria. This factor was not only cause for careful consideration about just what organisms were being described as species, but also about the genetic composition of the mixed cultures.

Four exotic *Eretmocerus* species have been recovered from *Bemisia* (*tabaci* group) collected in Texas and identified by the authors. These are *Eret. emiratus*, *Eret. hayati*, *Eret. melanoscutus*, and *Eret. mundus*.

Field recoveries of *Eret. emiratus*, *Eret. hayati*, *Eret. mundus*, and *Eret. staufferi* released against *Bemisia* (*tabaci* group) in California have been identified by the authors. A fifth population, *Eretmocerus* sp. from Ethiopia, was also released in Arizona and California. The Ethiopian population has been characterized by the authors following recent provision of additional early voucher specimens by E. Andress (USDA-APHIS). It has been recovered in Arizona, and appears well established and nearly completely dominant in the California Imperial Valley based upon 2001–2002 *Bemisia* (*tabaci* group) recovery samples provided by the California Department of Food and Agriculture (CDFA) Biological Control Program and USDA-APHIS (Chapter 13). Recovery samples from 2003 confirmed the widespread establishment of *Eretmocerus* sp. (Ethiopia) in AZ (Goolsby et al. 2005).

It must be emphasized that during the period 1998–2000, the authors found that a number of exotic *Eretmocerus* specimens reared by W. Roltsch (CDFA Biological Control Program) from *Bemisia* (*tabaci* group) from the Imperial Valley could not be placed because specific morphological characters were intermediate and therefore not clear. We surmised this was possibly due to hybridization based upon the close relationships of the various exotic populations that were released in the same locales. Preliminary crossing attempts undertaken by J. Goolsby (USDA Mission Biological Control Laboratory, unpublished data), where the exotic species released in California were initially imported and reared (Goolsby et al. 1998) and preliminary crossing tests undertaken by C. Pickett (CDFA Biological Control Program, unpublished data) between what was then called M93005 from India [now *Eret. hayati*] and M94023 from Thailand [now *Eret. melanoscutus*] indicated the possibility of hybridization. The authors have specimens from the California tests under examination.

Furthermore, examinations of specimens sent by P. DeBarro and J. Goolsby from their Queensland, Australia, laboratory culture of *Eret. hayati* that originated from the Lower Rio Grande Valley of Texas showed that although most specimens met the morphological criteria for *Eret. hayati*, some specimens from the same culture bore funicular segmentation similar to *Eret. mundus* (see figures in Zolnerowich and Rose 1998). Specimens reared from *Bemisia* (*tabaci* group) from the California Central Valley have been provided by C. Pickett to the authors for examination because both *Eret. hayati* and *Eret. mundus* were released in that area.

Currently, P. DeBarro and J. Goolsby (personal communication) are utilizing molecular techniques to examine possible hybridization between these two exotic species. This affords a great opportunity to apply morphologically based systematics to specimens from material that may be shown molecularly to include hybrids, and vice versa.

Given the preceding, the overall dominance of the Ethiopian population in the California Imperial Valley and its presence in Arizona are extremely interesting. It will be exciting to follow the ongoing dynamics of the various exotic and naturally occurring *Eretmocerus* populations in these areas.

Two additional exotic *Eretmocerus* populations were released against *Bemisia* (*tabaci* group) in Florida. These populations originated from Hong Kong and Sudan and were introduced to Florida and released there by Nguyen and Bennett (1995). McAuslane and Nguyen (1996) discussed the biology of the uniparental species from Hong Kong, which was described as *Eret. rui* by Zolnerowich and Rose (2004). *Eretmocerus rui* was recovered from *Bemisia* (*tabaci* group) in Florida by R. Nguyen (Florida Department of Agriculture & Consumer Services, personal communication). Both Nguyen and P. Stansly (personal communication) (University Florida) have recovered the Sudan population in Florida from *Bemisia* (*tabaci* group). The Sudan population has been characterized by the authors and is being described (Zolnerowich and Rose, in preparation).

Naturally occurring species of *Eretmocerus* parasitize *Bemisia* (*tabaci* group) in all the exotic species release areas; several of these species from Arizona, California, and Texas have been described by Rose and Zolnerowich (1997a). We use the term "naturally occurring" as a near synonym of "native" because we do not have meaningful historic evidence of exactly which species of *Eretmocerus*, and their whitefly and plant hosts, inhabited areas over time.

The authors are currently concentrating on examination of *Eretmocerus* from rearing samples of *Bemisia* (*tabaci* group) populations and other whitefly populations that occur within the bioclimatic ranges where exotic *Eretmocerus* species were colonized. Sampling has been and is still being conducted by the authors and a number of cooperators. Some other whitefly genera that have been and are being sampled include *Aleurocanthus, Aleuropleurocelus, Aleurothrixus, Dialeurodes, Paraleyrodes, Parabemisia, Tetraleurodes* and *Trialeurodes*.

At this time the authors are describing new species from various whitefly hosts from California, such as *Eret. picketti* (Rose and Zolnerowich 2003). Other manuscripts are in review or preparation. This research is designed to characterize and identify *Eretmocerus* species, and species complexes, utilizing various whitefly species as hosts in and near areas where exotic *Eretmocerus* species have been successfully colonized, and in some cases, apparently established. The authors are addressing the much discussed "critical issue" of non-target host utilization, and seek to learn about possible competition for host resources, and the outcomes of this possible competition (see Hoelmer and Rose 1999).

As an example, *Eretmocerus picketti* has been reared from *Tetraleurodes acaciae* (Quaintance) on carob, *Ceratonia siliqua* L. (Leguminosae), in El Centro in the southern California Imperial Valley. Species of *Tetraleurodes* in the *acaciae* group were discussed by Nakahara (1995); the proposed group consists of six described species and *T. acaciae*. All species in the whitefly host group are of American origin, thus, *Eret. picketti* is assumed to be a naturally occurring species. Carob is an evergreen tree native to the Mediterranean area that bears *T. acaciae* year round in the Imperial Valley. *Tetraleurodes acaciae* and *Eretmocerus picketti* are also found on other desert legumes. As yet, there is no evidence of utilization of *T. acaciae* by any exotic *Eretmocerus* species, and no evidence of utilization of *Bemisia* (*tabaci* group) by *Eret. picketti* from *T. acaciae*.

A naturally occurring and undescribed species of *Eretmocerus* from *Aleuropleurocelus* (Homoptera: Aleyrodidae: Aleyrodinae) on *Hymenoclea salsola* (Compositae: Heliantheae) also lives in the California Imperial Valley (Rose and Zolnerowich, submitted for publication). The host plant, *H. salsola*, known as cheesebrush, cheesebush, or burrobush is a member of a western American genus that inhabits sandy desert washes (Wiggins 1980; Payne 2002). Cheesebrush is a common desert plant in the Imperial Valley that bears year-round foliage, is interspersed with agricultural plantings in Imperial County, and is found as well in nonagricultural desert environments. *Aleuropleurocelus* has black larvae that are readily distinguished from the pale *Bemisia* (*tabaci* group) larvae that also are found on cheesebrush throughout this distribution.

As is the case with *Eret. picketti* from *T. acaciae*, as yet there is no evidence that this species of *Eretmocerus* from *Aleuropleurocelus* utilizes *Bemisia* (*tabaci* group), nor is there evidence that any of the exotic *Eretmocerus* species utilize *Aleuropleurocelus* sp. The authors plan to continue studies on the possibility of non-target host utilization and competition for host resources that include these species and others that are being described (Chapter 18).

We are studying possible non-target host utilization by parasitic Hymenoptera, in this case several introduced species of *Eretmocerus,* which were imported to combat *Bemisia* (*tabaci* group) in a classical biological control program. This is because the practice of biological control by natural enemies, a proven effective method to regulate pest insect populations and concurrently reduce insecticide/pesticide use (see DeBach 1974), has been painted with a broad brush of claims that introduced species could cause environmental damage, particularly because of possible non-target host utilization and subsequent competition for host resources that could eliminate native enemies (Simberloff and Stiling 1996). We believe the *Eretmocerus* complex is an excellent system to monitor potential non-target utilization of native whitefly hosts, and would be a means to address this issue. Long-term sampling to characterize and identify *Eretmocerus* utilizing various whitefly species in and around areas where exotic species have been introduced and/or established should continue.

To glean some answers about possible non-target host utilization and competition for these resources, the following are some of the most salient questions that we hope will be continue to be addressed in future research:

1. What are the species of *Eretmocerus* attacking *Bemisia* (*tabaci* group) on different plant species in states where exotic *Eretmocerus* were released?
2. Which species of *Eretmocerus* attacking *Bemisia* (*tabaci* group) in various locations and plantings should be evaluated, conserved, augmented and transferred to new locations?
3. Are naturally occurring *Eretmocerus* species still recovered from *Bemisia* (*tabaci* group) in areas of the USA where they were known to occur prior to release of exotic species?
4. Are alternate whitefly hosts utilized by naturally occurring *Eretmocerus* species that reproduce in *Bemisia* (*tabaci* group) in areas where exotic species are recovered?
5. What naturally occurring *Eretmocerus* species are found in non-target hosts in areas where exotic species are recovered?
6. Are naturally occurring and exotic *Eretmocerus* species competing for various whitefly host resources?
7. What are the outcomes of such competition?

Acknowledgements We thank our colleagues Martha Hunter, Charles Pickett, Bill Roltsch, Earl Andress, Matt Ciomperlik, John Goolsby, Kim Hoelmer, Walker Jones, Steve Naranjo, Fred Bennett, Greg Evans, Ru Nguyen, Phil Stansly, John LaSalle, Paul Debarro, and Marc Coombs for providing specimens for study. We express our appreciation to Robert Hendrickson, Jr., for curation of microslides; Jack McShea, Cinnabar Mac, Livingston, Montana, for writing the real-time data acquisition program (Rosebud II); and Ben Shaw, Lucid Art, for drawings. This chapter is Contribution No. 04–341-B from the Kansas Agricultural Experiment Station (KAES) and was supported in part by KAES Hatch Project No. 583, Insect Systematics.

References

Bellows, Jr., T.S., T.M. Perring, R.J. Gill, and D.H. Headrick. 1994. Description of a species of *Bemisia* (Homoptera: Aleyrodidae). Annals of the Entomological Society of America 87: 195–206.

Brown, J.K., D.R. Frohlich, and R.C. Rosell. 1995. The sweetpotato or silverleaf whiteflies: Biotypes of *Bemisia tabaci* or a species complex? Annual Review of Entomology 40: 511–534.

Brown, J.K, T.M. Perring, A.D Cooper, I.D. Bedford, and P.G. Markham. 2000. Genetic analysis of *Bemisia* (Hemiptera: Aleyrodidae) populations by isoelectric focusing electrophoresis. Biochemical Genetics 38: 13–18.

Cameron, P. 1912. On a collection of parasitic Hymenoptera (chiefly bred), made by Mr. Walter W. Froggatt, F.L.S., in New South Wales, with descriptions of new genera and species. Part iii. Proceedings of the Linnean Society of New South Wales 37: 172–216.

Clausen, C.P. and P.A. Berry. 1932. The citrus blackfly in Asia and importation of its natural enemies into Tropical America. US Department of Agriculture Technical Bulletin 320.

Compere, H. 1936. Notes on the classification of the Aphelinidae, with descriptions of new species. University of California, Publications in Entomology 6: 277–322.

Compere, H. 1961. The red scale and its insect enemies. Hilgardia 31: 173–278.

DeBach, P. 1960. The importance of taxonomy to biological control as illustrated by the cryptic history of *Aphytis holoxanthus* n. sp. (Hymenoptera: Aphelinidae), a parasite of *Chrysomphalus anonidum,* and *Aphytis coheni* n. sp., a parasite of *Aonidiella aurantii.* Annals of the Entomological Society of America 53: 701–705.

DeBach, P. 1969. Uniparental, sibling and semi-species in relation to taxonomy and biological control. Israel Journal of Entomology 4: 11–28.

DeBach, P. 1974. Biological Control By Natural Enemies, Cambridge University Press, Cambridge.

DeBarro, P.J., F. Driver, I.D. Naumann, S. Schmidt, G.M. Clarke, and J. Curran. 2000. Descriptions of three species of *Eretmocerus* Haldeman (Hymenoptera: Aphelinidae) parasitizing *Bemisia tabaci* (Gennadius) (Hemiptera: Aleyrodidae) and *Trialeurodes vaporariorum* (Westwood) (Hemiptera: Aleyrodidae) in Australia based on morphological and molecular data. Australian Journal of Entomology 39: 259–269.

DeSantis, L. 1946. Taxonomia de la familia Aphelinidae (Hymenoptera, Chalcidoidea). Revista del Museo de La Plata (Nueva Serie) (Zoologia) 5: 1–21.

DeSantis, L. 1948. Estudio monográfico de los afelínidos de la Republica Argentina (Hymenoptera, Chalcidoidea). Revista del Museo de La Plata (Neuve Serie) (Zoologia) 5: 23–280.

DeSantis, L. 1967. Catalogo de los Himenopteros Calcidoideos de America al Sur de los Estados Unidos. Primer Suplemento. Revista Peruana de Entomologia 21: 1–38.

Dozier, H.L. 1932. The identity of certain whitefly parasites of the genus *Eretmocerus* Hald., with descriptions of new species (Hymenoptera: Aphelinidae). Proceedings of the Entomological Society of Washington 34: 112–122.

Evans, G.A. and F.D. Bennett. 1996. A new *Eretmocerus* (Hymenoptera: Aphelinidae) species reared from *Dialeurodes kirkaldyi* (Homoptera: Aleyrodidae). Florida Entomologist 79: 579–582.

Gerling, D. 1969. Comments on *Eretmocerus serius* Silvestri (Hymenoptera: Aphelinidae) with a description of two new species. Bollettino de Laboratorio di Entomologia Agraria 'Filippo Silvestri', Portici 27: 79–88.

Gerling, D. 1970. Two African species of *Eretmocerus* Haldeman (Hymenoptera: Aphelinidae). Journal of the Entomological Society of Southern Africa 33: 325–329.

Gerling, D. 1972. Notes on three species of *Eretmocerus* Haldeman occurring in Israel with a description of a new species. Entomologische Berichten 32: 156–161.

Girault, A.A. 1921. Miscellaneous species of chalcid-flies from Australia (Hymenoptera: Chalcidae). Insecutor Inscitiae Menstruus 9: 1–3.

Girault, A.A. 1930. New Pests from Australia, IX. Privately Published. *See*: Memoirs of the American Entomological Institute 28: 278.

Girault, A.A. 1932. New lower Hymenoptera from Australia and India. Privately Published, Brisbane.

Goolsby, J.A., M.A. Ciomperlik, B.C. Legaspi, Jr., J.C. Legaspi, and L.E. Wendel. 1998. Laboratory and field evaluation of parasitoids of exotic parasitoids of *Bemisia tabaci* (Gennadius) (Biotype "B") (Homoptera: Aleyrodidae) in the Lower Rio Grande Valley of Texas. Biological Control. 12: 127–135.

Goolsby, J.A., M.A. Ciomperlik, A.A. Kirk, W.A. Jones, B.C. Legaspi, Jr., J.C. Legaspi, R.A. Ruiz, D.C. Vacek, and L.E. Wendel. 2000. Predictive and empirical evaluation for parasitoids of *Bemisia tabaci* (Biotype "B"), based on morphological and molecular systematics. In A. Austin and M. Dowton, eds., Hymenoptera: Evolution, Biodiversity, and Biological Control, pp. 347–358. International Hymenopterists Conference (4th, January 6–11,1999, Canberra, A.C.T., Australia) CSIRO Publishing, Collingwood, Victoria.

Goolsby, J.A., P. De Barro, A.A. Kirk, R. Sutherst, M. Ciomperlik, P. Ellsworth, J. Gould, K. Hoelmer, S. Naranjo, M. Rose, W. Roltsch, R. Ruiz, C. Pickett, and D. Vacek. 2005. Post-release evaluation of the biological control of *Bemisia tabaci* biotype "B" in the USA and the development of predictive tools to guide introductions for other countries. Biological Control 32: 70–77.

Haldeman, S.S. 1850. On four new species of Hemiptera of the genera *Ploiaria, Chermes, Aleurodes,* and two new Hymenoptera, parasitic on the last named genus. American Journal of Science IX: 108–111.

Hayat, M. 1972. The species of *Eretmocerus* Haldeman, 1850 (Hymenoptera: Aphelinidae) from India. Entomophaga 17: 99–106.

Hayat, M. 1976. Two new species of Aphelinidae parasitic on *Acaudaleyrodes rhachipora* from India. Entomophaga 21: 160–162.

Hayat, M. 1998. Aphelinidae of India (Hymenoptera: Chalcidoidea): A Taxonomic Revision. Memoirs on Entomology, International 13.

Heilmann, L. 1997. Development of species specific DNA probes for *Eretmocerus* species. In T.J. Henneberry, N.C. Toscano, T.M. Perring, and R.M. Faust, eds., Silverleaf Whitefly: National Research, Action and Technology Transfer Plan, 1997–2001: Fifth Annual Review, San Diego, CA, January 28–30, 1997. US Department of Agriculture ARS 1997–01, p. 87.

Hempel, A. 1904. Notas sobre dois inimagos de laranjeira. Familia Aleurodidae. Boletin de Agricultura (Sao Paulo) 5: 10–21.

Heraty, J.M. and M.E. Schauff. 1998. Mandibular teeth in Chalcidoidea: function and phylogeny. Journal of Natural History 32: 1227–1244.

Hoelmer, K.A. and M. Rose. 1999. Predicting and assessing non-target impacts of parasitic Hymenoptera attacking *Bemisia* (Homoptera: Aleyrodidae) in the southwestern USA. In Proceedings, International Symposium of the Global IOBC: Evaluating Indirect Ecological Effects of Biological Control, October 17–20, 1999, Montpellier, France. International Organization of Biological Control/Western Palearctic Regional Section Bulletin 22(2): 28.

Howard, L.O. 1895. Revision of the Aphelinidae of North America. US Department of Agriculture, Technical Series 1.

Howard, L.O. 1908. On two species of parasites of Aleyrodidae. Proceedings of the Entomological Society of Washington 10: 63–65.

Hulden, L. 1986. The whiteflies (Homoptera, Aleyrodidae) and their parasites in Finland. Notulae Entomologicae 66: 19–20.

Hunter, M.S., M.F. Antolin, and M. Rose. 1996. Courtship behavior, reproductive relationships, and allozyme patterns of three North American populations of *Eretmocerus* nr. *californicus* (Hymenoptera: Aphelinidae) parasitizing the whitefly *Bemisia* sp., *tabaci* complex (Homoptera: Aleyrodidae). Proceedings of the Entomological Society of Washington 98: 126–137.

Husain, T. and M.M. Agarwal. 1982. Studies on some aphelinid (Hym. Chalcidoidea) parasites of hemipterous insects of India. Journal of the Bombay Natural History Society 79: 155–162.

Ishi, T. 1938. Descriptions of six new species belonging to the Aphelinidae from Japan. Kontyû 12: 27–32.

Krishnan, B. and B.V. David. 1996. Records and descriptions of some aphelinid parasitoids of Aleyrodidae (Homoptera: Insecta) from India. Jai Research Foundation, Valvada, Gujarat.

Mani, M.S. 1941. Studies on Indian parasitic Hymenoptera. Indian Journal of Entomology 3: 25–36.

McAuslane, H.J. and R. Nguyen. 1996. Reproductive biology and behavior of a thelytokous species of *Eretmocerus* (Hymenoptera: Aphelinidae) parasitizing *Bemisia argentifolii* (Homoptera: Aleyrodidae). Annals of the Entomological Society of America 89: 686–693.

McKenzie, H.L. 1956. The Armored Scale Insects of California. Bulletin of the California Insect Survey, Vol. 5. University of California Press, Berkeley.

Mercet, R.G. 1931. Notas sobre Afelinidos (Hym. Chalc.), 4.ª nota. EOS 7: 395–399.

Mound, L.A. and S.H. Halsey. 1978. Whitefly of the World. A Systematic Catalogue of the Aleyrodidae (Homoptera) With Host Plant and Natural Enemy Data. British Museum (Natural History) and Wiley, Chichester.

Myartseva, S.N. 1973. *Eretmocerus nikolskajae* sp. n. (Hymenoptera, Chalcidoidea) – new parasite of *Bulgarialeurodes cotesi* Mask. (Homoptera, Aleurodoidea) in Turkmenistan. Izvestiya Akademii Nauk Turkmenskoi SSR Seriya Biologicheskikh Nauk 4: 81–85.

Nakahara, S. 1995. Taxonomic studies of the genus *Tetraleurodes* (Homoptera: Aleyrodidae). Insecta Mundi 9: 105–150.

Nguyen, R. and F.D. Bennett. 1995. Importation and field release of parasites against silverleaf whitefly, *Bemisia argentifolii* (Bellows and Perring) in Florida from 1990–1994. Proceedings of the Florida State Horticultural Society 108: 43–47.

Payne, W. W. 2002. *Hymenoclea*, Winged Ragweed. In B.G. Baldwin, S. Boyd, B.J. Ertter, R.W. Patterson, T.J. Rosatti, and D.H. Wilken, eds., The Jepson Desert Manual, Vascular Plants of Southeastern California, p. 159. University of California Press, Berkeley.

Pinto, J.D. 1998. Systematics of the North American species of *Trichogramma* Westwood (Hymenoptera: Trichogrammatidae). Memoirs of the Entomological Society of Washington Number 22: 1–287.

Rao, S.V. and P. DeBach. 1969a. Experimental studies on hybridization and sexual isolation between some *Aphytis* species (Hymenoptera: Aphelinidae). I. Experimental hybridization and an interpretation of evolutionary relationships among species. Hilgardia 39: 515–553.

Rao, S.V. and P. DeBach 1969b. Experimental studies on hybridization and sexual isolation between some *Aphytis* species (Hymenoptera: Aphelinidae). II. Experiments on sexual isolation. Hilgardia 39: 555–567.

Rao, S.V. and P. DeBach 1969c. Experimental studies on hybridization and sexual isolation between some *Aphytis* species (Hymenoptera: Aphelinidae). III. The significance of sexual isolation between interspecific hybrids and parental species. Evolution 23: 525–533.

Risbec, J. 1951. Les Chalcidoïdes de l'Afrique occidentale française. Memoires de l'Institute Français d'Afrique Noire 13: 403–405.

Rose, M. 2000. *Eretmocerus* (Hymenoptera: Chalcidoidea: Aphelinidae) reared from *Aleurothrixus floccosus* (Maskell) (Homoptera: Aleyrodidae) in the Americas. Vedalia 7: 2–35.

Rose, M. and P. DeBach. 1991–1992. Biological control of *Parabemisia myricae* (Kuwana) (Homoptera: Aleyrodidae) in California. Israel Journal of Entomology 25–26: 73–95.

Rose, M. and D. Rosen. 1991–1992. *Eretmocerus debachi* n. sp. (Hymenoptera: Aphelinidae), an effective parasite of *Parabemisia myricae* (Homoptera: Aleyrodidae). Israel Journal of Entomology 25–26: 199–207.

Rose, M. and G. Zolnerowich. 1994. *Eretmocerus furuhashii* sp. n. (Hymenoptera, Aphelinidae), a parasite of *Parabemisia myricae* (Kuwana) (Homoptera, Aleyrodidae) in Japan. Japanese Journal of Entomology 62: 285–292.

Rose, M. and G. Zolnerowich. 1997a. *Eretmocerus* Haldeman (Hymenoptera: Aphelinidae) in the United States, with descriptions of new species attacking *Bemisia* (*tabaci* complex) (Homoptera: Aleyrodidae). Proceedings of the Entomological Society of Washington 99: 1–27.

Rose, M. and G. Zolnerowich. 1997b. The genus *Eretmocerus* (Hymenoptera: Aphelinidae): Parasites of whitefly (Homoptera: Aleyrodidae). California Department of Agriculture, Biological Control Project (Special Publication).

Rose, M. and G. Zolnerowich. 2003. *Eretmocerus picketti* new species (Hymenoptera: Chalcidoidea: Aphelinidae), reared from *Tetraleurodes acaciae* (Quaintance) (Homoptera: Aleyrodidae: Aleyrodinae). Pan Pacific Entomologist 79: 119–127.

Rose, M. and G. Zolnerowich. 2004. *Eretmocerus perseae* n. sp. (Hymenoptera: Chalcidoidea: Aphelinidae) reared from *Tetraleurodes perseae* Nakahara (Homoptera: Aleyrodidae: Aleyrodinae) living on avocado in Michoacan, Mexico. Vedalia 11: 45–52.

Rose, M., G. Zolnerowich, and M.S. Hunter. 1996. Systematics, *Eretmocerus*, and Biological Control. In D. Gerling, ed., *Bemisia* 1995: Taxonomy, Biology, Damage And Management. Intercept, Andover, Hants.

Rosell, R.C., I.D. Bedford, D.R. Frohlich, R.J. Gill, J.K. Brown, and P.G. Markham. 1997. Analysis of morphological variation in distinct populations of *Bemisia tabaci* (Homoptera: Aleyrodidae). Annals of the Entomological Society of America 90: 575–589.

Rosen, D. 1978. The importance of cryptic species and specific identifications as related to biological control. pp. 23–35 In J. Romberger, ed., Biosystematics in Agriculture. Allanheld, Osmun, Montclair, NJ.

Rosen, D. 1985. Biological Control. Chapter 13. In G.A. Kenicut and L.I Gilbert, eds., Insect Control, Comprehensive Insect Physiology, Biochemistry and Pharmacology, Vol. 12. Pergamon Press, Oxford.

Rosen, D. 1986. The role of taxonomy in effective biological control programs. Agriculture Ecosystems and Environment 15: 121–129.

Rosen, D. and P. DeBach. 1970. Notes on the genus *Marlattiella* Howard (Hymenoptera: Aphelinidae). Mushi 43: 39–44.

Rosen, D. and P. DeBach. 1976. Biosystematic studies on the species of *Aphytis* (Hymenoptera: Aphelinidae). Mushi 49: 1–17.

Rosen, D. and P. DeBach. 1979. Species of *Aphytis* of the World (Hymenoptera: Aphelinidae). Series Entomologica Vol. 17. DR. W. Junk BV Publishers, The Hague, and Israel Universities Press, Jerusalem.

Schauff, M.E. and J. LaSalle. 1998. The relevance of systematics to biological control: protecting the investment in research. In M.P. Zalucki, R. Drew, and G.G. White, eds., Pest Management - Future Challenges. Proceedings of the Sixth Australasian Applied Entomological Research Conference, Brisbane, 29 September–2 October 1998, Vol. 1. University of Queensland Press, Brisbane.

Silvestri, F. 1914. Descrizione di nuovi Imenotteri Calcididi africani. Bolletino del Laboratorio di Zoologia generale e agraria della Reale Scuole superiore d'agricoltura in Portici 9: 366–367.

Silvestri, F. 1928. Contribuzione alla conoscenza degli Aleurodidae (Insecta: Hemiptera) viventi su *Citrus* in Estremo Oriente e dei loro parasitti. Bolletino del Laboratorio di Zoologia generale e agraria della Reale Scuole superiore d'agricoltura in Portici 21: 1–60.

Silvestri, F. 1934. Ricerche sugli Hymenoptera Chalcidoidea. I. Le specie italiane del genere *Anthemus* How. (Hym. Encyrtidae) e nota descrittiva sull' *Eretmocerus masii* Silv. (Hym. Aphelinidae). Bolletino del Laboratorio di Entomologia Agraria Filippo Silvestri, Portici 23: 249–264.

Simberloff, D. and P. Stiling. 1996. How risky is biological control? Ecology 77: 1965–1974.

Subba Rao, B.R. 1984. Descriptions of new species of Oriental Mymaridae and Aphelinidae (Hymenoptera: Chalcidoidea). Proceedings of the Indian Academy of Science Animal Science 93: 251–262.

US Department of Agriculture. 1992. R.M. Faust, ed., Conference report and 5-year national research and action plan for the development of management and control methodology for the sweetpotato whitefly. US Department of Agriculture, Agricultural Research Service, ARS-107.

Viggiani, G. 1982. New species and host records of African aphelinids (Researches on Hymenoptera Chalcidoidea LXX). Journal of the Entomological Society of Southern Africa 45: 27–32.

Viggiani, G. and D. Battaglia. 1983. Le specie Italiane del genere *Eretmocerus* Hald. (Hymenoptera: Aphelinidae). Bolletino del Laboratorio di Entomologia Agraria Filippo Silvestri, Portici 40: 97–101.

Werren, J.H. 1997. Biology of *Wolbachia*. Annual Review of Entomology 42: 587–609.

Wharton, R.A., J.B. Woolley, and M. Rose. 1990. Relationship and importance of taxonomy to classical biological control. In D.H. Habeck, F.D. Bennett, and J.H. Frank, eds., Classical Biological Control In The Southern United States, pp. 11–15. Southern Cooperative Series Bulletin No. 355. IFAS Editorial, University of Florida, Gainesville, FL.

Wiggins, I.L. 1980. Flora of Baja California. Stanford University Press, Palo Alto.

Woolley, J.B., M. Rose, and P.C. Krauter. 1995. Morphometric comparisons of *Aphytis* species in the *lingnanensis* group (Hymenoptera: Aphelinidae). In D. Rosen, ed., Advances in The Study of *Aphytis* (Hymenoptera; Aphelinidae). Intercept, Andover.

Yashnosh, V.A. 1974. New species of *Eretmocerus* Haldeman (Hymenoptera: Aphelinidae) from Georgia. Soobshcheniya Akademii Nauk Gruzinskoy SSR 73: 713–715.

Zchori-Fein, E., O. Faktor, M. Zeidan, Y. Gottlieb, H. Czosnek, and D. Rosen. 1995. Parthenogenesis-inducing microorganisms in *Aphytis* (Hymenoptera: Aphelinidae). Insect Molecular Biology 4: 173–178.

Zchori-Fein, E., Y. Gottlieb, S.E. Kelley, J.K. Brown, J.M. Wilson, T.L. Karr, and M.S. Hunter. 2001. A newly discovered bacterium associated with parthenogenesis and a change in host selection behavior in parasitoid wasps. Proceedings of the National Academy of Science 98: 12555–12560.

Zolnerowich, G. and M. Rose. 1998. *Eretmocerus* Haldeman (Hymenoptera: Aphelinidae) imported and released in the United States for control of *Bemisia* (*tabaci* complex) (Homoptera: Aleyrodidae). Proceedings of the Entomological Society of Washington 100: 310–323.

Zolnerowich, G. and M. Rose. 2004. *Eretmocerus rui* n. sp. (Hymenoptera: Chalcidoidea: Aphelinidae), an exotic natural enemy of *Bemisia* (*tabaci* group) (Homoptera: Aleyrodidae) released in Florida. Florida Entomologist 87: 283–287.

Endnote

Current species of *Eretmocerus*, followed by reported hosts. Note that species of *Eretmocerus* are commonly misidentified, or the host may be misidentified. Hosts for some species of *Eretmocerus* remain unknown.

Eret. adustiscutum, Krishnan and David 1996; *Lipaleyrodes euphorbiae* David and Subramaniam

Eret. aleurobi Ishi 1938; *Aleurolobus marlatti* (Quaintance)

Eret. aleurothrixus Rose 2000; *Aleurothrixus floccosus* (Maskell)

Eret. aleyrodesii (Cameron) 1912; *Aleyrodes atriplex* Froggatt

Eret. aleyrodiphaga (Risbec) 1951; *Aleurotrachelus socialis* Bondar

Eret. ampliatus Rose 2000; *Aleurothrixus floccosus* (Maskell)

Eret. australis Girault 1921

Eret. bisetae Hayat 1998; *Dialeurodes*

Eret. breviclavus Subba Rao 1984

Eret. cadabae Viggiani 1982; *Aleuroplatus cadabae* Priesner and Hosney

Eret. californicus Howard 1895; *Aleurothrixus floccosus* (Maskell), *Aleyrodes*, *Aleyrodes spiraeoides* Quaintance, *Bemisia* (*tabaci* group), *Trialeurodes vaporariorum* (Westwood)
Eret. clauseni Compere 1936
Eret. comperei Rose 2000; *Aleurothrixus floccosus* (Maskell)
Eret. corni Haldeman 1850; *Aleurocanthus*, *Aleurothrixus porteri* Quaintance and Baker, *Aleyrodes*, *Bemisia* (*tabaci* group), *Dialeurolonga fici* David and Subramaniam, *Pealius quercus* (Signoret), *Singhius hibisci* (Kotinsky), *Trialeurodes packardi* (Morrill), *Trialeurodes vaporariorum* (Westwood)
Eret. debachi Rose and Rosen 1991-92; *Parabemisia myricae* (Kuwana)
Eret. delhiensis Mani 1941; *Aleyrodes*, *Neomaskellia bergii* (Signoret)
Eret. desantisi Rose 2000; *Aleurothrixus floccosus* (Maskell)
Eret. dialeurolonga Krishnan and David 1996
Eret. diversiciliatus Silvestri 1914; *Acaudaleyrodes citri* (Priesner and Hosni), *Bemisia* (*tabaci*) group, *Siphoninus phillyreae* (Haliday)
Eret. dozieri Rose 2000; *Aleurothrixus floccosus* (Maskell)
Eret. eremicus Rose and Zolnerowich 1997; *Bemisia* (*tabaci* group), *Trialeurodes abutiloneus* (Haldeman), *Trialeurodes vaporariorum* (Westwood)
Eret. emiratus Zolnerowich and Rose 1998; *Bemisia* (*tabaci* group)
Eret. exilis Rose 2000; *Aleurothrixus floccosus* (Maskell)
Eret. flavus Krishnan and David 1996; *Lipaleyrodes euphorbiae* David and Subramaniam
Eret. furuhashii Rose and Zolnerowich 1994; *Parabemisia myricae* (Kuwana)
Eret. gracilis Rose 2000; *Aleurothrixus floccosus* (Maskell)
Eret. gunturiensis Hayat 1972; *Aleurocanthus woglumi* Ashby
Eret. haldemani Howard 1908; *Aleurolobus*, *Aleuroplatus coronata* (Quaintance), *Aleurothrixus floccosus* (Maskell), *Aleyrodes*, *Aleyrodes spiraeoides* Quaintance, *Bemisia* (*tabaci* group), *Trialeurodes abutiloneus* (Haldeman), *Trialeurodes vaporariorum* (Westwood)
Eret. hayati Zolnerowich and Rose 1998; *Bemisia* (*tabaci* group)
Eret. hydrabadensis Husain and Agarwal 1982
Eret. illinoisensis Dozier 1932; host unknown
Eret. indicus Hayat 1972; *Aleurocanthus woglumi* Ashby, *Aleurolobus*
Eret. jimenezi Rose 2000; *Aleurothrixus floccosus* (Maskell)
Eret. joeballi Rose and Zolnerowich 1997a; *Bemisia* (*tabaci* group)
Eret. lativentris Girault 1932
Eret. longicornis Viggiani and Battaglia 1983; *Aleurolobus wunni* (Ryberg), *Aleurotrachelus jelinekii* (Frauenfeld)
Eret. longipes Compere 1936; *Aleurotuberculatus takahashii* David and Subramaniam, *Tetraleurodes mori* (Quaintance)
Eret. longiscapus Hayat 1998; *Aleurolobus*
Eret. longiterebrus Rose 2000; *Aleurothrixus floccosus* (Maskell)
Eret. melanoscutus Zolnerowich and Rose 1998; *Bemisia* (*tabaci* group)
Eret. mundus Mercet 1931; *Acaudaleyrodes citri* (Priesner & Hosny), *Aleuroplatus cadabae* Priesner and Hosney, *Aleyrodes proletella* (Linnaeus), *Asterobemisia carpini* (Koch), *Bemisia afer* (Priesner & Hosny), *Bemisia ovata* (Goux), *Bemisia* (*tabaci* group), *Dialeurodes kirkaldyi* (Kotinsky), *Neomaskellia bergii* (Signoret), *Siphoninus phillyreae* (Haliday), *Trialeurodes ricini* (Misra), *Trialeurodes vaporariorum* (Westwood)
Eret. nairobii Gerling 1970; *Aleurocanthus hansfordi* Corbett, *Aleurocanthus zizyphi* Priesner and Hosny
Eret. nativus Girault 1930
Eret. neobemisiae Yashnosh 1974; *Asterobemisia atraphaxius* (Danzig)
Eret. nikolskajae Myartseva 1973; *Bulgarialeurodes cotesii* Maskell, *Tetralicia erianthi* Danzig
Eret. orientalis Gerling 1969; *Aleurocanthus inceratus* Silvestri, *Aleurocanthus spiniferus* (Quaintance), *Bemisia* (*tabaci* group)
Eret. pallidus Dozier 1932; *Tetraleurodes*
Eret. paulistus Hempel 1904; *Aleurothrixus floccosus* (Maskell), *Neomaskellia bergii* (Signoret)

Eret. perseae Rose and Zolnerowich 2004; *Tetraleurodes perseae* Nakahara
Eret. picketti Rose and Zolnerowich 2003; *Tetraleurodes acaciae* (Quaintance)
Eret. portoricensis Dozier 1932; *Aleurothrixus floccosus* (Maskell), *Tetraleurodes acaciae* (Quaintance)
Eret. queenslandensis Naumann and Schmidt 2000; *Bemisia* (*tabaci* group)
Eret. rajasthanicus Hayat 1976; *Acaudaleyrodes rachipora* (Singh)
Eret. rosei Evans and Bennett 1996; *Dialeurodes kirkaldyi* (Kotinsky)
Eret. roseni Gerling 1972; *Acaudaleyrodes citri* (Priesner & Hosny), *Acaudaleyrodes rachipora* (Singh), *Bemisia afer* (Priesner and Hosny), *Bemisia hancocki* Corbett
Eret. rui Zolnerowich and Rose 2004; *Bemisia* (*tabaci* group)
Eret. sculpturatus Hayat 1998
Eret. serius Silvestri 1928; *Aleurocanthus citriperdus* Quaintance and Baker, *Aleurocanthus rugosa* Singh, *Aleurocanthus spiniferus* (Quaintance), *Aleurocanthus woglumi* Ashby, *Bemisia* (*tabaci* group), *Neomaskellia bergii* (Signoret)
Eret. silvestrii Gerling 1969
Eret. siphonini Viggiani and Battaglia 1983; *Aleurolobus niloticus* Priesner and Hosny, *Siphoninus phillyreae* (Haliday)
Eret. staufferi Rose and Zolnerowich 1997a; *Bemisia* (*tabaci* group), *Trialeurodes abutiloneus* (Haldeman)
Eret. tejanus Rose and Zolnerowich 1997a; *Bemisia* (*tabaci* group)
Eret. trialeurodis Hayat 1998; *Trialeurodes ricini* (Misra)
Eret. warrae Naumann and Schmidt (in DeBarro et al. 2000); *Trialeurodes vaporariorum* (Westwood)
Eret. zippanguiphagus Hulden 1986: *Aleurotuberculatus similis* Takahashi, *Pealius quercus* (Signoret)

Chapter 6
Molecular Characterization with RAPD-PCR: Application of Genetic Diagnostics to Biological Control of the Sweetpotato Whitefly

Don C. Vacek[1], Raul A. Ruiz[1], Matthew A. Ciomperlik[1], and John A. Goolsby[2]

Abstract The application of genetic diagnostics under the umbrella of classical taxonomy was imperative for successful development and delivery of the biological control program against the sweet potato whitefly, *Bemisia tabaci* (Gennadius) biotype B (= silverleaf whitefly, *B. argentifolii* Bellows and Perring). In 1990, conventional genetic methodologies were not feasible because of the limited quantity of DNA contained in the tiny parasitoids imported for use against *B. tabaci*. At that time, a novel and rapid genetic method, RAPD-PCR, was adapted to categorize morphologically similar parasitoids emerging from parasitized whiteflies imported from foreign countries. Individuals of *Eretmocerus* and *Encarsia* from each unique culture in quarantine were quickly characterized using RAPD-PCR with selected DNA primers. Assigning unique RAPD banding patterns to native and foreign parasitoid collections set a precedent by capturing the maximum amount of parasitoid species diversity during quarantine processing; minimizing the number of duplicate cultures; assuring quality control of production colonies; and facilitating ecological studies and field evaluations.

6.1 Introduction

In the practice of biological control, knowing the identity of species in a collection from a particular geographic locality is important. Also critical is the identification of native and exotic cryptic species in a group of individuals that are nearly identical morphologically. Cryptic species can have significantly different biological

---

[1] USDA-APHIS-PPQ Center for Plant Health Science and Technology Laboratory, Moore Air Base, Bldg. 6414, 22675 N. Moorefield Rd., Edinburg, TX 78541

[2] USDA-ARS, Kika de la Garza Subtropical Agricultural Research Center, Beneficial Insects Research Unit, 2413 East Highway 83, Weslaco, TX 78596

J. Gould et al. (eds.), *Classical Biological Control of* Bemisia tabaci *in the United States.*

attributes that influence their success as biological control agents, yet may not be separable using known morphological characters. For example, is a collection of morphologically identical adults from a specific location different from previous collections by explorers at the same locale? On examination of voucher specimens classical taxonomists often say they are identical, yet molecular methods may confirm differences that underlie unique biological attributes.

Genetic technology had a significant impact on the development and delivery of the biological control program against the sweet potato whitefly, *Bemisia tabaci* (Gennadius) biotype B (= silverleaf whitefly, *B. argentifolii* Bellows and Perring). In 1991, challenges in identification of native and imported natural enemies were met with a specific molecular diagnostic tool, randomly amplified polymorphic DNA-polymerase chain reaction (RAPD-PCR). Subsequent taxonomic studies corroborated the data from the molecular assays. Genetic patterns assigned to world-wide natural enemy collections set a precedent for identification of cryptic species, facilitation of mass rearing, quality control, ecological studies, and field evaluation of natural enemies.

In 1990 available methodologies, isozyme electrophoresis and standard DNA extraction, coupled with restriction enzymes, provided differentiation, but were not feasible because of the limited quantity of DNA contained in the tiny parasitoids imported for use against *B. tabaci* (Legaspi et al. 1996). At that time, labor-intensive cloning and sequencing required large amounts of up-front time and funding, two resources that have historically limited research on biological control of insect pests. RAPD-PCR was the only technique readily available that could give immediate (48 h) discriminating DNA patterns for identification of insects with very low DNA content (Black et al. 1992). Consequently, early challenges in identification of unique cultures (defined in Chapter 7) were met with RAPD-PCR (Legaspi et al. 1996; Vacek et al. 1996). Later taxonomic (Rose and Zolnerowich 1997; Zolnerowich and Rose 1998), behavioral (Goolsby et al. 1996, 1998), satellite DNA (Heilmann 2001), and sequencing studies (Babcock and Heraty 2000; Debarro et al. 2000) corroborated the RAPD data.

6.2 Importation and Preparation of Specimens

Parasitized whiteflies on host material were imported from foreign countries to the USDA-APHIS-PPQ, Mission Biological Control Laboratory (MBCL) in Mission, Texas, now known as the Center for Plant Health Science and Technology Laboratory. Entomologists at the Arthropod Quarantine Facility of the MBCL received 235 cultures of natural enemies and reared 55 populations of *Encarsia* spp. and *Eretmocerus* spp. collected in Africa, Asia, Europe, and South America, some of which contained new species (Legaspi et al. 1996; Zolnerowich and Rose 1998; see Chapters 2, 4, 5, and 7).

In order to facilitate the molecular analysis, the *Eretmocerus* and *Encarsia* parasitoids were separated into distinct groups using the morphology of the pupae and adult females. Individual specimens from foreign collections were categorized by whitefly host plant, specific collection site, and macro-characters of the whiteflies and parasitic Hymenoptera. Individuals representing each unique combination of site, plant type, and macro-characters (see Chapter 7) were received fresh frozen from entomologists at the Arthropod Quarantine Facility (Goolsby et al. 2000) and assayed for RAPD patterns by the Genetics Unit. Periodically, samples of individuals from production colonies were submitted to the Genetics Unit for quality assessment.

Parasitoids were categorized with morphologically based systematics and RAPD-PCR. Cohorts of the original parental material were sent to cooperating systematists, who were able to characterize them while the original parental cohort was still alive. Typically, material was characterized using both genetic and morphological methods within 2–3 days after acceptance into quarantine.

6.3 RAPD-PCR Technology

The molecular genetic technique, RAPD-PCR (modified by Black et al. 1992), was applied and adjusted as necessary for the characterization of the parasitoids. Template DNA extracted from individual wasps was amplified in a GenAmp PCR System 9600 (Perkin Elmer) programmable thermal cycler utilizing *Taq* DNA Polymerase (Promega Corp.) and various Operon Technologies primers from 10-mer primer kits (OPA and OPC), and custom manufactured primers.

The RAPD-PCR process used 2.0 µl of DNA preparation containing the specimen's DNA (template) which was added to a 48.0 µl cocktail of reaction mix that contained dNTPs, $MgCl_2$, buffers, primer, and polymerase. The PCR program (Black et al. 1992) consisted of temperature profiles programmed into the thermal cycler as follows: (1) 80°C for 25 min; (2) 94°C for 1 min; (3) 92°C for 1 min; (4) 35°C for 1 min; (5) slope to 72°C at 1°C steps every 8 s (approximately 4:54); (6) 72°C for 2 min; (7) cycle to step 3, 45 times; (8) 72°C for 2 min; (9) hold at 4°C indefinitely.

The resulting DNA amplification products were separated by agarose gel electrophoresis (Sambrook et al. 1989). PCR products were loaded onto a 1.2% agarose gel and run for a period of time to allow the amplicons to separate into characteristic banding patterns. The electrophoresis gels were pre-stained with ethidium bromide and visualized on a UV transilluminator. All gels were digitally scanned and archived with an analytical imaging system (Millipore BioImage).

The imaging software, which captures the gel image and detects bands, assigned a base pair value to each band produced by the RAPD from individual insects. Base pair values are calculated from a molecular weight standard (PhiX174-*Hae* III) containing DNA fragments of known base pair size loaded alongside samples being investigated. Individual bands may vary in intensity, and range in length from 100 base pairs (bp) to 2,000 bp. Each sample was labeled with the pertinent information and stored on an optical disk. BioImage software was used to compare the RAPD banding pattern from one gel lane to the patterns in all lanes on that gel and to all gel images produced by previous diagnostic runs. Computer comparisons facilitated the identification of new banding patterns and determination of similarities among individuals and populations. A pattern key booklet was generated using computer images from electrophoresis gel photographs to allow the characterization of individuals as having a specific, known RAPD pattern or a new pattern. The information pertaining to each individual allowed the determination of RAPD variation within and among collections. Once the pattern was examined, a unique identification code (RAPD pattern number) was assigned to each insect or collection and entered into data base archives.

Reproducible banding patterns were determined by comparing RAPD patterns within and between templates of a single insect. Characteristic banding patterns and all bands unique to individual collections were labeled as possible DNA marker bands, which were then used as discrete differences to group a collection into a genetic type. Template standards were incorporated during the analysis to guarantee that the reagents and technique met acceptable standards, and positive and negative controls were included during each run as internal control standards. Repeatability was validated by assessment of each batch of PCR reaction mix with known DNA templates and expected RAPD bands were repeatable from year to year.

A RAPD pattern number or code was used to identify an individual parasitoid species or genetically distinctive population. As cultures of exotic parasitoids were received from overseas collectors, they were reared in the Arthropod Quarantine Facility and identified by taxonomists at the ARS Systematic Entomology Laboratory and Texas A&M University. As the taxonomic, behavioral, and other biological data collected by these scientists corroborated that a particular RAPD pattern represented a genetically distinct group (i.e., a new species), that RAPD pattern was given a species designation. Images of assays were annotated with the BioImage software to reveal the base pair length for selected bands in each pattern and to illustrate key information used in the interpretation of results.

6.4 Characterization of RAPD Patterns

Individuals of *Eretmocerus* and *Encarsia* from each unique culture were immediately characterized using RAPD-PCR with primers OPC-04 and OPA-10 (Operon Technologies), the two most discriminating primers. Relative intensities of RAPD bands, while important, were generally not used as a significant character in determining the identity of individuals. Intense bands that were easily scored, repeatable, and present across all individuals in an accession or species were used as salient characters for designating a RAPD pattern. Figure 6.1 shows examples of *Encarsia* spp. patterns observed from primer OPC-04. With only visual inspection, the *Encarsia* RAPD patterns EN-1 and EN-2 are clearly different. There is little variation among the individuals within an accession because the patterns, for the most

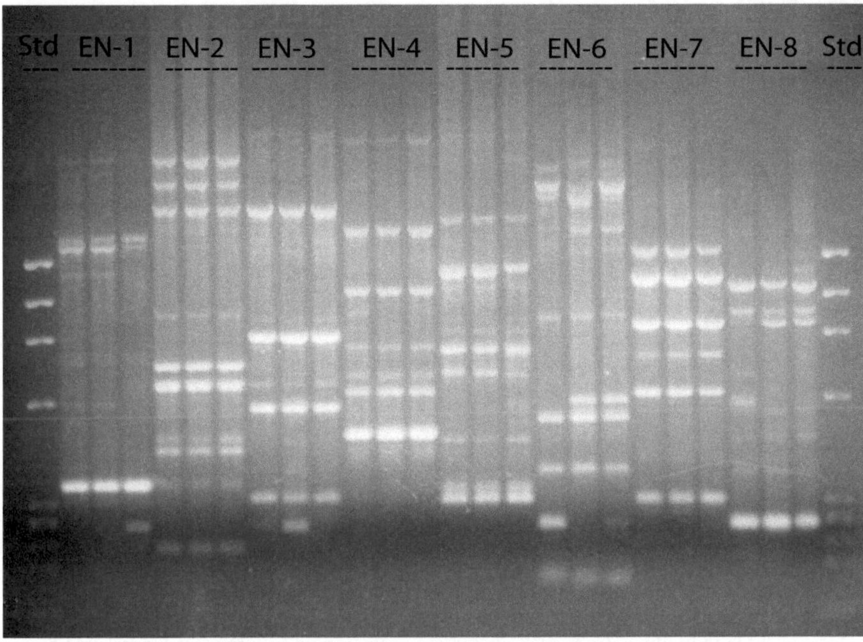

Fig. 6.1 *Encarsia* spp. RAPD patterns from amplification with primer OPC-04. Each vertical lane represents the RAPD banding pattern of a single amplified individual. The amplification products of three individuals from each quarantine accession or culture were loaded side by side on the electrophoretic gel. A DNA standard (PhiX174-*Hae* III) was loaded in the first and last lane. The species designation and origin for each of the eight patterns is as follows: EN-1, *Enc. bimaculata* (India) [formerly *Enc.* sp. nr. *strenua*]; EN-2, *Enc. formosa* (Greece and Egypt); EN-3 and 4, *Enc. sophia* (Taiwan); EN-5, *Enc. sophia* (Thailand, Malaysia, and Pakistan); EN-6, *Encarsia* sp. (Thailand); EN-7, *Enc. sophia* (Spain); EN-8, unknown. See Chapter 7, Goolsby et al. for more collection information and a complete list of *Encarsia* spp.

part, are identical among individuals within an accession or species. Figure 6.2 shows examples of *Eretmocerus* spp. patterns observed from primer OPC-04. Again, DNA banding patterns show clear differences that are consistent among different cultures. For example, a comparison of RAPD patterns ERET-1 and ERET-2 clearly shows that some bands are unique to ERET-1 while other bands are unique to ERET-2.

6.5 Impact of RAPD-PCR Technology

Assigning unique RAPD patterns to native and foreign parasitoid collections set a precedent by capturing the maximum amount of parasitoid species diversity during quarantine processing; minimizing the number of duplicate cultures; assuring quality control of production colonies; and facilitating ecological studies and field evaluations.

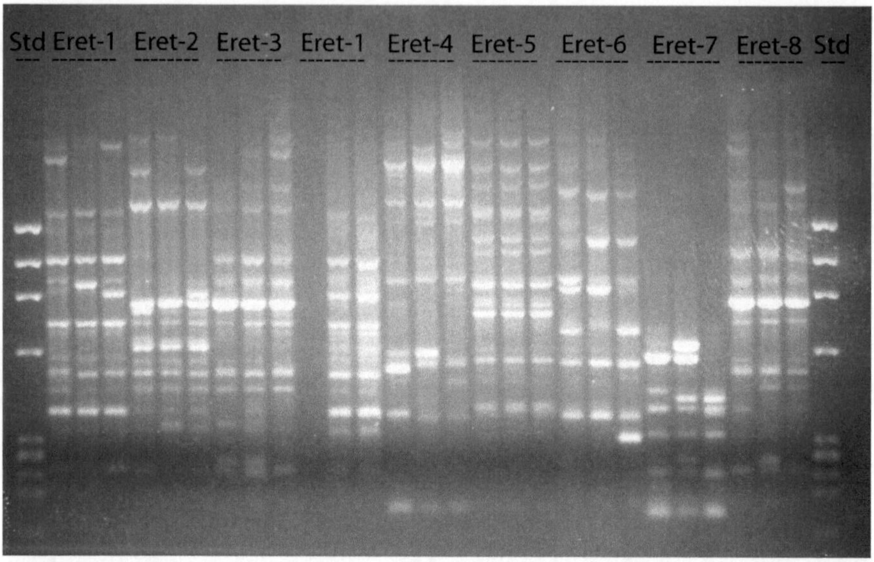

Fig. 6.2 *Eretmocerus* spp. RAPD patterns from amplification with primer OPC-04. Each vertical lane represents the RAPD banding pattern of a single amplified individual. The amplification products of three individuals from each quarantine accession or culture were loaded side by side on the electrophoretic gel. A DNA standard (PhiX174-*Hae* III) is loaded in the first and last lane. The species designation and origin for each of the eight patterns is as follows: ERET-1, *Eret. mundus* (Murcia, Spain; Padappai, India; Cairo, Egypt; Tainan, Taiwan; Frascati, Castel Gondolfo, and Testa Di Lespe, Italy; Gat, Golan Kibbutz, Golan Ma'Aleh Gamla, and Negev Desert, Israel; Santa Groce, Sicily, and Nicosia, Cyprus); ERET-2, *Eret. hayati* (Multan, Pakistan; Thirumala, India); ERET-3, *Eret. melanoscutus* (Chiang Mai and Kampang Sain, Thailand and Tainan, Taiwan); ERET-4, *Eret. eremicus* (Brawley, California, USA); ERET-5, *Eret. staufferi* (College Station, Texas, USA); ERET-6, *Eret. tejanus* (Mission, Texas, USA); ERET-7, unknown; ERET-8, *Eret. melanoscutus* (Sai Noi Klong Ha Roi, Thailand). See Chapter 7 for more collection information and a complete list of *Eretmocerus* spp.

In quarantine, RAPD analysis allowed for rapid population typing of parasitoid cultures. Based on RAPD patterns, unique parasitoid cultures (possible cryptic species) were set up in pure cultures and reared on the local *B. tabaci* biotype B. The integration of the morphology-based systematics and the molecular techniques optimized the characterization of natural enemies in quarantine. This integration allowed each new cryptic species to be fully evaluated and maintained as a pure culture, which cooperating systematists were able to access. Except for differences in RAPD patterns, new *Eretmocerus* species were only distinguishable from each other by minute differences in the first funicular antennal segment of the female (Rose and Zolnerowich 1997; Zolnerowich and Rose 1998). Some parasitoid types were simply not distinguishable based on morphological characters (for example, populations of *Encarsia sophia* exhibited biological differences). Screening with RAPD assays allowed us to distinguish such cryptic species at the time of culture initiation and during efficacy studies.

Multiple geographic cultures with redundant genotypes were eventually combined into a single culture, greatly conserving resources in quarantine. Without RAPD assays, multiple cultures of widely distributed species, such as *Eret. mundus*, would have consumed most of the quarantine space and personnel resources. Monthly quality assessments of production colonies of the parasitoids ensured that strain purity was maintained at an acceptable level for field delivery.

The techniques were equally valuable in field-testing. It was possible to release multiple species simultaneously in the field and determine the species in the recovery samples by their DNA profile. Combining the use of RAPD-PCR and classical systematics allowed for an increase in the numbers of individuals that could be characterized to the species level in the evaluation effort. Ultimately this led to the best possible determination of which species showed the greatest efficacy in the field. However, the molecular methods were valuable only because they were combined with classical systematics. Properly identified and curated specimens, coupled with frozen specimens, will provide the best permanent record of the species released and established. Representatives of all the cultures were cryogenically stored at MBCL and vouchered at the Texas A&M University, Department of Entomology Collection, College Station, Texas and the USDA-ARS, Systematic Entomology Laboratory, Washington, DC.

The cost of establishing RAPD-PCR capabilities at the MBCL was $258,000. If amortized for 10 years across all projects served, the implementation cost to the *B. tabaci* project was $26,000. Routine cost per specimen for identification with two RAPD primers was $10. Approximately 500 specimens were assayed per year for a total cost of $5,000 per year. There were considerable monetary and non-tangible benefits; savings of $415,000 were realized by the reduction in personnel, as well as lowered costs of supplies and utilities. In addition, project delivery time was considerably shorter, and precise identification for project evaluations was possible.

Integration of classical taxonomy and molecular diagnostics has proved useful in allowing the maximum amount of species diversity with a minimum amount of duplication in cultures. Because of limited funding and personnel, urgency of the program, and minuteness of the natural enemies, RAPD-PCR proved to be the most efficient and effective method to identify genetically unique accessions of exotic *Encarsia* spp. and *Eretmocerus* spp. in the early 1990s. In field evaluation efforts, the two methods were integrated for identifying indigenous and imported parasitoids (see Chapters 11–14).

This biological control program has demonstrated that the application of genetic diagnostics under the umbrella of classical taxonomy is imperative for a successful program. It is clear that there is a world wide decrease in classical systematics and simultaneously an increased investment with rapid advancement in molecular technologies. Future biological control programs must adequately fund basic taxonomy *and* rapid, high throughput, real-time PCR diagnostic methods to meet the needs of a large-scale biological control program.

Acknowledgements We are grateful to many individuals at the APHIS, PPQ, MBCL, Mission, TX: William Keale, Deanna Chapa, Owais Ahmed, and Edna Zavala for their technical support in RAPD-PCR; Lloyd Wendel, Ron Hennessey, and Elena Ibanez for managerial and logistical support. Tom Unruh, Agricultural Research Service, Yakima, Washington, DC, and Bill Black, Colorado State University, Ft Collins, Colorado, provided early training in the methods of electrophoresis and RAPD-PCR. Bill Roltsch and Charlie Pickett, California Department of Food and Agriculture, Sacramento, California and Greg Simmons and Juli Gould, Phoenix Plant Protection Laboratory, Phoenix, Arizona, provided important parasitoid materials from field evaluation studies. Larry Heilmann and Arland Oleson, North Dakota State University, Fargo, North Dakota, provided satellite DNA methodology.

References

Babcock, C.S. and J.M Heraty. 2000. Molecular markers distinguishing *Encarsia formosa* and *Encarsia luteola* (Hymenoptera: Aphelinidae). Annals of the Entomological Society of America 93 (4): 738–744.

Black, W.C. IV, N.M. DuTeau, G.J. Puterka, J.R. Nechols, and J.N. Pettorini. 1992. Use of the random amplified polymorphic DNA polymerase chain reaction (RAPD-PCR) to detect DNA polymorphisms in aphids (Homoptera: Aphididae). Bulletin of Entomological Research 82: 151–159.

DeBarro, P.J., F. Driver, I.D. Naumann, G.M. Clarke, and J. Curran. 2000. Descriptions of three species of *Eretmocerus* Haldeman (Hymenoptera: Aphelinidae) parasitising *Bemisia tabaci* (Gennadius)(Hemiptera: Aleyrodidae) and *Trialeurodes vaporariorum* (Westwood)(Hemiptera: Aleyrodidae) in Australia based on morphological and molecular data. Australian Journal of Entomology 39: 259–269.

Goolsby, J.A, J.C. Legaspi, and B.C. Legaspi, Jr. 1996. Quarantine evaluation of exotic parasites of the sweetpotato whitefly, *Bemisia tabaci* (Gennadius). Southwestern Entomologist 21: 13–21.

Goolsby, J.A., M.A. Ciomperlik, B.C., Legaspi, Jr., J.C. Legaspi, and L. E. Wendel. 1998. Laboratory and field evaluation of exotic parasitoids of *Bemisia tabaci* (Biotype ‘B’) in the Lower Rio Grande Valley of Texas. Biological Control 12: 127–135.

Goolsby, J.A., M.A. Ciomperlik, A.A. Kirk, W.A. Jones, B.C. Legaspi, R. A. Ruiz, D. C. Vacek, and L. E. Wendel. 2000. Predictive and empirical evaluation for parasitoids of *Bemisia tabaci* (Biotype “B”), based on morphological and molecular systematics. In A.A. Austin and M. Dowton, eds., Hymenoptera: Evolution, Biodiversity, and Biological Control, pp. 347–358. 4th International Hymenopterists Conference, 1999, Canberra, ACT, Australia, CSIRO Publishing, Collingwood, Victoria.

Heilmann, L.J, Jianzhong Zhang, E. R. Andress, A. E. Olson, D. R. Nelson, J. R. Gould, M.W. Ciomperlik, J.A. Goolsby, and D. C. Vacek. 2001. DNA based species identification test for parasitic wasps of the whitefly. In T.J. Henneberry, N.C. Toscano, R.M. Faust, and T.M. Perring, eds., Silverleaf Whitefly: 2001 Supplement to the 5-year National Research and Action Plan – second review. USDA-ARS, San Diego, CA.

Legaspi, J.C., B.C. Legaspi, Jr., R.I. Carruthers, J.A. Goolsby, W.A. Jones, A.A. Kirk, C. Moomaw, T.J. Poprawski, R.A. Ruiz, N.S. Talekar, and D.C. Vacek. 1996. Foreign exploration for natural enemies of *Bemisia tabaci* from Southeast Asia. Subtropical Plant Science 48: 48–53.

Rose, M. and G. Zolnerowich. 1997. *Eretmocerus* Haldeman (Hymenoptera: Aphelinidae) in the United States, with descriptions of new species attacking *Bemisia* (*tabaci* complex) (Homoptera: Aleyrodidae). Proceedings of the Entomological Society, Washington 99: 1–27.

Sambrook, J., E.F. Fritsch, and T. Maniatis. 1989. Molecular Cloning: A Laboratory Manual, 2nd Edition, Cold Spring Harbor Laboratory Press, New York.

Vacek, D.C., R.A. Ruiz, and L.E. Wendel. 1996. RAPD-PCR identification of natural enemies of SLWF. In Proceedings of the 4th Annual Progress Review. 5-Year National Research Action Plan Development, Management, and Control Methods for Silverleaf Whitefly, p. 149. February, 4–6 1996, USDA-ARS, San Antonio, TX. ARS 1996-01.

Zolnerowich, G. and M. Rose. 1998. *Eretmocerus* Haldeman (Hymenoptera: Aphelinidae) imported and released in the United States for control of *Bemisia* (*tabaci* complex) (Homoptera: Aleyrodidae). Proceedings of the Entomological Society, Washington 100: 31–323.

Chapter 7
Quarantine Evaluation of Parasitoids Imported into the USA for Biocontrol of *Bemisia tabaci* Biotype B

John A. Goolsby[1], Benjamin C. Legaspi, Jr.[2], and Jesusa C. Legaspi[3]

Abstract A total of 38 exotic and 2 native parasitoid populations of the sweetpotato whitefly, *Bemisia tabaci* biotype B, were evaluated in pre-release quarantine efficacy tests. Numbers of *B. tabaci* parasitized were counted in sleeve cages on cantaloupe melons (*Cucumis melo* 'Perlita'), cotton (*Gossypium hirsutum* 'Delta Pine 51'), and broccoli (*Brassica oleracea* 'Patriot'). Highest attack rates were found for *Encarsia* sp. nr. *pergandiella* (Brazil) and *Eretmocerus mundus* (Spain) on melons; *Eretmocerus hayati* (Pakistan) on cotton; and *Eretmocerus mundus* (Spain) on broccoli. In the laboratory, these three exotic parasitoids attacked significantly greater numbers of hosts than the native species of *Encarsia pergandiella* and *Eretmocerus tejanus*. This information was used to prioritize parasitoid species for mass rearing and release in biocontrol-based IPM programs against *B. tabaci*. The quarantine tests proved to be a good indicator of efficacy in the field.

7.1 Introduction

As the primary quarantine facility in the USA for the importation of exotic natural enemies of the sweetpotato whitefly, *Bemisia tabaci* (Genn.) strain B (also called silverleaf whitefly, *B. argentifolii* Bellows and Perring), the USDA-APHIS-PPQ Mission Biological Control Laboratory (MBCL) in Texas processed over 80 shipments of predators, parasitoids and pathogens sent by collectors world-wide from 1992 to 1998. These shipments resulted in the culture of 55 populations of *Encarsia* spp. and *Eretmocerus* spp. (both Aphelinidae) and several predator species. Parasitoids were categorized using RAPD-PCR and morphologically-based taxonomy (Chapters 4–6). Integration of the two techniques proved to be useful in capturing the maximum amount of species diversity with minimum duplication in cultures. Predictive

[1]USDA-ARS, Kika de la Garza Subtropical Agricultural Research Center, Beneficial Insects Research Unit, 2413 East Highway 83, Weslaco, TX 78596

[2]Florida Public Service Commission, 2540 Shumard Oak Blvd., Tallahassee, FL 32399

[3]USDA-ARS CMAVE – Florida A&M University, Center for Biological Control, 6383 Mahan Drive, Tallahassee, FL 32308

J. Gould et al. (eds.), *Classical Biological Control of* Bemisia tabaci *in the United States.*

evaluation methods were developed to analyze the performance of the imported parasitoid species under quarantine conditions (Goolsby et al. 1996). Promising parasitoid populations were then reared and released onto the same crops that were tested in quarantine in field cage or open-field evaluations in order to measure rates of parasitism under field conditions (Goolsby et al. 1998; Hoelmer 1998; Goolsby et al. 2000) (Chapters 8–10).

7.2 Quarantine Methods

Identification of new organisms being imported is one of the critical functions of quarantine work. Separation of natural enemies into distinct taxa should occur as cultures are initiated. Characterization of natural enemies in quarantine often requires a blend of observations of their morphology, mating behavior, host plant associations, etc. Taxonomic keys may be available, but are difficult to use with live material. At the start of the US *Bemisia* biological control program, up-to-date taxonomic keys to the *Eretmocerus* and *Encarsia* were not available for either the indigenous or imported species. *Eretmocerus* species indigenous to the southern and desert southwestern USA were poorly understood; based on biosystematic studies those *Eretmocerus* attacking *B. tabaci* were determined to be a complex of species (Hunter and Rose 1996). Foreign material arriving into quarantine showed extensive diversity, especially in the genus *Eretmocerus* (Legaspi et al. 1996a; Goolsby et al. 1998). To best handle the issues of cryptic species, species complexes, and the need to initiate pure cultures representing the maximum available diversity of *B. tabaci* natural enemies, a unique quarantine protocol was developed which integrated biosystematics and molecular techniques.

Foreign collections were categorized in quarantine according to their plant host, country or region of origin, and the macro-characters of the parasitic Hymenoptera and Aleyrodidae. Only parasitoids reared from individuals of the *B. tabaci* complex met the APHIS requirements for permitting, as stated in the Environmental Assessments of the genera *Eretmocerus* and *Encarsia* (USDA 1995a, b). Further, the imported species were required to have a uniparental, bi-parental, or autoparasitoid (whereby males typically develop as parasitoids of females of their own species) biology. Species which displayed obligate hyperparasitism (Hunter et al. 1996) of other taxa were not considered suitable for release. The requirements of the Environmental Assessments were intended to identify parasitoid species with the most specificity to the *B. tabaci* complex. Species that met these criteria were acceptable for further processing using our quarantine protocol.

Eretmocerus and *Encarsia* were separated into distinct subgeneric groups using the morphology of the pupae and adult females. Individuals from each unique accession were immediately characterized at the MBCL Genetics Laboratory using RAPD-PCR with primers CO4 and A10 (Black et al. 1992; Vacek et al. 1996). Detailed methodology and representative electrophoretic gel patterns for *Eretmocerus* and *Encarsia* parasitoids are contained in Legaspi et al. (1996a) and in Chapter 6. Cohorts of the original parental material were sent as specimens to cooperating taxonomists. This taxonomic and genetic information allowed the characterization of quarantine material while the original parental cohort was still alive. Typically, material was characterized using both methods within 2–3 days after acceptance into quarantine. Unique parasitoid accessions were set up as pure cultures reared on local Lower Rio Grande Valley (LRGV) populations of *B. tabaci* with hibiscus 'Kona Pink' (*Hibiscus rosasinensis* L.) as the plant host. Duplicate accessions were first evaluated to determine if they had similar attack rates and later combined or, in the later stages of the program, processed only for reference purposes. Representatives of all the accessions were cryogenically stored at the MBCL Genetics Laboratory, and voucher specimens were sent to the Texas A&M University, Department of Entomology Collection, College Station, Texas and the USDA-ARS, Systematic Entomology Laboratory, Washington, DC. All records of importations were recorded manually and electronically on the USDA-ARS Releases of Beneficial Organisms (ROBO) database (http://www.ars-grin.gov/cgi-bin/nigrp/robo/taxon.pl?333).

7.3 Quarantine Evaluations for Efficacy

Quarantine bioassays were developed as a preliminary means of rapidly screening the large number of accessions to identify which species or populations had high attack rates and which were relatively poor performers. Early experiences with the indigenous North American natural enemies showed that parasitism by a key native species, *Eretmocerus eremicus* Rose and Zolnerowich, was low on *B. tabaci* infesting fall/winter cole crops. This resulted in very low numbers of overwintering *Eret. eremicus* and outbreaks of *B. tabaci* on melons in the spring (Hoelmer 1998). The same drop in parasitism by the native *Encarsia* was not noted on winter cole crops. This was evidence of a tritrophic interaction between *Eret. eremicus, B. tabaci* and its cole crop host. To evaluate potential host plant effects on the imported parasitoids, their attack rates were evaluated on several key crop plants, including cotton (*Gossypium hirsutum* L), broccoli (*Brassica oleracea* L.), and cantaloupe melons (*Cucumis melo* L.), which are common hosts of *B. tabaci* in the USA. Parasitoids that performed well in the quarantine tests on these crop hosts were then prioritized in the mass rearing program which was quite extensive and utilized 20 environmental growth chambers, four greenhouses, and 20 field cages for parasitoid production. The quarantine tests provided a scientific rationale for prioritizing species for mass rearing, which was necessary to meet the needs of an action program for this serious pest of agriculture. High priority agents were then available for field evaluations in Arizona, Texas, and California (Chapters 8 and 9).

Parasitoid females selected at random from the quarantine rearing colonies were used in the tests. Parasitoids were confined in sleeve cages with *B. tabaci* for 2 days. *Eretmocerus* spp. were exposed to first and second instar nymphs of *B. tabaci*, whereas *Encarsia* were given more developed third and fourth instars.

The host plants used in the tests were cantaloupe melons ('Perlita'), cotton ('Delta Pine 51'), and broccoli ('Patriot'), common crop varieties in agriculture during the time of the experiments. The complement of species in culture varied over time as new material was collected and introduced into quarantine, and existing cultures determined to be duplicates were terminated or combined. Consequently not all species were screened simultaneously in this manner on all three host crops and some accessions were obtained too late in the program to be included in comparative screening.

7.4 Results and Discussion

7.4.1 Parasitoids

A total of 38 exotic and two native parasitoids were evaluated in laboratory experiments (Goolsby et al. 1996, 1998). Highest attack rates were found for *Encarsia* sp. nr. *pergandiella* (Brazil, M93055) and *Eret. mundus* Mercet (Spain, M92014) on melons, for *Eret. hayati* Zolnerowich and Rose (Pakistan, M95012) on cotton and for *Eret. mundus* (Spain) on broccoli (Fig. 7.1). These three exotic parasitoids attacked significantly greater numbers of hosts than the native species *Enc. pergandiella* Howard and *Eret. tejanus* Rose and Zolnerowich attacked in these laboratory tests. When results from tests on the three different crops were pooled, *Eret. mundus* (Spain) and *Eret. hayati* (Pakistan) performed significantly better than the other 36 populations tested (Table 7.1). The results also show significant tritrophic interactions; *Eretmocerus mundus* populations performed well on all three host plant species whereas *Eret. hayati* showed higher levels of attack on cotton than on melons or broccoli.

While the quarantine screening was being conducted, all the permitted species including the poor performers were being released into the field (Chapter 9). We found broad agreement between the results of the quarantine and field tests, indicating the utility of the quarantine evaluation program to help select the most promising species.

7.4.2 Predators

Three predator species were imported and cultured in quarantine (Table 7.2). *Serangium parcesetosum* Sicard was imported early in the program and was permitted and released in Arizona, California, and Texas (Chapter 8). In laboratory studies, *S. parcesetosum* preferred *B. tabaci* nymphs over other prey offered, with each beetle consuming about 600 individual prey every 24 h (Legaspi et al. 1996b). The second two species, *Acletoxenus formosus* Loew and *Serangium* sp.

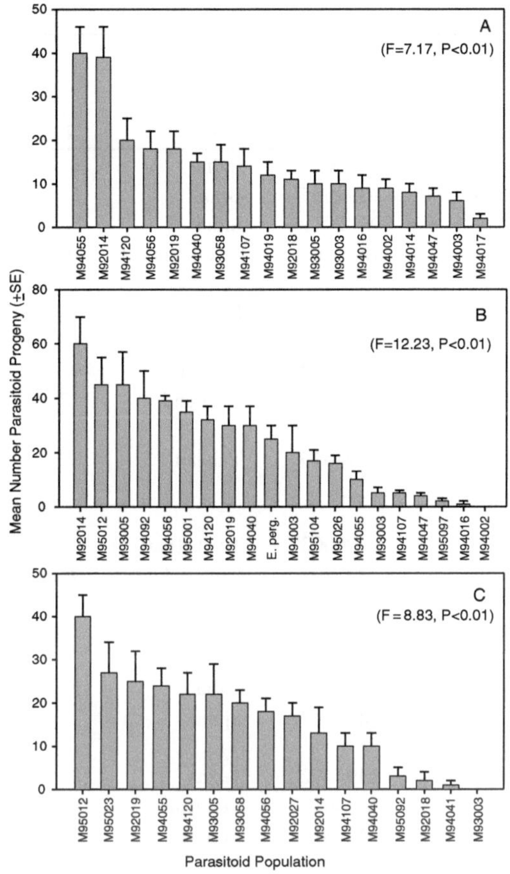

Fig. 7.1 Numbers of whitefly hosts attacked by different parasitoid populations in the laboratory on melons (**a**), broccoli (**b**), and cotton (**c**).

Table 7.1 Quarantine performance of *Bemisia tabaci* parasitoids ranked by number of *B. tabaci* nymphs attacked.

Species identification	MBCL ID culture	Collection origin	N	Mean no. *B. tabaci* nymphs attacked	LSD group
Eretmocerus mundus Mercet	M92014	Spain	10	44.7	a
Eretmocerus hayati Zolnerowich and Rose	M95012	Pakistan	10	41.8	a
Eretmocerus mundus	M94092	Italy	10	38.9	ab
Encarsia sp. (*parvella* group)	M95001	Dominican Republic	2	36.4	abc
Encarsia bimaculata Heraty and Polaszek	M95023	Thailand	9	27.4	bcd

(continued)

Table 7.1 (continued)

Species identification	MBCL ID culture	Collection origin	N	Mean no. *B. tabaci* nymphs attacked	LSD group
Eretmocerus hayati	M93005	India	12	26.4	cde
Encarsia sp. nr. *pergandiella*	M94055	Brazil	5	24.9	cde
Eretmocerus mundus	M92019	India	5	24.1	def
Encarsia pergandiella Howard	(none)	Texas (native)	10	24.0	cde
Encarsia sp. nr. *hispida*	M94056	Brazil	13	18.1	defg
Eretmocerus mundus	M92027	Egypt	1	17.3	defgh
Eretmocerus mundus	M93058	Taiwan	9	16.9	defghi
Eretmocerus melanoscutus Zolnerowich and Rose	M94040	Thailand	12	15.2	efghi
Encarsia sophia (Girault and Dodd)	M94019	Taiwan	8	13.1	fghij
Eretmocerus tejanus Rose and Zolnerowich	M94003	Texas (native)	4	12.8	fghij
Eretmocerus emiratus Zolnerowich and Rose	M95104	United Arab Emirates	9	11.9	ghijik
Eretmocerus sp. nr. *furahashi*	M94026	Taiwan	2	10.9	ghijk
Encarsia lutea Masi	M94107	Philippines	9	8.1	ghijk
Encarsia bimaculata	M92018	India	3	8.0	ghijk
Encarsia sophia	M93003	Spain	2	6.8	ghijk
Eretmocerus staufferi Rose and Zolnerowich	M94002	Texas	3	6.1	hijk
Encarsia sophia	M94047	Malaysia	6	5.8	hijk
Encarsia sophia	M94016	Taiwan	38	5.5	ijk
Eretmocerus mundus	M94092	Italy	1	2.9	jk
Eretmocerus sp.	M95097	Taiwan	1	2.1	jk
Encarsia sophia	M94017	Taiwan	1	1.6	jk
Encarsia sophia	M94041	Thailand	9	0.8	k

Note: Parasitoids are listed by name and unique identification number, country of collection, sample size (N = number of replicate leaves), numbers of hosts attacked, and LSD grouping (values in rows followed by common letters are not significantly different at $P = 0.05$).

underwent preliminary no-choice host range testing in quarantine using several common species of whitefly and armored scale. Both species fed and reproduced on many of the species tested. Based on concerns about releasing polyphagous predators with the potential to attack non-target species and the initial success of the more specific parasitoid species, both predator species were terminated in quarantine.

Table 7.2 Predators imported into quarantine at the Mission Biological Control Laboratory.

Predator Species	MBCL culture number	Molecular ID code	Country of origin	Collectors	Date	Original host plant
Serangium parcesetosum Sicard	M93008	SER-1	India, Podumbu	Kirk, Lacey, Kumar and Kris	I-93	castor bean
Serangium sp. nov.	M94050	SER-2	Malaysia, Kuala Lumpur	Kirk and Lacey	III-94	*Mussaenda* sp.
Acletoxenus formosus Loew	M97018	ACH-1	Spain, Canary Islands	Kirk	V-97	*Sonchus* sp.

7.5 Conclusions

One of the unique aspects of the *Bemisia* biological control program was the availability of numerous species of natural enemies collected during intensive, directed foreign explorations in combination with extensive mass rearing facilities. This unique situation provided the opportunity to evaluate on a large scale many different parasitoid species simultaneously, and predict which species would be the most effective. Many authors have proposed that biological control needs to become a more predictive science rather than an empirical method (Ehler 1990; Harris 1998). Despite considerable discussion of the topic, few biological control programs have attempted to evaluate prospective candidates in a predictive manner. In the *Bemisia* biological control program an attempt was made to predict and test the efficacy of the imported parasitoids. If the predictive methods used in a biological control program are accurate the time needed to identify and evaluate an effective natural enemy can be shortened which has value in conserving resources in a research program and reducing the short-term impact of the target pest.

It is now known retrospectively that four introduced species of *Eretmocerus* and one *Encarsia* have established in the USA (Chapters 11–14). This result is consistent with the pre-release quarantine evaluation which predicted that the *Eretmocerus* species would perform best in the field, adding valuable insights into the biological control program. Future programs should consider using pre-release evaluations as one of the tools available to select the best potential candidates for release. Current USDA Agricultural Research Service guidelines call for host range testing of arthropod biological control agents. Considering the time and resources needed to conduct these tests, pre-release attack rate studies could be used to select the agents that will be host range tested and ultimately considered for release.

References

Black, IV, W.C., N.M. DuTeau, G.J. Puterka, J.R. Nechols, and J.N. Pettorini. 1992. Use of the random amplified polymorphic DNA polymerase chain reaction (RAPD-PCR) to detect DNA polymorphisms in aphids (Homoptera: Aphididae). Bulletin Entomological Research 82: 151–159.

Ehler, L.E. 1990. Introduction strategies in biological control of insects. In M. Mackauer, L.E. Ehler, and J. Roland, eds., Critical Issues in Biological Control, pp. 111–130. Intercept, Andover, Hants.

Goolsby, J.A, J.C. Legaspi, and B.C. Legaspi, Jr. 1996. Quarantine evaluation of exotic parasites of the sweetpotato whitefly, *Bemisia tabaci* (Gennadius). Southwestern Entomologist 21: 13–21.

Goolsby, J.A., M.A. Ciomperlik, B.C. Legaspi, Jr., J.C. Legaspi, and L.E. Wendel. 1998. Laboratory and field evaluation of exotic parasitoids of *Bemisia tabaci* (Biotype 'B') in the Lower Rio Grande Valley of Texas. Biological Control 12: 27–135.

Goolsby, J.A., M.A. Ciomperlik, A.A. Kirk, W.A. Jones, B.C. Legaspi, R.A. Ruiz, D.C. Vacek, and L.E. Wendel. 2000. Predictive and empirical evaluation for parasitoids of *Bemisia tabaci* (Biotype "B"), based on morphological and molecular systematics. In A. Austin and M. Dowton, eds., Hymenoptera: Evolution, Biodiversity, and Biological Control, pp. 347–358. 4th International Hymenopterists Conference, 6–11 January 1999, Canberra ACT, Australia, CSIRO Publishing, Collingwood, Victoria.

Harris, P. 1998. Evolution of classical weed biocontrol: meeting survival challenges. Bulletin of the Entomological Society of Canada 30: 134–143.

Hoelmer, K.A. 1998. Comparative field cage evaluations of top-performing introduced parasitoids in desert cantaloupes. In T.J. Henneberry, N.C. Toscano, T.M. Perring, and R.M. Faust, eds., Silverleaf Whitefly National Research, Action and Technology Transfer Plan, 1997–2001. 1st Annual Review, Charleston, SC, 3–5 February 1998, p. 68. US Department of Agriculture ARS 1998–01.

Hunter, M.S., M.F. Antolin, and M. Rose. 1996. Courtship behavior, reproductive relationships, and allozyme patterns of three North American populations of *Eretmocerus* nr. *californicus* (Hymenoptera: Aphelinidae) parasitizing the whitefly *Bemisia* sp., *tabaci* complex (Homoptera: Aleyrodidae). Proceedings of the Entomological Society of Washington 98: 126–137.

Hunter, M.S., M. Rose, and A. Polazek. 1996. Divergent host relationships of males and females in the parasitoid *Encarsia porteri* (Hymenoptera: Aphelinidae). Annals of the Entomological Society of America 89: 667–675.

Legaspi, J.C., B.C. Legaspi, Jr., R.I. Carruthers, J.A. Goolsby, W.A. Jones, A.A. Kirk, C. Moomaw, T.J. Poprawski, R.A Ruiz, N.S. Talekar, and D. Vacek. 1996a. Foreign exploration for natural enemies of *Bemisia tabaci* from Southeast Asia. Subtropical Plant Science 48: 48–53.

Legaspi, J.C., B.C. Legaspi, Jr., R.L. Meagher, Jr., and M. A. Ciomperlik. 1996b. Evaluation of *Serangium parcesetosum* (Coleoptera: Coccinellidae) as a biological control agent of the silverleaf whitefly (Homoptera: Aleyrodidae). Environmental Entomology 25: 1421–1427.

Vacek, D.C., R.A. Ruiz, and L.E. Wendel. 1996. RAPD-PCR identification of natural enemies of SPWF. In T.J. Henneberry, N.C. Toscano, T.M. Perring, and R.M. Faust, eds., Silverleaf Whitefly: National Research, Action and Technology Transfer Plan, 1997–2001. Fourth Annual Review, 4–6 February 1996, San Antonio, TX, US Department of Agriculture ARS 1996–01.

US Department of Agriculture, Animal and Plant Health Inspection Service. 1995a. Field releases of nonindigenous parasitic wasps in the genus *Eretmocerus* (Hymenoptera: Aphelinidae) for biological control of whitefly pests (Homoptera: Aleyrodidae). Environmental Assessment, Riverdale, MD.

US Department of Agriculture, Animal and Plant Health Inspection Service. 1995b. Field releases of nonindigenous parasitic wasps in the genus *Encarsia* (Hymenoptera: Aphelinidae) for biological control of whitefly pests (Homoptera: Aleyrodidae). Environmental Assessment, Riverdale, MD.

Chapter 8
Evaluation of Exotic Parasitoids and Predators in Field Cages in California

Kim A. Hoelmer[1] **and William J. Roltsch**[2]

Abstract Field-cage evaluations of parasitoids in the genera *Eretmocerus* and *Encarsia* from 13 different countries were conducted from 1995 to 1997 to identify effective new species or geographic populations of natural enemies of *Bemisia tabaci* on key host crops of whiteflies in desert valleys in the southwestern USA. Top performing species and selected populations of *Eretmocerus* were then compared on cantaloupe, a key crop for spring whitefly increase, and on citrus, an overwintering host. The best-performing species included *Eret. emiratus* and/or *Eret.* sp. near *emiratus* from the United Arab Emirates and Ethiopia, several populations of *Eret. mundus,* and a population of *Enc. sophia* from Pakistan. These species originated in regions with climates very similar to the Imperial Valley, California.

8.1 Introduction

Outbreaks of the sweetpotato whitefly biotype B, *Bemisia tabaci* (Genn.) (= silverleaf whitefly, *B. argentifolii* Bellows and Perring) in the USA first began in the late 1980s in Florida and spread rapidly across the southern USA. The impact of the whitefly in agricultural areas in Arizona and California was especially severe, causing major economic losses in cotton (*Gossypium hirsutum* L.), alfalfa (*Medicago sativa* L.), melons (*Cucumis melo* L.), winter vegetables, and many other crops (Perring et al. 1991; Gonzalez et al. 1992). Although the development of new and highly active pesticides, such as imidacloprid, provided temporary relief to growers, classical biological control was seen as an important component of integrated management of *B. tabaci* (Faust 1992).

[1]USDA-ARS, Beneficial Insect Introduction Research Unit, 501 South Chapel Street, Newark, DE 19713-3814

[2]California Department of Food & Agriculture, Biological Control Program, 3288 Meadowview Rd., Sacramento, CA 95832

J. Gould et al. (eds.), *Classical Biological Control of* Bemisia tabaci *in the United States.*

The growth and impact of *B. tabaci* populations in irrigated desert valleys is determined largely by crop sequences and high mean seasonal temperatures. In areas where susceptible crops are in year-round production, such as California's Imperial Valley, *B. tabaci* populations persist at low levels during the winter on various hosts. Densities increase significantly on spring melon crops; as melon crops mature, whiteflies migrate to cotton, where populations continue to increase to very high levels. On fall crops young seedlings may be severely impacted before temperatures decline and whitefly populations diminish. Alfalfa, although not a preferred host for *B. tabaci*, is grown continuously on a major portion of cropland in the Imperial Valley, and thus has a significant effect on population dynamics. Populations are also influenced by the presence of agricultural weeds and a wide variety of native and introduced desert plants. Ornamental plantings in urban and rural home landscapes may also be heavily impacted by *B. tabaci* migrating from crops. These non-agricultural host plants are often good reservoirs for natural enemies of *B. tabaci* because they are not routinely treated with pesticides and many are perennials. Whitefly populations in other regions of the USA, such as the Lower Rio Grande Valley in Texas, are influenced by similar factors although summer temperatures are not as extreme as in the Southwest.

For natural enemies to have significant impact on whitefly populations in this environment they must be present and active in the spring when whiteflies begin to increase in agricultural crops, before the need for extensive use of chemical pesticides is felt by growers. Reservoirs must exist where natural enemies persist through the winter when whitefly populations on crops are minimal, and migrating natural enemies must be able to follow whiteflies from reservoirs into new crops.

Compared with other parts of the world where *B. tabaci* occurs, relatively few indigenous parasitoid species attack *B. tabaci* in the southwestern USA (see Chapter 18). Only three species occur here with regularity, in contrast to four to five species in the Rio Grande Valley in Texas, six or more in Florida, and as many as a dozen species reported in India and Pakistan. Of these, the indigenous *Eretmocerus eremicus* Rose and Zolnerowich (reported as *Eret. californicus* Howard in earlier literature) is the most abundant species, attacking *B. tabaci* on melons, cotton and on most non-crop host plants. Many indigenous polyphagous predators present in desert crops also attack *B. tabaci*. Some of these, such as big-eyed bugs (*Geocoris*) and lacewings (*Chrysoperla*), cause significant mortality to whitefly populations (Naranjo 2001).

The potential for increasing the species diversity and impact of parasitoids by introducing exotic parasitoids appeared to be very good from the outset of the National *B. tabaci* Research and Action Program. Surveys of parasitism of *B. tabaci* in desert crops showed that parasitism by the native *Eret. eremicus* on winter cole (crucifer) crops was very low compared to levels recorded in melons and cotton (authors' unpublished data). As a result, this parasitoid was not present in sufficient numbers in the spring to restrain whitefly population growth in spring melons. Laboratory studies also demonstrated that the presence of setae on melon leaves reduced the searching efficiency of *Eret. eremicus* on melons when compared with cotton and sweet potato (Headrick et al. 1996). A primary goal of importation of new natural enemies of *B. tabaci* was to identify species that would be more effective than the native parasitoids in both desert and non-desert regions of the USA. Field evaluations conducted in Texas, Arizona and California during 1994–1997 evaluated numerous geographic populations comprising eight species of *Eretmocerus* and seven species of *Encarsia* from 13 different countries. This chapter will discuss field cage evaluations conducted in the Imperial Valley, California.

Different species and different populations of the same species often vary in important characteristics such as host acceptance, fecundity, development time, etc. (see chapter 2, section 2.15 *in* Quicke 1997), which are often highly dependent on environmental conditions. Most species imported from foreign locations were reared and maintained successfully in cultures at the USDA-APHIS-PPQ Mission Biological Control Laboratory (MBCL, now called the Center for Plant Health Science and Technology Laboratory) quarantine in Mission, Texas in growth chambers using hibiscus, *Hibiscus rosa-sinensis* L., as a host plant, although some were more easily reared and maintained than others. It was anticipated that differences in climate among target release areas and host plant characteristics might result in differences in performance by imported natural enemies under varying field conditions that could significantly affect their establishment and impact on *B. tabaci*. Consequently, evaluations were conducted in two stages; the first screening was done on a variety of key crop host plants in the MBCL quarantine laboratory in Mission, Texas (Goolsby et al. 1996, 1998; see Chapter 7). The top performing cultures from the first screening were then evaluated under more realistic field conditions in California and Arizona on host crops of importance in desert agriculture. It was reasoned that cultures that failed to perform well under the favorable controlled conditions in quarantine laboratory screenings had a low probability of establishing and doing well in the field. Consequently, low performing cultures were not evaluated further in the field, although some were field-released at selected sites on the chance that the assumption was wrong (which was not the case, since none of the low performing species or geographic populations in quarantine screenings became established in the field).

Field cage tests conducted in the Imperial Valley from 1995 to 1997 evaluated numerous populations comprising eight species of *Eretmocerus* and seven species of *Encarsia* from 13 different countries. Each population was evaluated in comparative tests on one or more crops in quarantine screenings alone or in both quarantine and field evaluations. The measure of efficacy chosen for evaluations was the number of F_1 progeny produced per female. Because these cultures became available for study at different times, many of the best performers identified in the initial quarantine evaluations could not be compared on each crop concurrently. A final evaluation in 1997 of the top performing populations of *Eretmocerus* from the earlier field cage evaluations on various crops was performed on cantaloupe, which was selected because it was deemed the crop in which increased biological control would lead to the greatest overall reduction of whiteflies in desert regions with year-round production of whitefly hosts. One population of *Enc. sophia* (Girault and Dodd) from Pakistan performed exceptionally well on cotton, and was re-evaluated to validate the earlier results. In order to assess the ability of native and introduced parasitoids to suppress overwintering whitefly populations in citrus, an evaluation of top performing *Eretmocerus* was conducted in the fall of 1997. This evaluation was done as a result of studies by the California Department of Food and Agriculture (CDFA) and the University of California that identified citrus orchards (especially Washington navel and Valencia orange cultivars) as overwintering refuges for *B. tabaci* in the San Joaquin Valley which could reinfest field crops the following spring (see Chapter 14).

One species of Platygastridae, *Amitus bennetti* Viggiani and Evans, originally collected in Puerto Rico and released in Florida, was studied in the laboratory and in field cages at Riverside in 1995–1996, using cotton and beans as *B. tabaci* hosts (Joyce and Bellows 1997; Joyce et al. 1999). It was released in the Imperial Valley by University of California researchers (see Chapter 13). However, comparative evaluations between *A. bennetti* and the introduced aphelinid parasitoids were not done, and the species did not establish.

Relatively few predators were collected by foreign explorers and sent to the MBCL; these included the coccinellids *Clitostethus arcuatus* (Rossi) and *Serangium parcesetosum* Sicard, and the drosophilid *Acletoxenus formosus* Loew. Of these, only *S. parcesetosum* (APHIS culture M93008, Podumbu, India) was evaluated further. A number of literature reports discuss releases of *Serangium parcesetosum* against citrus whiteflies in Corsica, France and the Republic of Georgia, but the species in those studies was actually *S. montazerii* Fürsch (Booth and Polaszek 1996). Both species are apparently native to the Indian subcontinent. The identity of *S. parcesetosum* specimens cultured from the Podumbu collection and released in the USA against *B. tabaci* was verified by Dr. Roger Booth (personal communication). After its release from quarantine, *S. parcesetosum* was evaluated in the laboratory (Legaspi et al. 1996) and the greenhouse (Ellis et al. 2001), and field releases were made in Puerto Rico but without any indications of recovery (Pantoja et al. 2002). The beetle was also cultured at the APHIS Plant Protection Center in Phoenix, Arizona to support evaluations and releases in Arizona and California (discussed below).

8.2 *Encarsia* and *Eretmocerus*

8.2.1 Evaluation Methods

Field-cage evaluations of *Eretmocerus* and *Encarsia* were conducted in isolated fallow fields at the Irrigated Desert Research Station in Brawley, California beginning in fall of 1995 and concluding in the summer of 1997; evaluation on citrus was done in the fall of 1997. Parasitoids were evaluated against *B. tabaci* on cantaloupe in the spring, cotton in the summer, and broccoli in the early fall, which were the seasons when these crops were grown commercially in the Imperial Valley. Evaluation of *Eretmocerus* on alfalfa and citrus was also done in the fall, which is when they are typically invaded by whiteflies migrating from cotton.

Parasitoids for this study were provided by the USDA-APHIS Mission Biological Control Laboratory in Mission, Texas and the APHIS Imperial Valley Insectary in Imperial, California, the CDFA Biological Control Program in Sacramento, the University of Arizona insectary in Tucson, Arizona, and Novartis BCM/Bunting USA. The autoparasitic species *Enc. sophia* and *Enc. bimaculata* Heraty and Polaszek were compared initially with a suite of *Eretmocerus* species in the 1995 USDA evaluations on cantaloupe. Subsequently, *Eretmocerus* and uniparental *Encarsia* species were evaluated by USDA-APHIS and autoparasitoid *Encarsia* species (in which males are produced as secondary parasitoids on female primary parasitoids of whitefly) were evaluated separately by CDFA using similar methods (discussed below). The autoparasitic aspect of their biology made it necessary to manipulate these species somewhat differently from *Eretmocerus* species (in which both sexes are primary parasitoids of whitefly) in the evaluations.

Evaluations were conducted with releases of parasitoids into field cages (3–5 cage replicates per species or geographic population) constructed of UV resistant nylon monofilament mesh screening (52×52 threads per inch to prevent whitefly and parasitoid escape or entry) fitted over PVC tubing frames. Cages were either 1.85 m on each side (for *Eretmocerus* and uniparental *Encarsia*) or 1.25×1.85×1.85 m (for autoparasitic *Encarsia*). Cages were accessed through zippers sewn in one side of the cage, and all lower edges were buried in the soil. Temperature and RH was monitored inside and outside the cages using data loggers; these data were compared to data from an official weather station located ca. 0.5 km from the study plot for verification. Crops were planted as seedlings with two rows on raised beds per cage, in a manner comparable to production practice, i.e., 5–6 seedlings, or clusters of alfalfa stems per row. Furrow irrigation was provided within cages by pipes buried under cage frames.

For the evaluation of citrus, 10 potted citrus seedlings were placed inside cages in two rows. Cages were isolated from other plant hosts of *B. tabaci* by placement in the center of a fallow plot of land on the research station with the nearest vegetation at least several hundred meters distant, which was tilled regularly for weed suppression. Cages were entered only as needed to inoculate with whiteflies and female parasitoids and to check insect development. In addition, a fallow period of several weeks separated harvests from one set of tests from inoculations of the following set, with all crop residues removed and disposed of in order to minimize the chances of contaminating new cages with parasitoids from the previous tests. These measures effectively minimized contamination from other whiteflies and parasitoids in all evaluations except those on cotton, as was shown by the use of control cages that received no parasitoid releases so that any parasitism represented an estimate of background (non-release) contamination. Voucher specimens were collected from release stock before releasing parasitoids into cages to verify the identity of each culture released, and from their progeny to detect contamination by other species during the test.

Inoculation with equal numbers of whiteflies per replicate cage (numbers varied by crop) and of mated female parasitoids (100 per cage for *Eretmocerus* and uniparental *Encarsia*) made it possible to compare the number of progeny produced per female parasitoid in the F_1 generation. The lifespan of adult female *Eretmocerus* and *Encarsia* was much shorter than their generation time (authors' observations), so F_1 production did not overlap significantly with the next generation. For example, during an inspection of autoparasitic *Encarsia* cages prior to F_1 emergence in April 1996, an average of only 1.3 (SD ± 1.2) live females was found in each experimental cage and none were found in the control (non-release) cages. Autoparasitoid *Encarsia* females were initially introduced into cages in groups of 50, with 10 males included to ensure mating and female progeny production. Successive introductions consisting of 20 females and two to five males were made 14 and 21 days later to ensure mating of the F_1 generation and avoid highly skewed sex ratios in the F_2 and subsequent generations that conceivably might have arisen from a highly synchronous initial introduction. This ensured the presence of males for continued production of subsequent generations, while providing enough of a time lag between releases so that a clear assessment of production by the initial 50 females could be accomplished. Parasitoids were mated within 24 h of emergence and introduced into *Eretmocerus* cages when 2nd and 3rd instar whiteflies were present and into *Encarsia* cages when early 4th instars were present, according to their host stage preferences. The development of parasitoid progeny was monitored to determine when each cage was ready for sampling of the F_1 generation. The number of progeny in the F_1 generation was determined when the first adult parasitoids began to emerge from parasitized whiteflies, at which time the majority had developed to the late pupal stage. Immature (pupae or older larvae) parasitoids inside the whitefly host could be readily detected in the leaf samples, so samples collected at this point constituted a reliable measure of the F_1 generation. Samples consisted of one-half of each leaf bearing 4th instar whiteflies on all caged plants. The remaining leaf portions were left on plants to allow development of a successive generation. Samples were examined using a stereoscopic microscope and the progeny produced by each species or population recorded. We observed that a few cultures initially increased but failed to persist in subsequent generations. To detect these we looked for continued persistence and increase of each parasitoid culture after the F_1 generation by recording the F_2 progeny on samples consisting of 50 leaf discs taken from different leaves using a cork borer with a diameter of 1.85 cm (2.7 cm^2).

The final evaluation on cantaloupe compared the top performing *Eretmocerus* from all previous tests by measuring the total production of progeny in each field cages during the F_1 generation. Uniparental *Encarsia* were not included because none performed well in the initial field evaluations. At the conclusion of the F_1 generation all cantaloupe leaves bearing 4th instar and pupal whitefly were harvested and all parasitoids were counted (larval, pupal and emerged) in the laboratory. Similarly, in the evaluation on citrus, the F_1 generation was measured by counting all parasitoid progeny found on all leaves.

Species evaluated in these tests included populations of *Eretmocerus mundus* Mercet (from Spain: Murcia, India: Padappai, and Israel: Golan Ma'Aleh Samla), *Eretmocerus* sp. nr. *furuhashii* Rose and Zolnerowich (Taiwan: Tainan), *Eretmocerus melanoscutus* Zolnerowich and Rose (Thailand: Chiang Mai, Sai Noi Klong Ha

Roi and Kampang Saen, and Taiwan: Tainan), *Eretmocerus hayati* Zolnerowich and Rose (Pakistan: Multan), *Eretmocerus emiratus* Zolnerowich and Rose (United Arab Emirates, Strait of Hormuz), *Eretmocerus emiratus* or nr. *emiratus* (Ethiopia: Melka Weher), *Encarsia bimaculata* (India: Parbhani), *Encarsia sophia* (Thailand: Chiang Mai, Malaysia: Kuala Lumpur, Spain: Murcia, and Pakistan: Multan), *Encarsia* sp. nr. *pergandiella* Howard (Brazil: Sete Lagoas), *Encarsia* sp. nr. *hispida* De Santis (Brazil: Sete Lagoas), and an undescribed *Encarsia* sp. (*parvella* group) (Dominican Republic: Azua). Several other species were evaluated prior to the comparative tests described here, such as *Eretmocerus staufferi* Rose and Zolnerowich (a new species of unknown geographic origin first discovered parasitizing *B. tabaci* on tomatoes in greenhouses at Texas A&M University, College Station, Texas) and *Encarsia formosa* Gahan, but were found to be ineffective under Imperial Valley conditions and were not evaluated further.

8.2.2 *Evaluation Results: Eretmocerus and Uniparental Encarsia*

Table 8.1 summarizes the mean number of F_1 progeny produced by the indicated species or population of *Eretmocerus* and uniparental *Encarsia* on cantaloupe, cotton, broccoli, and alfalfa during 1995 and 1996.

8.2.2.1 Evaluation in Cantaloupe

In 1995, six species were compared: two populations of *Eretmocerus mundus* from Spain (APHIS quarantine culture M92014) and Israel (M94120), *Eret. melanoscutus* from Thailand (M94036), two autoparasitic *Encarsia, Enc. bimaculata* from India

Table 8.1 Mean F_1 progeny production in cage releases of *Eretmocerus* and uniparental *Encarsia* on key Imperial Valley whitefly host crops.

Species/culture	Origin	Mean progeny per female	Species/culture	Origin	Mean progeny per female
			Alfalfa		
Fall 1995			**Fall 1996**		
Enc. sp. nr. *pergandiella*	Brazil	0.58	*Enc.* sp. nr. *pergandiella*	Brazil	0.34
Enc. sp. nr. *hispida*	Brazil	0.08	*Enc.* sp. (*parvella* group)	Dominican Republic	0.13
Eret. mundus	Israel	0.29	*Eret. eremicus*	Imperial Valley, CA	0.74
Eret. hayati	Pakistan	0.09	*Eret. melanoscutus*	Taiwan	0.64
Eret. melanoscutus	Thailand	0.08	*Eret.* nr. *furuhashii*	Taiwan	0.86
			Eret. emiratus	UAE	0.60

(continued)

Table 8.1 (continued)

Species/culture	Origin	Mean progeny per female	Species/culture	Origin	Mean progeny per female
			Broccoli		
Fall 1995			**Fall 1996**		
Enc. sp. nr. *pergandiella*	Brazil	1.27	*Enc*. sp. (*parvella* group)	Dominican Republic	0.42
Enc. sp. nr. *hispida*	Brazil	3.04	*Eret. eremicus*	Imperial Valley, CA	0.43
Eret. mundus	Israel	8.81	*Eret. mundus*	Spain	1.36
Eret. hayati	Pakistan	3.94	*Eret. melanoscutus*	Taiwan	1.35
Eret. melanoscutus	Thailand	6.81	*Eret*. nr. *furuhashii*	Taiwan	1.95
			Eret. emiratus	UAE	1.08
			Cantaloupe		
Spring 1995			**Spring 1996**		
Enc. bimaculata[a]	India	7.50	*Enc*. sp. nr. *hispida*	Brazil	5.82
Enc. sophia[a]	Spain	12.00	*Enc*. sp. (*parvella* group)	Dominican Republic	0.95
Enc. sp. nr. *pergandiella*	Brazil	17.00	*Eret. hayati*	Pakistan	9.22
Eret. mundus	Spain	89.50	*Eret. melanoscutus*	Taiwan	18.14
Eret. melanoscutus	Thailand	127.00	*Eret*. nr. *furuhashii*	Taiwan	10.11
Eret. mundus	Israel	140.00	*Eret. emiratus*	UAE	20.77
			Cotton[b]		
Summer 1995			**Summer 1996**		
Enc. sp. nr. *pergandiella*	Brazil	2.82	*Enc*. sp. (*parvella* group)	Dominican Republic	10.47
Enc. sp. nr. *hispida*	Brazil	2.03	*Eret. mundus*	Israel	19.96[b]
Eret. mundus	Spain	15.86	*Eret. hayati*	Pakistan	101.33[b]
Eret. mundus	Israel	8.77	*Eret. melanoscutus*	Taiwan	78.36[b]
Eret. melanoscutus	Thailand	16.99	*Eret*. nr. *furuhashii*	Taiwan	115.66[b]
			Eret. emiratus	UAE	110.43[b]

[a] In spring 1995 only, two autoparasitoid *Encarsia* were evaluated in cantaloupe tests.
[b] 1995 and 1996 cotton cages became contaminated with *Eret. emiratus,* so the results show high levels of reproduction by unknown numbers of this species in addition to the species tested. Results for *Enc*.sp. (*parvella* group) are accurate.

(M92018) and *Enc. sophia* from Spain (M92003), and one uniparental *Encarsia, Enc.* near *pergandiella* (M94055) from Brazil.

In 1996, the following six species were evaluated: *Eret. hayati* from Pakistan (M95012), *Eret. melanoscutus* from Taiwan (M95097), *Eret.* sp. near *furuhashii* from Taiwan (M95098), *Eret. emiratus* from the United Arab Emirates (M95104), and two uniparental *Encarsia, Enc.* near *hispida* from Brazil (M95056), and the undescribed uniparental *Encarsia* species (*parvella* group) from the Dominican Republic (M95001). Overall, the *Eretmocerus* species appeared better adapted to cantaloupe and were more productive than uniparental *Encarsia* (or the two autoparasitoids). Higher levels of parasitism were achieved during the 1995 evaluations due to weather conditions more favorable to parasitoid activity. Despite the weather differences in the years tested, the superiority of *Eretmocerus* over *Encarsia* on this host was evident. The low productivity of the *Encarsia* species from the Dominican Republic was particularly surprising given its favorable results in quarantine evaluations.

8.2.2.2 Evaluation in Broccoli

Mean production of progeny on broccoli was much lower than on cantaloupe or cotton, for which there are several possible reasons. Cages with broccoli were inoculated during late fall and the progeny harvested during the early winter, during which time the whitefly population had declined. Because the parasitoids were less active at lower light levels and temperatures prevalent at that time they encountered a reduced proportion of the available hosts; a portion of acceptable hosts were on the older leaves which were shed, resulting in loss of some whiteflies and developing parasitoids.

In 1995, the tests compared five species: *Eret. mundus* from Israel (M94120), *Eret. hayati* from Pakistan (M95012), *Eret. melanoscutus* from Thailand (M94040), and the two uniparental *Encarsia, Enc.* sp. nr. *pergandiella* (M94055) and *Enc.* sp. nr. *hispida* (M95056) from Brazil. The greatest number of progeny was produced by *Eret. mundus* from Israel and *Eret. melanoscutus* from Thailand. The uniparental *Encarsia* species did not compare well, although not as poorly as in the cantaloupe tests.

In 1996, six species were evaluated in tests, including *Eret. mundus* from Spain (M92014), *Eret. emiratus* from the United Arab Emirates (M95104), *Eret. melanoscutus* from Taiwan (M95097), *Eret.* sp. nr. *furuhashii* from Taiwan (M95098), *Eret. eremicus* Rose and Zolnerowich from Imperial Valley (=*Eret.* sp. nr. *californicus* in earlier literature), and the uniparental *Encarsia* species (*parvella* group) from the Dominican Republic (M95001). Performance was in general lower than the previous year, possibly influenced by weather. The non-indigenous *Eretmocerus* were again slightly more productive than the native *Eretmocerus eremicus* and the uniparental *Encarsia* (M95001).

8.2.2.3 Evaluation in Alfalfa

Cages were inoculated during late fall and the F_1 progeny were harvested during the early winter. Evaluations conducted during late fall 1995 compared *Eretmocerus melanoscutus* (M94040, Thailand), *Eret. mundus* (M94120, Israel), *Eret. hayati*

(M95012, Pakistan) and the uniparental *Encarsia* sp. nr. *pergandiella* (M94055, Brazil) and *Enc.* sp. nr. *hispida* (M94056, Brazil). The most progeny were produced by *Enc.* nr. *pergandiella* and *Eretmocerus mundus*. Alfalfa was the only crop on which *Enc.* nr. *pergandiella* performed well relative to other species. Low numbers of whiteflies in cages was a problem in this test which may have adversely impacted the performance of some species in alfalfa.

Evaluations in fall 1996 compared *Eretmocerus emiratus* (M95104, United Arab Emirates), *Eret. melanoscutus* (M95097, Taiwan), *Eret.* nr. *furuhashii* (M95098, Taiwan), the native *Eret. eremicus* and the uniparental *Encarsia* M95001 (Dominican Republic) and M94055 (Brazil). All of the *Eretmocerus* species evaluated produced comparable numbers of progeny, while the uniparental *Encarsia* species were again less productive. The native *Eretmocerus eremicus* was comparable to all three non-indigenous *Eretmocerus* species on alfalfa.

In both sets of evaluations on alfalfa, all of the *Eretmocerus* species or populations examined produced comparable numbers of progeny, while the uniparental *Encarsia* species were less productive. Overall the number of offspring produced in alfalfa was very low for the same reasons discussed for broccoli. An additional contributing factor may be the more complex environment alfalfa presents to female parasitoids searching for hosts, with its multiple stems and numerous small and setose leaves.

8.2.2.4 Evaluation in Cotton

Results for cotton evaluations were confounded by high levels of contamination despite cage modifications and other procedures designed to avoid such contamination; it was the only crop thus affected. The contamination by *Eretmocerus* meant that none of the cotton cage results were truly indicative of the tested species' performance, except for *Encarsia,* whose larvae and pupae could be distinguished from *Eretmocerus*. Nearly all of the contamination in 1996 was due to *Eretmocerus emiratus* (*ex* UAE), as determined by vouchers of adults reared from samples of progeny. The source of these contaminating parasitoids may have been offspring from the preceding test on cantaloupe that became established in the vicinity, or parasitoids dispersing from crops or urban areas adjacent to the research station. High levels of searching and dispersal by contaminating parasitoids during the high summer temperatures may have quickly biased the results. For contamination to occur, the contaminating parasitoids would have had to escape from a closed cage of cotton and invade other closed cages in significant numbers, or more probably, to have invaded during those occasions when cages had to be entered, possibly by hitchhiking on clothing. Although this was an undesirable outcome for the evaluations on cotton, it demonstrated that *Eret. emiratus* possessed several qualities needed for an efficient biological control agent – dispersal and host-finding efficiency – and suggests that this species was well adapted to the extreme desert climate.

8.2.2.5 Final Comparison of Top Performing Species on Cantaloupe

The re-evaluation of the seven best-performing species from previous tests on all crops was conducted in the spring and early summer of 1997. This comparison was made to determine whether the exotic species performed significantly better than the native *Eret. eremicus*, and whether there were significant differences among the exotic species. Species selected for evaluation included three populations of *Eretmocerus mundus* (M92014, Spain; M94120, Israel; and M92019, India), *Eret. hayati* (M95012, Pakistan), *Eret. emiratus* (M95104, UAE), *Eret.* sp. near *emiratus* (M96076, Ethiopia) and the southwestern desert native *Eret. eremicus* as a control comparison. *Eret. melanoscutus* performed comparably with some of the cultures chosen but it was not available for evaluation during the fall of 1997.

The mean number of progeny produced per female of each culture is shown in Table 8.2. Both this and earlier cage tests showed a great deal of variation between some of the replicates despite inoculation of each cage with equal numbers of whiteflies and female parasitoids. The greatest numbers of progeny per female (66) were produced by *Eret. emiratus* from the United Arab Emirates and *Eret.* nr. *emiratus* from Ethiopia. *Eretmocerus mundus* cultures M92014 (Spain) and M94120 (Israel) were next most productive, with 55 and 51 progeny/female respectively. Native *Eret. eremicus* and *Eret. hayati* from Pakistan produced the least, with less than half the progeny of the top performing cultures. After the conclusion of this evaluation it was learned that the supply of *Eret. mundus* from the APHIS culture M92019 (Padappai, India) may have been contaminated during its early culture in quarantine or had been established as a mixed culture from the onset. This culture was therefore thought to be a mixture of *Eret. mundus, Eret. hayati* and possibly a third species indigenous to India that was never successfully cultured. However, RAPD-PCR tests indicated that *Eret. mundus* predominated in

Table 8.2 Mean F_1 progeny production in cage releases of *Eretmocerus* on cantaloupe in spring, 1997.

Species	Origin	Mean offspring per female[a]	Range
Eret. hayati	Pakistan	26.69 a	6.80–40.06
Eret. eremicus	Imperial Valley	32.06 ab	18.80–48.35
Eret. mundus (M92019)	India	45.07 ab	32.28–67.24
Eret. mundus (M92014)	Spain	50.68 b	32.42–71.68
Eret. mundus (M94120)	Israel	54.83 b	39.31–77.24
Eret. emiratus	UAE	66.07 b	39.28–102.04
Eret. nr. *emiratus*	Ethiopia	66.53 b	39.89–134.92

[a] ANOVA based on transformed data (log + 1.0). Values in table are shown back transformed following statistical analysis. Means within a test followed by the same letter are not significantly different, LSD means comparison.

General model F = 2.22, df = 6, 18; P = 0.0892.

the mixture, and specimens were usually identified as this species based on morphological characters. Also, the culture of *Eretmocerus emiratus* or nr. *emiratus* (Ethiopia) could not be separated from *Eret. emiratus* from the UAE by morphological means, but were distinguishable by RAPD-PCR.

Log + 1 transformed data were subjected to one-way analysis of variance with orthogonal contrasts of means. When each of the seven cultures was analyzed as a separate experimental treatment, significant differences among pairs of cultures included the native *Eret. eremicus* and *Eret. hayati* ($P = 0.0151$), *Eret. hayati* and *Eret. emiratus* (UAE) ($P = 0.0432$), *Eret. hayati* and *Eret. emiratus*/nr. *emiratus* (Ethiopia) ($P = 0.0098$), and *Eret. hayati* and *Eret. mundus* (India) ($P = 0.0243$). Similar results were obtained when the 3 geographic cultures of *Eret. mundus* (or predominantly *Eret. mundus*) were combined and the cultures of *Eret. emiratus* and nr. *emiratus* were combined and each group treated as one species instead of five distinct populations. In this case, the native *Eret. eremicus* differed significantly from *Eret. hayati* ($P = 0.0096$), and *Eret. mundus* and *Eret. hayati* each differed from *Eret. emiratus*/nr. *emiratus* ($P = 0.0272$ for *mundus:emiratus;* $P = 0.0025$ for *hayati:emiratus*). Although the variation in numbers of progeny within each cage was high, the results of this comparison supported the earlier field evaluation results in which *Eret. mundus* and *Eret. emiratus*/nr. *emiratus* outperformed native *Eret. eremicus* and *Eret. hayati*.

8.2.2.6 Evaluation in Citrus

Washington navel orange 'Lane Late' seedlings were used as the host plants for *B. tabaci*. Flushes of new growth of this cultivar are susceptible to high populations of *B. tabaci* that migrate from senescing cotton fields in the Central Valley of California during the fall. Species evaluated included *Eretmocerus mundus* ex Spain (M92014), *Eret. hayati*, *Eret. emiratus*, *Eret.* sp. *emiratus* or nr. *emiratus* (Ethiopia), and the indigenous *Eret. eremicus*. Sampling to measure F1 progeny production was conducted in late December of 1997. High levels of predation by ants in all citrus cages precluded leaving a portion of the whitefly population for measurement of overwintering survival as planned, so all foliage containing *B. tabaci* immatures was sampled for measurement.

Table 8.3 Mean F_1 progeny production in cage releases of *Eretmocerus* on citrus in fall, 1997.

Species	Origin	Mean offspring per female[a]	Range
Eret. eremicus	Imperial Valley	3.58 a	1.39–8.28
Eret. mundus	Spain	5.14 a	0.27–8.37
Eret. hayati	Pakistan	2.11 a	0.44–4.28
Eret. emiratus	UAE	5.87 a	1.14–14.06
Eret. nr. *emiratus*	Ethiopia	3.17 a	0.00–6.08

[a]ANOVA based on transformed data (log + 1.0). Values in table are shown back transformed following statistical analysis. Means within a test followed by the same letter are not significantly different, LSD means comparison. General model $F = 0.74$, $df = 4, 14$; $P = 0.5778$.

All five *Eretmocerus* species/geographic populations evaluated attacked and developed in *B. tabaci* nymphs on navel orange (Table 8.3). However, the average number of progeny per female ranged from 2.1 to 5.9, considerably less than production on other hosts such as cantaloupe in warmer weather. Mean numbers of progeny were analyzed by ANOVA with orthogonal contrasts as above, but there were no significant differences among cultures. In absence of ant predation the number of progeny might have been higher. As was the case with other fall–winter crops tested, decreasing temperatures slowed the development of whitefly and immature parasitoids, and probably reduced the activity of the adult parasitoids.

8.2.3 Evaluation Results: Autoparasitic Encarsia

Each test of autoparasitic *Encarsia* was subjected to a one-way analysis of variance following a Log + 1 transformation of data. Mean separations were performed using the Waller-Duncan *k* ratio procedure.

Encarsia sophia populations from Spain (M93003), Thailand (M94041) and Malaysia (M94047) were evaluated simultaneously on broccoli in fall 1995 and again on cantaloupe in spring 1996. The populations were morphologically indistinguishable but the Spanish population was separable from the other two by a different PCR pattern (see Chapter 13). On each host the Spanish population produced the greatest number of progeny (Table 8.4). In cantaloupe, the Malaysian

Table 8.4 Mean F_1 progeny production in cage releases of autoparasitic *Encarsia.*

Host crop	Date	Species and origin	Mean progeny per female[a]	Range
Broccoli	Fall 1995	*Enc. sophia*, Spain	12.8 a	10.3–15.5
		Enc. sophia, Thailand	4.0 b	2.7–4.8
		Enc. sophia, Malaysia	6.0 b	4.2–7.5
	Fall 1996	*Enc. sophia*, Spain	12.7 a	10.3–14.8
		Enc. sophia, Pakistan	4.6 b	2.0–9.6
		Enc. bimaculata, India	1.6 b	0.4–2.7
Cantaloupe	Spring 1996	*Enc. sophia*, Spain	16.7 a	10.6–24.7
		Enc. sophia, Thailand	13.8 a	9.2–22.7
		Enc. sophia, Malaysia	1.4 b	0.3–2.4
	Spring 1997	*Enc. sophia*, Pakistan	14.0 a	10.3–21.4
		Enc. bimaculata, India	7.2 b	6.6–7.7
Cotton	Summer 1996	*Enc. sophia*, Spain	12.2 a	9.0–40.3
		Enc. sophia, Pakistan	52.5 a	20.3–185.2
		Enc. bimaculata, India	9.9 a	6.9–8.1
	Summer 1997	*Enc. sophia*, Pakistan	159.2	36.3–253.2

[a]ANOVA based on transformed data (log + 1.0). All mean values (except for cotton 1997) back transformed following statistical analysis. Means within a test followed by the same letter are not significantly different. Broccoli 1995: $F = 14.5$, df = 2, 6; $P < 0.05$; Cantaloupe 1996: $F = 18.1$, df = 2, 6; $P < 0.05$; Cotton 1996: $F = 2.38$, df = 2, 6; df = 2, 6; $P = 0.17$; Broccoli 1996: $F = 8.08$, df = 2, 6; $P < 0.05$; Cantaloupe 1997: $F = 8.5$, df = 1, 4; $P < 0.05$.

population performed very poorly (1.4 progeny/female) compared to the other two. Production of a second generation of all three populations on broccoli was weak, with production falling below F_1 levels. Further testing of *Enc. sophia* populations from Thailand and Malaysia was not continued on other crops because of their poor performance compared to the Spanish population.

Testing was continued by comparing *Enc. sophia* from Spain (M93003) with another population from Pakistan (M95107) and with *Enc. bimaculata* from India (M92018). These species and geographic populations were compared on cotton in the summer and on broccoli in the fall of 1996. In the cotton tests, the *Enc. sophia* from Pakistan stood out from the others with a mean production of over 50 progeny/female. Because of unusually high variation among treatment replicates, treatments were not statistically different in this test. High contamination levels by *Eretmocerus* spp. in cotton replicates (as noted for the *Eretmocerus* tests during the summer) may have been responsible for a portion of the variability. The fall test on broccoli showed again that the Spanish *Enc. sophia* population performed best on this host plant. The mean production for the Spanish population was essentially identical to that recorded in the 1995 test. During the spring of 1997, evaluations in cantaloupe included *Enc. sophia* from Pakistan (M95107) and *Enc. bimaculata* (Table 8.4). By this time, the Spanish population of *Enc. sophia* had been mass reared and released and was not available in culture any longer.

The evaluation on cotton was repeated with the Pakistan population in summer 1997 resulting in an even higher mean production of 159 progeny per female (Table 8.4) despite levels of *Eretmocerus* contamination that were approximately the same as in 1996. Consequently, this geographic population of *Enc. sophia* was chosen by CDFA for widespread releases in Imperial Valley, where it has become established (see Chapter 13).

The consistency of the results obtained with certain species, such as *Enc. sophia* from Spain across all host crops tested, and by *Enc. sophia* from Pakistan on cotton, when compared with the disparity of results with other geographic populations of the same species (i.e., from Malaysia on cantaloupe) or other species (such as *Enc. bimaculata* on broccoli) suggests that there are significant biological differences among these *Encarsia* populations. These may be due to host plant effects or behavioral responses under differing climatic conditions. The population of *Enc. sophia* from Multan, Pakistan, is clearly better adapted for cotton than broccoli, perhaps reflecting an adaptation to hot climates.

8.3 *Serangium parcesetosum*

8.3.1 Evaluation Methods and Results

Field evaluations were conducted in 1994 and 1995 at the APHIS Plant Protection Center in Phoenix and in commercial fields in Parker, Arizona, and in Imperial Valley, California, at the former USDA Irrigated Desert Research Station in Brawley,

California. To supply beetles for caged field trials the beetle was cultured by APHIS in Phoenix using eggs of pink bollworm, *Pectinophora gossypiella* (Saunders), for food. Tests in Arizona were conducted using cages of the same construction as described for *Eretmocerus* that were placed over rows of melons in commercial fields.

Beetle adults and larvae reduced whitefly in the cages sufficiently to result in healthy plants and fruit, as compared with dead or nearly dead plants in control cages without releases of beetles. Further experiments using several release rates in open field releases in Arizona demonstrated significant differences in whitefly populations only at high release rates (APHIS, unpublished data). In Brawley, caged releases were made in an experimental field of broccoli in November 1994. Coverings of spun polyester insect exclusion netting (1×2.5 m) supported at intervals with wire hoops were placed over raised beds of young broccoli plants with the edges buried with soil to secure the cages. *Bemisia tabaci* adults were introduced into the enclosures for oviposition, followed 1 week later by *S. parcesetosum* adults. One week after the introduction of beetles, the netting was removed and the broccoli leaves examined for evidence of beetle feeding and oviposition. Feeding on whitefly eggs and nymphs was visible, but predation was not extensive and few *S. parcesetosum* eggs or larvae were found. After the cage covers were removed most adult beetles quickly dispersed from the plants, and few remained in the experimental plot after 24 h. No increase in the predator population occurred in the broccoli plot. Further evaluations of *S. parcesetosum* were not continued since no sign of the beetle was detected in any subsequent surveys in Imperial Valley following its release from the caged plots in Brawley or at selected release sites in three communities in Imperial Valley, and Yuma, Arizona, in fall 1994.

8.4 Summary

Considering that the introduced *Eretmocerus* and *Encarsia* cultures compared in this study were field-evaluated because they performed well in quarantine evaluations (Chapter 7), it is noteworthy that the best-performing species or geographic populations in this desert Southwest trial were those that originated from similar climatic regions: the Arabian Peninsula, arid northeastern Africa, and hot, dry regions bordering the Mediterranean. The *Eretmocerus* species that have established in the desert valleys, *Eret. emiratus* and/or near *emiratus* (see Chapter 13), came from particularly hot arid regions. Although recovery surveys showed that *Eret. hayati* did not establish in the desert Southwest, its establishment and present dominance in the Lower Rio Grande Valley of Texas is not surprising given that the climate in this region of Texas closely matches its source in Pakistan. Likewise, the *Enc. sophia* population that became established in Imperial Valley originated in the hot interior valley of central Pakistan near Multan, while a population previously obtained from a cooler northern region of Pakistan near Rawalpindi and released in Imperial Valley in 1988 did not establish (see Chapter 18). Laboratory studies indicated that the Rawalpindi population did not reproduce well above 32° C (G. Nuessly, personal communication). Three different uniparental *Encarsia* species performed poorly under desert field conditions; perhaps limited genetic diversity restricts them from adapting to a harsh environment different from their origin. As a widespread indigenous species, *Eretmocerus eremicus* is presumably well adapted to the arid southwestern USA where it readily attacks *Bemisia,* but is perhaps better adapted as a parasitoid of native species of *Trialeurodes* which occur in the same habitats as the introduced *B. tabaci* and which are most likely its natural hosts in the region.

The failure of the coccinellid predator *Serangium parcesetosum* to establish following releases in several regions of North America may be related to habitat preference. In natural populations it appears to be associated with whiteflies in arboreal habitats rather than field crops and weeds. Although a climatic match is lacking between Podumbu, India, and the southwestern deserts of the USA, the climate of the Lower Rio Grande Valley in Texas is similar enough to expect the beetle might establish, yet it did not.

The establishment of one or more top performing exotic parasitoids in *B. tabaci* on several key crops grown during the spring and summer when whitefly populations are increasing should improve the impact of biological control in these important agricultural areas. Likewise, the gap in activity of the native *Eret. eremicus* on cole crops during the winter months should be at least partly reduced by the establishment of additional *Eretmocerus* species and *Encarsia sophia.*

References

Booth, R.G. and A. Polaszek. 1996. The identities of ladybird beetle predators used for whitefly control, with notes on some whitefly parasitoids, in Europe. Brighton Crop Protection Conference – Pests & Diseases, Brighton England, November 18–21, 1996. Proceedings Vol. 2B-2: 69–74. British Crop Protection Council, Alton Hampshire, UK.

Ellis, D.R., R.J. McAvoy, L. Abu Ayyash, M. Flanagan, and M.A. Ciomperlik. 2001. Evaluation of *Serangium parcesetosum* (Coleoptera: Coccinellidae) for biological control of silverleaf whitefly, *Bemisia argentifolii* (Homoptera: Aleyrodidae), on poinsettia. Florida Entomologist 84: 215–221.

Faust, R.M. 1992. Conference report and 5-year national research and action plan for development of management and control methodology for the sweetpotato whitefly. US Department of Agriculture, Agricultural Research Service, ARS-107.

Goolsby, J.A., M.A. Ciomperlik, B.C. Legaspi, J.C. Legaspi, and L.E. Wendel. 1998. Laboratory and field evaluation of exotic parasitoids of *Bemisia tabaci* (Biotype 'B') in the Lower Rio Grande Valley of Texas. Biological Control 12: 127–135.

Goolsby, J., J.C. Legaspi, and B.C. Legaspi. 1996. Quarantine evaluation of exotic parasitoids of the sweetpotato whitefly, *Bemisia tabaci* (Gennadius). Southwestern Entomologist 21: 13–21.

Gonzalez, R.A., G.E. Goldman, E.T. Natwick, H.R. Rosenberg., J.I. Grieshop, S.R. Sutter, T. Funakoshi, and S. Davila-Garcia. 1992. Whitefly invasion in Imperial Valley costs growers, workers millions in losses. California Agriculture 46 (Sept.–Oct.): 7–8.

Headrick, D.H., T.S. Bellows, Jr., and T.M. Perring. 1996. Behaviors of female *Eretmocerus* sp. nr. *californicus* (Hymenoptera: Aphelinidae) attacking *Bemisia argentifolii* (Homoptera:

Aleyrodidae) on cotton, *Gossypium hirsutum* (Malvaceae) and melon, *Cucumis melo* (Cucurbitaceae). Biological Control 6: 64–75.

Joyce, A. and T.S. Bellows. 1997. Field cage evaluation of *Amitus bennetti* (Hymenoptera: Platygasteridae), a parasitoid of *Bemisia argentifolii*. In T.J. Henneberry, N.C. Toscano, T.M. Perring, and R.M. Faust, eds., Silverleaf Whitefly: 1997 Supplement, 5-Year National Research and Action Plan, 5th Annual Review, San Diego, CA, January 28–30, 1997. US Department of Agriculture, ARS, Washington, DC. ARS 1997–02, p.153.

Joyce, A.L., T.S. Bellows, and D.H. Headrick. 1999. Reproductive biology and search behavior of *Amitus bennetti* (Hymenoptera: Platygasteridae), a parasitoid of *Bemisia argentifolii* (Homoptera: Aleyrodidae). Environmental Entomology 28: 282–289.

Legaspi, J.C., B.C. Legaspi, Jr., R.L. Meagher, Jr., and M.A. Ciomperlik. 1996. Evaluation of *Serangium parcesetosum* (Coleoptera: Coccinellidae) as a biological control agent of the silverleaf whitefly (Homoptera: Aleyrodidae). Environmental Entomology 25: 1421–1427.

Naranjo, S.E. 2001. Conservation and evaluation of natural enemies in IPM systems for *Bemisia tabaci*. Crop Protection 20: 835–852.

Pantoja, A., M.A. Ciomperlik, N. Gabriel, P. Vasquez, L. Wiscovitch, and W. Robles. 2002. Evaluation of the establishment and distribution of exotic biological control agents of silverleaf whitefly in Puerto Rico. In T.J.Henneberry, R.M. Faust, W.A. Jones, and T.M. Perring, eds., Silverleaf Whitefly: National Research, Action and Technology Transfer Plan (Formerly Sweetpotato Whitefly, Strain B): 4th Annual Review of the Second 5-Year Plan and Final Report for 1992–2002. San Diego, CA, February 10–12, 2002. US Department of Agriculture, ARS, Washington, DC, p.165.

Perring, T.M., A. Cooper, D.J. Kazmer, C. Shields, and J. Shields. 1991. New strain of sweetpotato whitefly invades California vegetables. California Agriculture 45 (Nov.–Dec.): 10–12.

Quicke, D.L.J. 1997. Parasitic Wasps. Chapman & Hall, London.

Chapter 9
Field Evaluation of *Bemisia* Parasitoids in Texas

Matthew A. Ciomperlik[1] and John A. Goolsby[2]

Abstract Two methods were employed to assess the potential of candidate parasitoid species/strains to parasitize *B. tabaci* under field conditions in Texas. Sleeve cage evaluations were conducted in kale, cantaloupe melons, and cotton in 1994–1995. In kale, the highest parasitism rates were observed for two strains of *Eret. mundus* (Spain M92014, and India M92019). Similar evaluations in cantaloupe melons identified *Eret. mundus* (Spain M92014) and *Enc.* sp. nr. *pergandiella* (Brazil M94055) as parasitizing the greatest numbers of *B. tabaci*. Sleeve cage evaluations in cotton showed that *Eret. hayati* (Pakistan M5012) and two *Eret. mundus* strains (Spain M92014, and India M92019) parasitized the most whitefly hosts.

Open field release evaluations were conducted in cotton, broccoli and cantaloupe in 1995–1996 on species or strains of parasitoids that performed well in sleeve cage evaluations. Field release evaluations in cotton showed that *Eret. mundus* (Spain M2014) and *Eret. hayati* (India M93005) were regularly recovered, but they did not parasitize more hosts than the native *Eret. tejanus* or *Enc. pergandiella*. This may have been due to an 'incumbent advantage' of greater numbers of native parasitoid present during the releases of exotic parasitoids. Field releases of *Eret. mundus* (Spain M92014) in winter broccoli showed higher parasitism rates than either of the native parasitoids. Evaluations of *Eret. mundus* (Spain M92014) and *Eret. hayati* (Pakistan M95012) in cantaloupe showed higher parasitism rates than the native *Eret. tejanus* and *Enc. pergandiella*.

Subsequent parasitoid recovery studies conducted in 2001 through 2004 showed repeated collection of *Eret. hayati* (Pakistan M95012) across all crop types sampled. A small number of *Eret. mundus* (Spain M92014) were recovered as well, principally from cole crops. During collections made in 2005, *Eret. hayati* (Pakistan M95012) was again collected in large numbers for export to combat *B. tabaci* in Australia. These findings indicate that *Eret hayati* has become well established in the LRGV of Texas.

[1]USDA-APHIS-PPQ, Center for Plant Health Science and Technology Laboratory, 22675 N. Moorefield Rd., Bldg. 6414, Edinburg, TX 78541

[2]USDA-ARS, Kika de la Garza Subtropical Agricultural Research Center, Beneficial Insects Research Unit, 2413 East Highway 83, Weslaco, TX 78596

J. Gould et al. (eds.), *Classical Biological Control of* Bemisia tabaci *in the United States.*

9.1 Introduction

In the past, biological control programs have been implemented in an empirical fashion where species are released either one at a time or in groups, with the intent that they will establish and control the target pest. The sweetpotato whitefly, *Bemisia tabaci* (Genn.) biotype B (also referred to as silverleaf whitefly, *B. argentifolii* Bellows and Perring) program in the Lower Rio Grande Valley of Texas (LRGV) was implemented with a more predictive approach (Goolsby et al. 2000) in which geographic strains/species were evaluated in quarantine for attributes thought to hold potential for field establishment, and ultimately, whitefly control. Those attributes of the parasitoid species included one or more of the following: originally collected from a region closely matched in climate to the Lower Rio Grande Valley, a demonstrated high fecundity on the pest, a unique genotype, or lab data indicating ability to efficiently parasitize hosts on a specific host plant. Imported exotic agents were initially screened for their performance on several key crops while in quarantine culture (Goolsby et al. 1996, 1998; Chapter 7); successful candidate natural enemies were then further evaluated for their potential to parasitize *B. tabaci* under field conditions. Ultimately, data gathered from these tests were used to justify mass rearing and widespread field release of selected biological control agents (Chapters 10–14).

9.2 Sleeve Cage Evaluations 1994–1995

The first level of field evaluation for the candidate natural enemies used sleeve cages. These tests allowed natural enemies to be evaluated for their potential to parasitize *B. tabaci* under field conditions on crops that were widely affected by the pest and economically important in the LRGV (Goolsby et al. 1998). Leaves of the crop of interest were visually inspected for contamination by Lepidoptera eggs or larvae, aphids, and for the presence of approximately 500–750 whitefly eggs. Preliminary trials indicated that leaves containing Lepidoptera eggs, Lepidoptera larvae, or aphids often were consumed or developed problems with sooty mold from aphid honey-dew exudates. Once the leaves were determined to be suitable for use (i.e., sufficient numbers of whitefly (WF) eggs and no contamination), a nylon organza sleeve cage was placed over the leaf and tied at both ends with a twist tie. Individual caged leaves were checked every 4 days to gauge the development of immature whiteflies. When the whitefly immatures reached mixed populations of second and third instars, mated female *Eretmocerus* spp. were introduced into the sleeve cages. *Encarsia* spp. were released in the sleeve cages when the whitefly immatures reached the third to fourth instar.

The parasitoid species selected for study were collected from laboratory cultures as parasitized pupae and held in small (25 cm^3) Plexiglas screened cages. Adult parasitoids were allowed to emerge in the cages and mate at random. All adult parasitoids were aspirated from the cage every 24 h in order to ensure a uniform age of adult parasitoids used in the field evaluation. Parasitoids used in this experiment were separated by placing a 1/4 dram vial over the individual until it walked into the vial, and the vial was sealed with a cotton plug so the parasite could not escape. Each parasitoid was identified by sex; females were held for use in the experiment and males were discarded. Forty female parasites of each of the strains were isolated only hours prior to release into the sleeve cages. Two females were introduced into each sleeve cage. Twenty replicates were set up per parasitoid species/strain and an additional twenty replicates without any parasitoids added served as controls. The female parasitoids were allowed to oviposit for 15 days, at which point the experiment was terminated. Each sleeved leaf was cut from the remaining foliage and the entire sleeve cage was returned to the laboratory, where counts of whitefly exuviae, and parasitized pupae were made.

In all field evaluations regardless of plant host, the proportions of parasitized hosts were compared among the parasitoid species/strains. The data were arcsine transformed prior to analysis of variance. An analysis of variance test (SAS 1995) was used to determine significant differences in the mean proportion of parasitized whitefly nymphs among the parasitoid species/strains tested. Significantly different means were separated, where appropriate, using Tukey's HSD multiple range test.

9.2.1 Kale, Winter 1994–1995

Eretmocerus mundus Mercet (Spain, culture M92014 and India, M92019), *Encarsia sophia* (Girault and Dodd) (Spain, M93003), *Enc. lutea* (Masi) (Israel, M94107) and *Enc.* nr. *pergandiella* (Brazil, M94055) were evaluated on kale plants (*Brassica oleracea* var. 'Acephala') in the field in the winter of 1995. An analysis of variance showed highly significant differences in the mean number of parasitized pupae among the parasite strains ($F = 9.80$; $d.f. = 4,69$; $P < 0.0001$). Of the five species tested, *Eret. mundus* parasitized the greatest proportion of whitefly nymphs followed by *Eret. hayati* (Table 9.1). None of the three remaining species parasitized more than 10% of the available whitefly nymphs.

The results of the field sleeve cage evaluations suggested that the *Eret. mundus* (M92014 and M92019) were better adapted to parasitizing *B. tabaci* on kale than the *Encarsia* spp. tested. This could be due to the effect of the host plant on the foraging parasitoid. For example, *Eretmocerus eremicus* Rose and Zolnerowich, native to the desert valleys of California and Arizona, does not parasitize significant numbers of *B. tabaci* on cole crops (K. Hoelmer, personal communication). Host plant effects on the *Encarsia* spp. were also observed in mass rearing efforts on kale. Low rates of parasitism were found for *Enc.* nr. *pergandiella* (Brazil M94055) and *Enc. sophia* (Spain M93003) when they were reared on kale and collards in the field insectary (Goolsby, unpublished data). In conclusion, the two *Eretmocerus* spp. tested appeared to be the best candidates for further field release and evaluation in cole crops.

Table 9.1 Results of sleeve cage evaluations of selected parasitoid species/strains under field conditions in the Lower Rio Grande Valley, Texas, during 1994–1995.

			Mean no. parasitized nymphs		
Species	Origin	Culture #	Kale	Cantaloupe	Cotton
Eret. mundus	Spain	M92014	14.9 ± 4.9 a	17.6 ±4.5 a	17.9 ± 5.1 a
Enc. bimaculata	India	M92018		13.6 ±4.4 a	
Eret. mundus	India	M92019	13.9 ± 3.4 ab	10.9 ± 2.7 a	11.9 ± 3.1 ab
Enc. lutea	Israel	M94107	4.0 ± 1.4 bc		
Enc. nr. *pergandiella*	Brazil	M94055	1.3 ± 0.7 c	17.5 ± 6.7 a	
Enc. sophia	Spain	M93033	0.7 ± 0.3 c	8.4 ± 3.0 a	
Enc. lutea	Cyprus	M93064		14.0 ± 3.8 a	
Eret. mundus	Israel	M94120		10.3 ± 2.0 a	4.5 ± 1.8 b
Eret. melanoscutus	Thailand	M94023		5.8 ± 1.2 a	
Eret. melanoscutus	Thailand	M94036			5.6 ± 3.1 a
Eret. hayati	Pakistan	M95012			18.3 ± 5.0 a

Note: Means (+SEM) followed by the same letter within each column are not significantly different (Tukey's HSD, $P > 0.05$).

9.2.2 *Cantaloupe Melon, Spring 1995*

In the spring of 1995, eight species/strains were evaluated on cantaloupe melons (*Cucumis melo* L. 'Perlita'): *Eretmocerus mundus* (Spain M92014) and *Eret. mundus* (Israel M94120 and India M92019), *Enc.* sp. nr. *strenua* (India M92018), *Enc.* nr. *pergandiella* (Brazil M94055), *Encarsia sophia* (Spain M93003), *Encarsia lutea* (Cyprus M93064), and *Eret. melanoscutus* Zolnerowich and Rose (Thailand M94023). Previous quarantine screening tests (Goolsby et al. 1996) showed several species to be promising candidates for field testing on melons. No significant differences in percent parasitism were found among the eight species/strains tested (Table 9.1; $F = 0.9$; df = 6, 90; $P < 0.47$).

We predicted that *Encarsia* nr. *pergandiella* (Brazil M94055) would perform better than the other species tested on melons based on its performance in earlier quarantine screening experiments (Goolsby et al. 1996; Chapter 7). The potential of this species and the others tested may have been unrealized due to experimental conditions. Whitefly numbers were at or near damaging levels very early on in the melon crop; they produced copious amounts of honeydew within the sleeve cage, which may have made the environment inside the sleeve too sticky for the parasitoids to search effectively. In addition, melon plants died shortly after the test due to high whitefly pressure. For these reasons, these sleeve cage evaluations on melons may not have accurately depicted the potential of the exotic parasitoids against *B. tabaci* on melons. It is also likely that control of whitefly on melons will require the integration of additional control agents, and/or compatible insecticides to be successful (Chapter 16).

9.2.3 Cotton, Summer 1995

Five *Eretmocerus* strains were evaluated to determine their ability to find and parasitize *B. tabaci* on cotton (*Gossypium hirsutum* L. 'DPL50') in the summer of 1995: *Eretmocerus mundus* (Spain M92014 and India M92019), *Eret. hayati* (Pakistan M95012), *Eret. mundus* (Israel M94120), and *Eret. melanoscutus* (Thailand M94036). An analysis of variance showed significant differences in the mean proportion of parasitized pupae among the parasite strains (Table 9.1; $F = 4.40$; df. = 4,84; $P = 0.0029$). Of those strains tested, *Eretmocerus hayati* (Pakistan M95012) and *Eret. mundus* (Spain M92014) parasitized significantly more hosts than the other strains. Parasitism rates for these two strains reached approximately 18%, while the *Eretmocerus hayati* from India parasitized about 12%. The remaining two species, *Eret. melanoscutus* (M94036) from Thailand and *Eret. mundus* (M94120) from Israel, parasitized less than 10% of the available hosts.

Data gathered during the field evaluations of the exotic parasitoid species and strains in kale, melons and cotton suggest that the *Eret. mundus* (Spain M92014 and M92019) performed well in all crop types. *Eretmocerus hayati* (M95012) from Pakistan performed as well as the species mentioned above in cotton; however, further testing is needed to determine its performance on kale and melons. *Encarsia* sp. nr. *pergandiella* from Brazil (M94055) performed well in melons, but did not perform well in the kale evaluations. Further testing of this species is needed in cotton, so that accurate comparisons can be made to the several *Eretmocerus* spp. tested.

Results of these field sleeve cage evaluations indicated that the *Eretmocerus* spp. from Spain and Pakistan and *Enc.* sp. nr. *pergandiella* from Brazil should be mass reared for field release in the future. Parasitism rates during the evaluations never exceeded 25% by any one parasitoid species. However, the experiments were continued for only one generation in the field and were likely not entirely characteristic of potential control of *B. tabaci* over multiple generations. Additional experiments to test the ability of the various parasitoid strains to attack whitefly under open field conditions and across multiple crop types appear to be warranted. Differences observed in attack rates across host type indicate that individual species may readily attack *B. tabaci* on one host plant but infrequently attack them on others.

Riley and Ciomperlik (1997) determined the seasonal abundance patterns of *B. tabaci* in relationship to crop phenology of the Lower Rio Grande Valley of Texas. This study showed that whitefly maintained year-round infestations by seasonal migration from crop to crop utilizing annual row crops grown in the area. The most economically important row crops included cole crops (cabbage and kale), cucurbits (melons and cucumbers) and cotton. Therefore, the ability of parasitoid strains/species to attack *B. tabaci* on multiple host plants, especially those of economic importance in the growing region, is vital to the establishment of parasitoids for biological control of the pest.

9.3 Open Field Evaluations of Parasitoids of *B. tabaci*

Species/strains of parasitoids that showed promise in either quarantine or sleeve cage evaluations were mass reared in field insectary cages at the Mission Biological Control Laboratory for further testing. The next level of evaluation was aimed at determining the efficacy of individual species as parasitoids of *B. tabaci* under field conditions in crops commonly grown in the LRGV of Texas. In these evaluations, adult parasitoids were released from large Petri dishes at multiple points within the plot. Open field evaluations determined whether the released parasitoid species was undergoing within-season reproduction and what proportion of hosts were attacked. Comparison of attack rates between native and exotic parasitoid species provided a relative measure of efficacy so that future rearing efforts could be focused on effective candidate species.

9.3.1 *Organic Cotton, Summer 1995*

Six 0.4 ha plots of cotton ('DPL-50') were planted by farmers in cooperation with the US Fish and Wildlife Service (USFWS) on refuge lands in La Joya, Texas. The cotton plots were organically managed, without the use of herbicides or insecticides. Five of the six plots received two releases of 5,000 parasitoids per month per plot, which were released over an 8-week period. There were two replicates each of *Eretmocerus mundus* (Spain M92014) and *Eret. hayati* (India M93005), a single plot for release of *Eret. staufferi* Rose and Zolnerowich (Texas M94002), and one plot served as a control. Release plots were separated from each other by a distance of approximately 1 km.

Samples of parasitized *B. tabaci* nymphs were collected before each release, and sampling continued for 9 weeks after releases ceased. The samples were examined to identify parasitized and non-parasitized nymphs. Individual parasitized nymphs were separated from leaf material and isolated in ¼ dram glass vials; emerging adult parasitoids were separated by species and counted to determine the proportion of nymphs parasitized by native parasites and by released exotic species.

Figure 9.1 summarizes data gathered during the field evaluations of cotton at La Joya, Texas. Whitefly densities were relatively low throughout the study period, with peak populations averaging less than 20 late instar nymphs per leaf. Whitefly populations generally reach much higher population levels in commercially grown cotton (Riley and Ciomperlik 1997) than those observed in this study. The study plots were located within USFWS property boundaries, and were isolated from large commercial agriculture production areas, which may explain why whitefly populations remained at low levels throughout the duration of the study. Two native species of parasitoids, *Enc. pergandiella* Howard and *Eret. tejanus* Rose and Zolnerowich greatly exceeded the number of exotic individuals recovered in our samples. Further, *Enc. pergandiella* consistently outnumbered the native *Eret.*

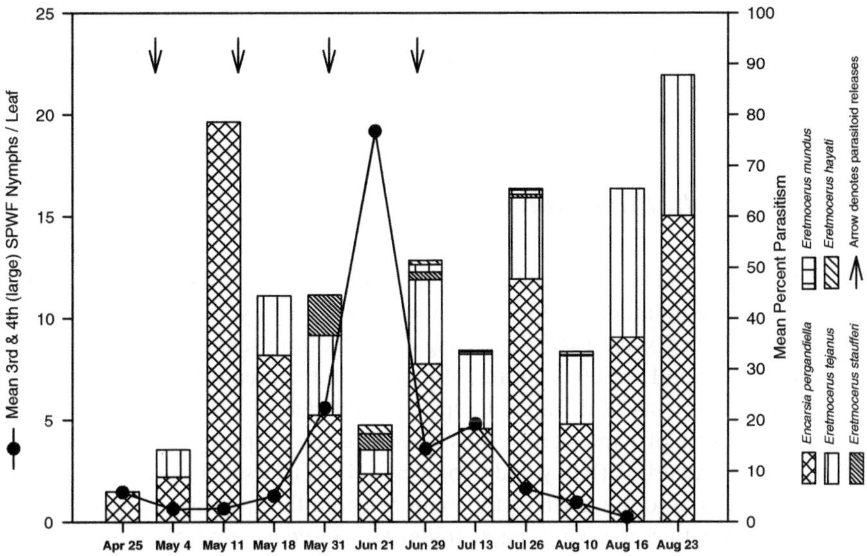

Fig. 9.1 Parasitism rates of *Bemisia tabaci* by native and released exotic parasitoids in a wildlife refuge planting of cotton in summer of 1995. Arrows indicate releases of exotic parasitoids.

tejanus. *Eretmocerus staufferi* was collected on several sample dates, and reached its greatest densities, approximately 5% of the overall parasitism of *B. tabaci* nymphs, about 2 weeks after the last release. *Eretmocerus mundus* and *Eret. hayati* were occasionally recovered from field samples, but never exceeded more than 5% parasitism.

Samples were collected 10 weeks after the final release; they contained proportionately small numbers of the exotic species. Sub-samples of those parasites were sent to Mike Rose at Texas A&M University, who confirmed the recovery of all three species from the cotton plots by standard morphological identification, and to Don Vacek at MBCL, whose genetic verification by RAPD-PCR confirmed the taxonomic identification (Chapter 6). Further analysis of the data showed that *Eretmocerus mundus* occurred in samples from other plots indicating that it had dispersed a distance greater than 1 km from its original release location.

The vast majority of parasitoids collected in the study were native to the area, and were most likely well established within the cotton plots prior to the release of the exotic parasitoids. The native parasitoids probably had an 'incumbent advantage' over the exotics by establishing mating populations well before the exotic species were released in the study. This 'incumbent advantage' poses a significant hurdle to the successful establishment of exotic parasitoids, especially when attempting to establish those populations early in the growing season. Competition for suitable hosts to parasitize and difficulty in locating mates among larger populations of native species may limit the successful introduction of exotic parasitoids, especially if relatively small numbers are released. Early season releases of exotics, as soon as suitable host stages are present, may help overcome the incumbent advantage held by native parasitoids.

9.3.2 Broccoli, Fall/Winter 1995

We established an 8.1 ha research farm at the Mission Biological Control Laboratory that allowed us to maintain a continuous annual cropping cycle of whitefly infested crops typical to that of the LRGV growing area. Planting schedules, growing practices, crop cycles and varieties commonly grown in the area were used to simulate typical agroecosystems of the area. Crops were planted in 2.0 ha blocks according to the LRGV cycle of winter cole crops (*Brassica* spp.), spring cucurbits, summer cotton, and fall cucurbits (Riley and Ciomperlik 1997). A separate 1.2 ha farm was established 0.75 km to the north of the main research farm for comparison of parasitoid release treatments (exotic parasitoids) to no release controls (native parasitoids) and to provide separation to minimize migration. Insect pests other than whiteflies were either treated with biorational insecticides (such as *Bacillus thuringensis*) or damage was tolerated so that parasitoids released in the experiments were not adversely impacted. The experiments reported hereafter were conducted on these research farms.

Broccoli ('Sultan') was planted at both farm locations and monitored for the presence of whitefly. Once whitefly densities reached 0.5 nymphs per cm^2, approximately 3,000 *Eret. mundus* (Spain M92014) adults were released per 0.4 ha on 18 November at the south farm location (release plot). No parasitoids were released at the north farm (no release plot) in order to measure parasitism by the native parasitoids. Thirty basal leaves selected at random were collected every 2–4 weeks to determine the proportion of parasitized to non-parasitized large nymphs. Figure 9.2 compares the parasitism rates of *Eret. mundus* (Spain M92014) in broccoli in the release and no release plots. Although whitefly nymph densities were less than one nymph per cm^2, moderate rates of parasitism were observed. As whitefly densities decreased throughout the growing season, parasitism rates declined as well. The highest parasitism rates occurred 4 weeks after the release of the exotic parasitoids. *Eretmocerus mundus* parasitized approximately 40% of the large nymphs, while the native parasitoids *Eretmocerus tejanus* and *Encarsia pergandiella* (combined) parasitized less than 25% of the available hosts.

The net benefit of releasing exotic parasitoids into winter cole crops, in this case broccoli, is twofold. First, overwintering whitefly numbers are further reduced so that fewer adult whiteflies are present to migrate to spring crops. Secondly, immature parasitoids that overwinter in whitefly nymphs will emerge in the spring and can attack whitefly in spring crops. Whitefly numbers are relatively low during the winter months, and consequently the numbers of overwintering parasitoids are low as well. Identifying exotic parasitoid species like *Eret. mundus* that readily attack *B. tabaci* on cole crops grown in winter may be key to preventing large spring

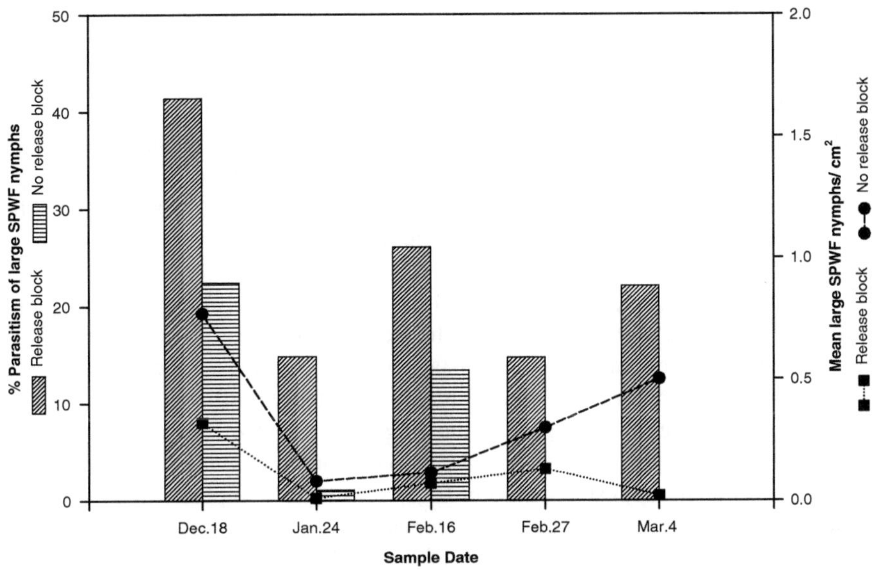

Fig. 9.2 Parasitism of *Bemisia tabaci* in broccoli on the MBCL research farm in the winter of 1995. *Eret. mundus* (Spain M92014) was released on 18 November at a rate of approximately 3,000 adult parasitoids per 0.4 ha.

populations of the pest. In these field trials, *Eret. mundus* parasitized greater numbers of *B. tabaci* nymphs than the two most prominent native parasitoids combined. It also performed well in sleeve cage evaluations on melons and cotton, and in previous quarantine screening efforts (Goolsby et al. 1996, 1998). As a result, mass rearing and area-wide release efforts focused on this exotic parasitoid species appeared to be warranted (Chapter 11).

9.3.3 Cantaloupe Melons, Spring 1996

Imidacloprid (Admire® 2) was registered for use against *B. tabaci* in cucurbits in 1995, and many growers in the LRGV began using the product. The efficacy of the parasitoids and their compatibility with the use of imidacloprid in cantaloupe melons was tested in the spring of 1996. Two *Eretmocerus* spp. (*Eret. mundus* and *Eret. hayati*) that performed well in sleeve cage evaluations on kale, cantaloupe melons, and cotton were evaluated in these tests.

Two experimental plots, 2.0 ha and 1.2 ha plots separated by 0.75 km, were prepared for field trials. Cantaloupe melons were direct seeded to 80 in. beds on 21 February in a series of four rows, each separated by two rows of sorghum x sudan hybrid grass as a wind break. Irrigation, fertigation and chemigation were administered through underground T-tape® drip irrigation lines. Four weeks after planting

(19 March), a single treatment of Admire® 2 flowable systemic insecticide was applied via the drip irrigation lines at 16 oz (0.5 lb ai) per 0.4 ha to both test plots.

Two exotic parasitoid species, *Eretmocerus mundus* (Spain M92014), and *Eret. hayati* (Pakistan M95012) were released at 2-week intervals at approximately 3,000 adult parasitoids per release into the 2 ha plot, while no releases were made in the 1.2 ha plot so that comparisons to naturally occurring native parasitoid species could be made. Adult parasitoids were released from Petri dishes placed at the base of the plant canopy as soon as suitable nymphal instars were detected in the release plot.

On each sample date, one leaf was collected randomly from each of 30 plants from a specific node on the main stem. The leaf node selected was the one with the largest population of early to mid-instar whitefly nymphs. Once a node was selected, leaves from that node were sampled repeatedly until significant numbers of pupal exuviae begin to appear, indicating maturation of the nymphal cohort associated with that leaf node. Upon detection of pupal exuviae, a younger node, again harboring the largest population of mixed-instar nymphs, was selected for subsequent sampling, and the cycle was repeated.

Figure 9.3 shows the mean number of large (third and fourth instar) whitefly nymphs per cm^2 on selected main-stem leaves sampled from the cantaloupe melons. Early to mid-season nymphal populations on the cantaloupe melons were very low, not exceeding 1 large nymph per cm^2 and remained low until the melons were ready for harvesting on 7 June. A final sample on 19 June to assess post-harvest nymph densities was taken from the 15th main stem leaf node due to the declining plant

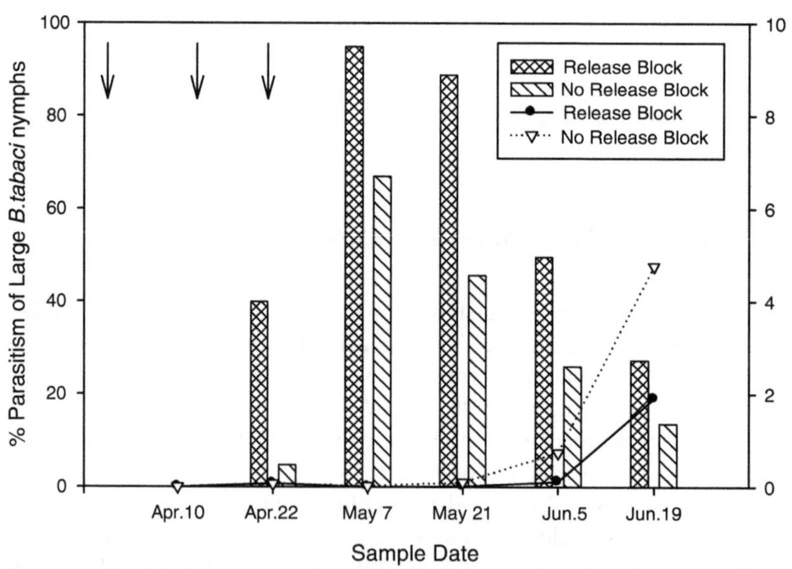

Fig. 9.3 Parasitism of *Bemisia tabaci* in cantaloupe melons on the MBCL research farm in the spring of 1996. Arrows indicate the release of exotic parasitoids.

quality. Nymph densities were higher than those observed at any time previously during the study. Whitefly nymph densities averaged 1.8 nymphs per cm^2 in the exotic parasitoid release plot and 4.8 nymphs in the no release plot.

Whitefly densities were much lower than previously recorded in our field trials (Ciomperlik, unpublished data). The application of imidacloprid, made 4 weeks after planting, appeared to limit early season population build up of whiteflies, and the subsequent release of the exotic parasitoids on 27 March appears to have been well timed. Even though whitefly density was very low, high rates of parasitism were observed 2–3 weeks after the final parasitoid release. No attempt was made to separate parasitism caused by the exotic and native *Eretmocerus* species in the release plots. The combined parasitism rates for all *Eretmocerus* spp. in the release plots exceeded 95% of the available SPWF nymphs. The native *Eret. tejanus* and *Enc. pergandiella* parasitoids attacked about 65% and 43% of the whitefly nymphs in the no release plots 2 and 3 weeks after release.

The combination of imidacloprid and releases of exotic parasitoids maintained whitefly at approximately half the population level as that of imidacloprid and native parasitoids in the final two samples. This study illustrates the compatibility of using systemic applications of imidacloprid and parasitoids to control *B. tabaci* (see also Chapter 16). Prior experience in conducting similar field trials suggest that imidacloprid and parasitoid releases used separately do not control the pest as well as using the two control measures in combination. The length of time that imidacloprid is effective against *B. tabaci* ranges from 6 to 8 weeks in cantaloupe melons (S. Fraser, Bayer Corp., personal communication). From date of application, the range in effective control for imidacloprid should have continued long enough for two additional generations of whitefly to be completed before the crop was harvested. However, the native and exotic parasitoids maintained whitefly populations below 1 nymph per cm^2.

Since *Eretmocerus mundus* (Spain) and *E. hayati* (Pakistan and India) repeatedly outperformed other exotic and native parasitoid species in our field evaluations, these two species were given precedence in mass rearing efforts for area-wide release against *B. tabaci*. All of the quarantine screening, laboratory, and field evaluation data collected indicated that these two species would contribute significantly to suppression of *B. tabaci* in the lower Rio Grande Valley of Texas.

9.3.4 Parasitoid Recoveries 2001–2004

Samples of parasitized *B. tabaci* were collected from multiple locations throughout Hidalgo County, Texas in the fall of 2000 and spring of 2001 from cabbage, collards and cucumbers. Parasitized pupae were removed from the leaves, held in ¼ dram glass vials until the adult parasitoid emerged and the adult parasitoids were separated into groups of either *Eretmocerus* spp. or *Encarsia* spp. prior to being identified by RAPD-PCR.

RAPD-PCR analyses revealed that 98.3% (n = 135) of the banding patterns produced by the *Eretmocerus* spp. matched banding patterns produced by *Eret. hayati* (Pakistan M95012), which were collected across all the crop types sampled. A small number (less than 1% of the sample) of *Eret. mundus* (Spain M92014) were recovered and only from the cole crops. One individual from the *Encarsia* spp. sample (5.5%) matched banding patterns produced by *Enc. sophia* (Pakistan M95107), while the remaining individuals were consistent with native *Enc. pergandiella* patterns.

During 2003, field collections were made in cultivated sunflower (*Helianthus annuus* L. var. black oil seed) infested with large numbers of bandedwing whitefly (*Trialeurodes abutiloneus* (Haldeman)) in order to ship *Eretmocerus hayati* to Australia for culture and release to control *Bemisia tabaci*. Emergence of the adult parasitoids showed that they were heavily parasitized by the native *Eret. tejanus* and *Enc. pergandiella*, and that the exotic *Eret. hayati* was not present in the sample. The collection of additional samples the following spring from cantaloupe melon infested with *B. tabaci* showed more than 99% parasitism by *Eret. hayati*, and provided more than 500 individuals for export (Goolsby et al. 2005). The native *Encarsia pergandiella* was also collected from the samples in large numbers, but were not retained.

The native parasitoids *Eret. tejanus* and *Enc. pergandiella* readily attack the native bandedwing whitefly and apparently moderate populations of that pest insect. It appears that the native parasitoids have the ability to utilize *B. tabaci* biotype B as a host (Riley and Ciomperlik 1997), which probably represents a new host association, but the resulting population suppression is insufficient to keep populations of *B. tabaci* below economically damaging levels. Recoveries of parasitoids from field surveys indicate that *Eret. hayati* (Pakistan M95012) is well established and contributes to the control of *B. tabaci* across multiple host crops in the LRGV of Texas. Small numbers of *Eret. mundus* (Spain M92014) have been recovered only from cole crops, suggesting that this parasitoid plays a minor role in *Bemisia* control.

References

Goolsby, J.A., J.C. Legaspi, and B.C. Legaspi, Jr. 1996. Quarantine evaluation of exotic parasitoids of the sweetpotato whitefly, *Bemisia tabaci* (Gennadius). Southwestern Entomologist 21: 13–21.

Goolsby, J.A., M.A. Ciomperlik, B.C. Legaspi, Jr., J.C. Legaspi, and L.E. Wendel. 1998. Laboratory and field evaluation of exotic parasitoids of *Bemisia tabaci* (Gennadius) (Biotype "B") (Hymenoptera: Aleyrodidae) in the Lower Rio Grande Valley of Texas. Biological Control 12: 127–135.

Goolsby, J.A., M.A. Ciomperlik, A.A. Kirk, W.A. Jones, B.C. Legaspi, R.A. Ruiz, D.C. Vacek, and L.E. Wendel. 2000. Predictive and empirical evaluation for parasitoids of *Bemisia tabaci* (Biotype "B"), based on morphological and molecular systematics. In A. Austin and M. Dowton, eds., Hymenoptera: Evolution, Biodiversity, and Biological Control, pp. 347–358.

Fourth International Hymenopterists Conference, 6–11 January 1999, Canberra, ACT, Australia, CSIRO Publishing, Collingwood, Victoria.

Goolsby, J.A., P. J. De Barro, A.A. Kirk, R. Sutherst, L. Canas, M. Ciomperlik, P. Ellsworth, J. Gould, K.A. Hoelmer, S. J. Naranjo, M. Rose, W. Roltsch, R. Ruiz, C. Pickett, and D. Vacek. 2005. Post-release evaluation of the biological control of *Bemisia tabaci* biotype "B" in the USA and the development of predictive tools to guide introductions for other countries. Biological Control 32: 70–77.

Riley, D.G. and M.A. Ciomperlik. 1997. Whitefly (Homoptera: Aleyrodidae) and associated parasitoid (Hymenoptera: Aphelinidae) population dynamics in the Lower Rio Grande Valley of Texas. Environmental Entomology 26: 1049–1055.

SAS Institute, Inc. 1995. SAS/STAT users guide. Statistical Analysis System Institute, Cary, NC.

Chapter 10
Mass-Rearing *Bemisia* Parasitoids for Support of Classical and Augmentative Biological Control Programs

Gregory S. Simmons[1], Charles Pickett[2], John Goolsby[3], James Brown[2], Juli Gould[4], Kim Hoelmer[5], and Albino Chavarria[6]

Abstract Three systems for production of *Bemisia* parasitoids in the genera *Eretmocerus* and *Encarsia* are described: a smaller scale system for initial production and evaluation of the numerous cultures collected during the foreign exploration effort and two systems for larger scale production to support augmentative biological control demonstration projects and establishment of new species for classical biological control. Efficient production systems depended on providing high-quality host plants that were free of pests, good environmental control, and careful control and monitoring of the whitefly host population. When these conditions were met the greenhouse based production systems could produce millions of parasitoids per week, and production levels as high as 172,000 parasitoids/m^2/generation were achieved. After harvesting, processing techniques allowed for separation of unparasitized whitefly and removal of debris, producing a very clean product with as little as 0.5% whitefly contamination.

[1] USDA-APHIS-PPQ Center for Plant Health Science and Technology Laboratory, 3645 East Wier Avenue, Phoenix, AZ 85040

[2] California Department of Food & Agriculture, Biological Control Program, 3288 Meadowview Rd., Sacramento, CA 95832

[3] USDA-ARS, Kika de la Garza Subtropical Agricultural Research Center, Beneficial Insects Research Unit, 2413 East Highway 83, Weslaco, TX 78596

[4] USDA-APHIS-PPQ Center for Plant Health Science and Technology Laboratory, Building 1398, Otis ANGB, MA 02542

[5] USDA-ARS Beneficial Insects Introduction Research Unit, 501 South Chapel Street, Newark, DE, 19713-3814

[6] USDA-APHIS-PPQ Center for Plant Health Science and Technology Laboratory, 22675 N Moorefield Rd., Building 6414, Edinburg, TX, 78541

J. Gould et al. (eds.), *Classical Biological Control of* Bemisia tabaci *in the United States.*

10.1 Introduction

The development of efficient mass-rearing systems for *Bemisia* parasitoids was crucial for the implementation of the classical and augmentative biological control programs. Early work relied on adapting methods for the production of *Encarsia formosa* (Gahan) on the greenhouse whitefly, *Trialeurodes vaporariorum* (Westwood). Production systems for the Nearctic parasitoid *Eretmocerus eremicus* Rose and Zolnerowich were developed using this system, but new production techniques were needed for the exotic Palearctic parasitoids such as *Eretmocerus mundus* Mercet, *Eret. emiratus* Zolnerowich and Rose and *Eret. hayati* Zolnerowich and Rose, which are unable to develop successfully in *Trialeurodes* species (K. Hoelmer, unpublished data).

In this chapter, we describe several rearing techniques developed to support augmentative biological control demonstration projects in spring melons; for large-scale releases in California, Arizona, and Texas to establish new species of *Eretmocerus*; and for greenhouse biological control evaluations in several states. All of the *Eretmocerus* species that were collected in the worldwide effort to find more effective *Bemisia* parasitoids (Chapter 2) were reared on *Bemisia tabaci* (Gennadius) biotype 'B' using the systems described here. These include some of the most effective Palearctic and North American species: *Eretmocerus emiratus*, *Eret. mundus*, *Eret. hayati*, and *Eret. eremicus*.

10.2 Natural Enemy and Whitefly Production Methods Used by the USDA-APHIS Mission Biological Control Laboratory

The initial rearing methods for production of *B. tabaci* and its imported natural enemies were developed at the USDA-APHIS-PPQ Mission Biological Control Laboratory (MBCL) in Mission, Texas, using hibiscus and eggplant systems, and were smaller in scale than those described in the second part of this chapter. The rearing operations supported parasitoid colony establishment and maintenance, and quarantine evaluations (Goolsby et al. 1996, 1998) of new exotic *B. tabaci* cultures (Chapter 2). Parasitoids were reared in laboratory colonies to provide material for field cage evaluations in Arizona, California and Texas (Chapters 7 and 8); for greenhouse biological control evaluations in Colorado, Mississippi, and New York; to provide seed material for the rearing operations in California; and for direct field releases throughout the USA.

Fifty-five populations and/or species of parasitoids were reared at MBCL from 1992 to 1999 (Chapter 2, Table 10.1). Initially, all cultures in quarantine were colonized on whitefly infested hibiscus (*Hibiscus rosa-sinensis* L.). Following release from quarantine, parasitoid cultures were increased in the laboratory production facility on both whitefly-infested hibiscus and/or eggplant (*Solanum melongena* L.)

Table 10.1 Problems encountered during rearing of whiteflies and parasitoids and the solutions that proved most effective.

Problem/need	Solution	Application rate	Comments
Fertilization of Plants	20-20-20, Peter's Liquid Fertilizer, Allentown, PA	16:1 ratio of water to fertilizer	TX and CA
	Romero 20-10-20 soluble fertilizer	16:1 ratio of water to fertilizer	CDFA
Soil-free potting mixture	Redi-Earth, Scotts, Marysville, OH		CA
Sand for potting mixture	Play Sand, Quikrete, Atlanta, GA		CA
Soil-free potting mixture	Redigro Corp., Sacramento, CA	48% peat moss, 30% perlite, 15% vermiculite, and 7% washed sand	
Fertilize eggplant plants	Osmocote slow release fertilizer (Scotts, Marysville, OH (14-14-14, 3–4 month formulation)	6 oz/ft^3	CA
Pesticide Applications			
Aphelinid parasitoids, greenhouse whitefly, and spider mites infesting whitefly colony	Acephate (Orthene 75 S, Valent)	0.77 ml/l	Whiteflies were resistant to acephate so only the pests were killed
Green peach aphid, *Myzus persicae*; cotton aphid, *Aphis gossypii*; *B. tabaci*; long-tailed mealybug	Pyrethrin-rotenone (Pyrellin EC, Webb Wright Corp.)	2.5 ml/l	
Pseudococcus longispinus; and western flower thrips, *Frankliniella occidentalis*	Pyrethrin-piperonyl butoxide (Concern, Miracle-Gro Lawn Products, Inc)	2.5 ml/l	
	Insecticidal soap (M-Pede, Mycogen Corp.)	19.2 ml/l	
	Cyfluthrin (Tempo 2EC, Bayer Corp.)	0.3 ml/l	
	Azadirachtin (BioNeem, Safer)	23 ml/l	
Broad mite, *Polyphagotarsonemus latus*	Avermectin (Avid 0.15 ec, Syngenta Inc.)	0.6 ml/l	
Argentine ants, *Linepithema humile*	Hydramethylon ant bait (Amdro Pro, BASF)	10–15 g in a small petri dish in the corner of a cage	Ants tended aphids and preyed on parasitoid pupae

(continued)

Table 10.1 (continued)

Problem/need	Solution	Application rate	Comments
Pythium and Phytophthora root rot	metalaxyl (Subdue 2E, Syngenta)	0. 31 ml/l	Applied weekly alternating with fosetyl aluminum
	fosetyl aluminum (Aliette 80WP, Bayer)	1.0 ml/l	
Natural Enemy Applications			
Green peach aphid, *Myzus persicae*	*Aphidius colemani* Viereck (parasitoid)	2–5 per plant	
Long-tailed mealybug, *Pseudococcus longispinus*	*Cryptolaemus montrouzieri* Mulsant (the mealybug destroyer)	1–2 beetles per plant	
Western flower thrips, *Frankliniella occidentalis*	*Amblyseius cucumeris* Oudemans (predatory mite)	5–20 mites per plant	
	Minute pirate bug *Orius insidiosus* (Say)	1–5 bugs per plant	Also predators of whiteflies so only released in the plant production greenhouse
Two-spotted spider mite, *Tetranychus urticae*	Predaceous mites *Phytoseiulus persimilis*	500 predators per 25 plants	Control in TX greenhouses
	Mixed release of *Phytoseiulus persimilis, Mesoseiulus longipes,* and *Neoseiulus californicus*	10–20 mites per plant	Control in CA greenhouses
Fungus gnats, *Bradysia* spp	*Bacillus thuringiensis* var *israelensis* (Gnatrol,Valent)	Weekly soil drenches with 10.0 ml/l	

plants. Production in the laboratory facility allowed for stable year round rearing, permitting inoculation of outdoor rearing cages designed for economical rearing of multiple species for field release. Together the use of these two systems resulted in the production of several million parasitoids from the 58 species and/or populations of exotic natural enemies.

10.2.1 Laboratory Culture on Hibiscus at the USDA-APHIS Mission Biological Control Laboratory

10.2.1.1 Plant Culture

All parasitoid and predator cultures were reared in environmental growth chambers on whitefly infested *Hibiscus rosa-sinensis* L. 'Brilliant Red'. This variety produced large glossy leaves that were ideal for oviposition by *B. tabaci*. Hibiscus cuttings or seedlings were grown in 15.2 cm diameter pots and mature plants were ready for whitefly oviposition approximately 2 months after potting. The plants were reused twice and then discarded. More than 500 plants were held in rotation to supply adequate numbers of plants for the rearing colonies. Before use in the insect colonies, newly expanding leaves were removed to force oviposition onto the large mature leaves that were more efficiently handled in the rearing process. Plants were watered weekly and completely hand wiped to remove aphids, mites and other pests.

10.2.1.2 Whitefly Oviposition

A population of approximately 10^5 adult *B. tabaci* collected from McAllen, Texas, was used to initiate the whitefly colony. To ensure long-term viability of the colony, a new colony was restarted every 2 years, always replacing it with field collected whitefly from the local agricultural area near Mission, Texas. Mature hibiscus plants were placed in an environmental growth chamber set at 27°C and a 14:10 L:D photoperiod. Each cage had overhead illumination provided by two each of Vitalite® and Chroma-50® 40 watt bulbs to produce full spectrum light, which was important for good plant growth during the winter months. Inside the environmental chamber, plants were placed inside a 0.5 × 0.5 × 1 m aluminum frame fitted with a white organza shroud to confine the emerging adult whiteflies to the hibiscus plants. Approximately 100 mature clipped hibiscus leaves bearing 4th instar whiteflies were stapled to paper towels and placed in the cage for emergence and subsequent oviposition onto new plants. The plants were kept in the whitefly infested cage for 2 days, at which point egg density reached approximately 50 eggs per cm^2. After oviposition the whiteflies were removed with a hand held vacuum. The short oviposition period produced plants with a strict cohort of whiteflies, which were then held at 27°C for 7 days, by which time most had developed into 2nd instar nymphs suitable for parasitization. Seven percent of the plants were held for maturation of the whiteflies to produce adults for the next cycle of oviposition. Spider mite outbreaks were prevented by quarterly inoculative releases of the phytoseiid predator mites (Table 10.1).

10.2.1.3 Parasitoid Culture

Parasitoids were reared in environmental growth chambers operated under the same conditions as those described in the previous section. Whitefly infested plants were placed inside 0.25 × 0.25 × 0.5m cages that had three sides made of white translucent Plexiglas and top and front from clear Plexiglas® that were designed to prevent escape of the extremely small aphelinids. A 10cm hole on the top was covered with fine mesh organza for ventilation; the right side had a similar hole with a 20cm long organza sleeve, which allowed access for watering and parasitoid collection. The left side had a 30cm oblong hole with large sleeve, which was large enough to allow for addition hibiscus plants. The cages could hold a total of four hibiscus plants.

One hundred adult *Eretmocerus* parasitoids of mixed sexes were aspirated into vials and released into each cage, which contained two to four hibiscus plants infested with mostly 1st and 2nd instar whiteflies. Plants with a larger proportion of 3rd and 4th instar whitefly were used for *Encarsia* species. For autoparasitic *Encarsia*, i.e., *Enc. sophia*, a third plant containing previously parasitized whiteflies was added to allow for production of male progeny. Plants with parasitized whiteflies were held for 14 days until the majority of the parasitoids had pupated. At this point the leaves were clipped, scanned for contamination and stapled to paper towels and hung in emergence cages. As parasitoids emerged they were collected with aspirators and used to inoculate the next series of cages or for field releases or evaluation studies. One top corner of production and emergence cages was streaked with honey, which significantly increased the longevity and fecundity of the parasitoids.

Congeneric *Eretmocerus* species are difficult or impossible to distinguish using macro-morphological characters; therefore only one species of *Eretmocerus* was reared per chamber. To minimize contamination of a colony with another parasitoid genus, each growth chamber contained no more than one *Eretmocerus* and one *Encarsia* culture. *Eretmocerus* and *Encarsia* have morphologically different adults and pupae and could easily be identified by production workers to detect early stages of contamination. Periodic samples from each parasitoid colony were checked using RAPD-PCR (Chapter 6) to ensure purity. If contamination was detected, a new culture was started from ten new lines developed from 10 single females that matched the original parental RAPD-PCR pattern. As a result of these efforts by the insect production and genetics teams, no populations were lost due to contamination during the 9 years of insect rearing.

10.2.2 Outdoor Field Cage Production at Mission Biological Control Laboratory

10.2.2.1 Plant Culture

All parasitoid cultures were reared on whitefly infested eggplant, *Solanum melongena* L. 'Ichiban', a variety of eggplant that produces large leaves ideal for oviposition by *B. tabaci*. Kale (*Brassica oleracea acephala*) was used from

November to February because of its tolerance of cool winter temperatures. Plants were grown from seed in 15.2 cm diameter pots and were ready for whitefly oviposition approximately 3 months after potting. More than 1,000 plants were kept in the production cycle with approximately 200 plants started each week in order to maintain the supply of young vigorous plants.

10.2.2.2 Whitefly Oviposition

The *B. tabaci* colony in the field cage was initiated and maintained as described for the laboratory colony. Mature plants were placed in 3.1 × 3.1 × 1.8 m field cages covered in 52 × 52 mesh Lumite Saran™ (Lumite Industries, Gainesville, Georgia). Cages were pinned to the ground using 25 × 1.23 cm soil auger screws and were sealed along the bottom edges with weights made of sand-filled lengths of 4-in fire hose. Plants and whiteflies were reared under ambient outdoor temperatures, except during November to February when the mean minimum and maximum temperatures for McAllen fell to 7.6°C and 20.1°C respectively. During the winter, cages were covered with clear greenhouse plastic and small electric portable heaters were used to increase the temperature about 5°C above ambient temperature. Approximately 10,000 whiteflies were placed in each of eight outdoor cages containing 40 eggplants plants/cage; this was sufficient to provide a continuous supply of whiteflies to inoculate parasitoid production cages. All the plants were watered daily using an automated irrigation timer; liquid fertilizer was added to the irrigation water monthly or as needed (Table 10.1). The whitefly production facility was 1.6 km from the parasitoid production facility to reduce the possibility of contamination from the parasitoid colonies. As whitefly adults emerged, they were vacuumed from the plants using a high volume, low speed ventilation box fan. The adults were collected in a fine mesh cotton organza sock for transfer to the parasitoid production facility or used to inoculate new whitefly production cages. The plants were monitored until whitefly egg density reached 50 eggs per cm^2, at which point adult whiteflies were removed by vacuuming. The short oviposition period produced a defined whitefly cohort that was held for 7–21 days by which time they were mostly 2nd instar whitefly nymphs suitable for parasitization. Spider mite and cotton aphid outbreaks were prevented by quarterly inoculative releases of the phytoseiid predator mite, *Phytoseiulus persimilis* Athias-Henriot, and *Aphelinus gossypii* Timberlake, respectively.

10.2.2.3 Parasitoid Culture

Parasitoids were reared in the same cages with the same environmental parameters as those described above. Eggplant (or in winter, kale) plants (n = 120) growing in 15.2 cm diameter pots were placed in cages and allowed to mature to the 6-leaf stage. Adult whiteflies were then added using the methods described above. Parasitoid cages had an additional Velcro® flap to prevent escape or contamination of the parasitoid colony. An on-site weather station provided degree-day information used to predict when parasitoids would be ready for harvest (von Arx 1983).

Two to six thousand *Eretmocerus* parasitoids of mixed sexes were added to each cage, which contained plants infested with mostly 1st and 2nd instar whitefly; later instars were provided for production of *Encarsia*. Plants with parasitized whiteflies were held from 2 weeks to 2 months (depending on outdoor temperatures) until the majority of the unparasitized whitefly had emerged and the parasitoids were in the pupal stage. At this point the leaves were clipped, scanned for contamination and stapled to paper towels and hung up in indoor laboratory emergence cages maintained at 25°C. Emergence cages (1 × 1.5 × 1 m) were covered with black shrouds except for two round openings designed for placement of 10 cm Petri dish bottoms at the cage top, facing the fluorescent lights. After the emerged parasitoids gathered on the Petri dish bottoms, they were removed and the Petri dish tops, which were streaked with honey, replaced. The numbers of parasitoids were counted using a sub-sample on a fixed grid. Petri-dishes were sealed with Parafilm® M and placed in an environmental chamber at 16°C with a 14:10 L:D cycle. Dishes containing adult parasitoids were shipped interstate to cooperators, held for inoculation of cages, or used for field release.

The outdoor cage production facility provided a low cost, low input method for multiplying parasitoids, although production was somewhat unpredictable during cooler winter months. Development of the parasitoids stopped if the mean temperature dropped below 16°C in the cage. It was not unusual for the parasitoid generation time to take up to 2.5 months during the winter. The use of heated greenhouses instead of the outdoor cages made mass production more predictable, but the cost for the structures was considerably higher than for outdoor cages. The outdoor cage method should be considered as an alternative method for multiplication of whitefly parasitoids during warmer months, especially if budgets are limited.

10.3 Rearing Native and Exotic *Eretmocerus* Species in the Desert Southwest Using a Greenhouse-Based System

A joint project operated by the California Department of Food and Agriculture (CDFA), USDA-APHIS-PPQ, the Imperial County Agricultural Commissioners office, and private industry was launched in Imperial County, California to mass-rear the most effective exotic whitefly parasitoid species. The Imperial Valley insectary was in operation from 1994 to 1998 to support the classical biological control releases and augmentation demonstration projects in California and Arizona. One project supported by the rearing program was to increase biological control of *B. tabaci* biotype B in spring melons by releasing several species of whitefly parasitoids. Parasitism of whiteflies by native species (*Eretmocerus eremicus* and *Encarsia* spp.) is generally low in the spring melon crop in desert production areas of California and Arizona, where whitefly populations first start to rapidly increase, leading to high regional populations that progress to infest cotton and alfalfa after the melon harvest. In addition, the mass rearing supported the classical biological control program with the goal of establishing new species of parasitoids. Benefits of this program included the release of more than 60 million exotic whitefly parasitoids into the Imperial Valley, which helped establish several new species; the transfer of rearing technology; the introduction of exotic, more effective, whitefly parasitoids to the beneficial insect industry; and cooperation with industry to improve the use of mechanized beneficial insect delivery systems.

10.3.1 Plant Production

In the Imperial Valley, plants were grown year-round in a greenhouse without artificial lights. Eggplant was chosen as a host plant because it is a good host of *B. tabaci*, is quick growing and has relatively few pests, and its large leaves are easy to process for parasitoid collection. The main eggplant varieties used were 'Black Beauty' and 'Whopper' although good results were also achieved with Asian varieties 'Sennari' and 'Ping Tung'. Eggplant seeds were planted in one gallon plastic pots filled with a 60:40 mixture of peat-based soil mixed with coarse grained sand and Osmocote™ slow release fertilizer or a similar generic brand (Table 10.1). The soil mix was sterilized with an electric box sterilizer raising the temperature of the soil to 77°C for ½h before use. The greenhouse had both heating and cooling capabilities so that plants were grown at temperatures between 22°C and 33°C throughout the year; due to the southern location, artificial lights were unnecessary. Plants were provided additional fertilizer through the irrigation system with a soluble fertilizer (Table 10.1). Under these conditions plants suitable for egg deposition were produced within 5–8 weeks depending on the season. Plants ready for production were growing robustly to a height of 0.6–1.0m and had 25–40 large leaves up to 19cm wide that were suitable for production. Good plant health was the single most important factor for good parasitoid production.

10.3.2 Insect and Disease Control

Insect infestations caused problems in several ways. Eggplant plants infested with insect pests were weaker and less able to tolerate the high density of whitefly nymphs needed for high parasitoid production; they grew more slowly, became deformed and shed leaves before the parasitoids were mature. Infestations of pests also reduced oviposition activity by both whitefly and parasitoids making the plants less suitable for feeding and oviposition and by creating competition for leaf space.

Infestations of *Bemisia* in the clean plant production greenhouse caused problems when they became established on a cohort of plants prior to exposure to whiteflies in the production system. Early whitefly infestation led to contamination by *Encarsia* spp. (especially *Encarsia hispida* De Santis), which had become a pest in the greenhouses after it had been in production. Whiteflies that established too early also emerged as adults in the parasitoid exposure cages, resulting in heavy feeding and honeydew excretion that weakened plants and interfered with parasitoid oviposition.

For all of these reasons, it was important to keep the plants in the plant production greenhouse as free of pests as possible. To reduce pest entry, greenhouse vents and doors were screened with 52 × 52 mesh Lumite Saran™. A combination of pesticides and natural enemy releases were used as needed to keep plants as clean as possible (Table 10.1). Treatment of pests in the insect colonies was mostly limited to releases of natural enemies, especially spider mite predators and aphid parasitoids. Pesticides with no or short residual times were used so that treated plants could be placed into the parasitoid production system with little delay. Pesticides were frequently rotated to avoid problems with resistance. Pesticides were tested for phytotoxicity to eggplant leaves before use; in some cases application rates were lower than recommended label rates to avoid damage to the leaves. Young seedlings were the most susceptible to damage from spraying and, if possible, plants younger than 2–3 weeks were not sprayed. Most applications of pesticides to the plants in production were made before their introduction into the whitefly and parasitoid colonies to avoid applying pesticides directly to the insect cultures.

Pesticides were applied to control several common pest species (Table 10.1) and were either used alone or as a tank mix with insecticidal soap. Most applications were made for control of *B. tabaci* in the eggplant production greenhouse as the other pests were occasional problems that could be controlled by both releases of natural enemies as well as the pesticides. Treatments for whitefly were made on average once a week during May–September and usually consisted of insecticidal soap alternated with insecticidal soap mixed with a pyrethrin compound.

Occasionally the whitefly colony would become infested with parasitoids (especially *Encarsia* spp.), greenhouse whitefly, or spider mites. Because a badly infested colony could actively produce whiteflies for 4–8 weeks and it was impractical to start over we developed a technique to clean these colonies. Imperial Valley populations of *B. tabaci* were nearly 100% resistant to acephate (Steve Castle, USDA-ARS, Phoenix, Arizona, personal communication) while aphelinid parasitoids, greenhouse whitefly, and spider mites were completely susceptible. We were able to take advantage of the selectivity of this pesticide to treat an infested whitefly colony with acephate (Table 10.1) to eliminate everything except the whitefly population.

Within the whitefly colony and in the parasitoid rearing, plants were occasionally affected by root rot diseases, mainly *Pythium* and *Phytophthora* species. These diseases were generally only a problem during the hottest months when the greenhouse evaporative cooling system was less efficient at keeping the greenhouse at temperatures below 38°C. The combination of heat stress, high densities of feeding whiteflies and exposure to plant pathogens sometimes caused plants to be diseased, leading to premature leaf-drop and the loss of many developing parasitoids. These problems were primarily controlled by careful watering, which reduced favorable conditions for root rot diseases, and by applying a preventive application of fungicides as needed. When disease conditions were prevalent, drench applications of metalaxyl or fosetyl aluminum (Table 10.1) were made on alternate weeks.

10.3.3 Whitefly Colony and Inoculation of Host Plants

10.3.3.1 Whitefly Colony

To start a new colony, whitefly adults were collected from field locations in the Imperial Valley and transferred to colony cages containing potted eggplants. The first colonies were initiated in the late winter with whiteflies collected from cole crops such as cabbages, broccoli and cauliflower. Whitefly adults were collected with a gas powered vacuum (D-VAC Vacuum Insect Net Model 24, Rincon-Vitova Insectaries, Inc., Ventura, California) with a nylon organdy bag fitted over the end of the collection hose. Whitefly adults were collected for 5 min after which the net bags were tied off with a piece of twine and transferred to an ice chest maintained at 13–15°C for transport to the laboratory.

Whiteflies were released inside a 0.9×0.5×0.4 m high wooden cage with a slant-glass top in the laboratory. A fluorescent light source was placed over the cage to attract the whiteflies to the top of the glass and to aid in separation from other species that had been collected. Whitefly adults were collected with an insect aspirator (0.64 cm plastic tubing attached to 0.64 cm aluminum tubing fitted into a 9 dram vial with a rubber plug) connected to an electric vacuum pump (Model G582DX-S55NXMLD-6711, 1/3 hp, 5.5A, 115v, Gast, Mfg. Corp, Benton Harbor, Michigan) with 0.64 cm plastic tubing. The collection of whitefly adults was inspected for contaminating species and stored at 13–15°C for transfer between greenhouses. If there were other insect contaminants in the collection, it was necessary to transfer the whiteflies to another cage and recollect the whiteflies to obtain a clean collection.

10.3.3.2 Inoculating Plants with Whiteflies for Parasitization

Four to 6 week old plants were washed with a fine spray of water to remove dust and pesticide residues, damaged and yellow leaves were removed, and plants were placed into 1.8×1.8×1.8 m whitefly oviposition cages inside a 6×12 m greenhouse.

The greenhouse temperature was maintained between 20 and 38°C by heating and cooling supplied as needed throughout the year. Production in the late winter/early spring for the first releases on spring melons was an important time of the year for rearing. It was observed that when temperatures fell below about 20°C production levels of both whitefly and parasitoids was poor and that at temperatures above 38°C, production declined since eggplant plants appeared to be less tolerant of high levels of whitefly feeding at these temperatures. At high temperatures many leaves became yellow and died before parasitoids could mature.

Depending on the size of the plants, 14–20 plants were placed in each cage. The cages were made of Lumite Saran™ (52 × 52 mesh) with doors secured by zippers covered with a Velcro™ fabric flap to keep small insects from entering around the zipper teeth. Plants were connected to a drip irrigation system and were watered 2–3 times a week as needed and fertilized weekly with a soluble fertilizer (Table 10.1).

Whitefly adults were collected from the colony using a battery powered electric vacuum device with a fine mesh organdy bag fit over the collection nozzle (Modified CDC Backpack Aspirator Model 1412, John Hock Co. Gainesville, Florida) for 2–3 min, and then transferred to cages at the rate of 0.12 l of adult whiteflies per 20 plants. Plants were exposed to the whiteflies until the density of whiteflies on the leaves was greater than 15 whitefly per cm^2; this took 24–48 h depending on temperature. Plants were monitored over the next 1–3 days by collecting ten 2.54 cm^2 leaf discs per cage and counting the number of eggs. Once the target density was reached, all adult whitefly were removed by vacuuming as described above.

After initial whitefly removal, plants were monitored every 2–3 days to remove any remaining whitefly and any damaged or yellowing leaves, and wash off accumulated honeydew. One week after whitefly introduction, the plants were monitored to determine the stage of whitefly development by examining the leaves of several plants with a ×10 hand lens. *Eretmocerus* spp. prefer to oviposit into 2nd instar nymphs (Headrick et al. 1996); whitefly nymphs typically reached this stage of development in 7–8 days during warmer summer months and about 12–13 days during cooler periods of the year. When half of the whitefly nymphs were 2nd instars or older, the cage was readied for parasitoid introduction.

10.3.3.3 Parasitoid Production

Parasitoids were released as adult wasps or as pupae. To protect parasitoids from the heat they were transported to the rearing cages in styrofoam ice chests cooled to 13–16°C. Cool adults were also less likely to fly to the roof of the cage rather than stay on the plants with the whitefly nymphs. Adult parasitoids were released at a rate of 667 adults per plant from 20 dram vials or 100 cm Petri dishes containing no more than about 1,000 wasps per dish and placed on the soil at the base of the plants. Parasitoid pupae were released in 60 cm Petri dishes, which were also placed at the base of the plant, at a rate of 1,334 pupae per plant. Because emergence rates were less than 100%, more parasitoid pupae than adult wasps were released.

One week after release the undersides of several leaves from the upper and middle portions of several plants were inspected for searching parasitoids. If no adult parasitoids were seen, any emerging adult whiteflies that escaped parasitism were removed by vacuuming to reduce feeding stress on the plants during parasitoid pupal development; this was continued every few days, or as needed until harvest. It was important to carefully balance whitefly egg density and parasitism rates to ensure that plants thrived long enough to produce a good crop of parasitoids before other stresses caused the plants to decline. When whitefly eggs were too numerous or low parasitism rates occurred, prolonged adult whitefly feeding caused the plant to drop leaves before parasitoid development was complete.

Ten days after parasitoid introduction, developing parasitoid pupae were monitored by examining fourth instar whitefly nymphs with a ×10 hand lens for signs of parasitism such as displaced mycetomes, development of an amber colored cuticle, parasitoid eye development, and the appearance of a margin on the developing parasitoid pupae within the whitefly (Roltsch and Mayhew - Whitefly and *Eretmocerus* parasitoid life stages. http://www.cdfa.ca.gov/phpps/ipc/biocontrol/pdf/insects/19eretmocerus-lifestages.pdf). Older leaves sometimes yellowed and dropped from the plant at this time. Because these leaves typically had the oldest and most developed parasitoids, they were collected to begin the parasitoid harvest.

Harvesting began when 50% or more of parasitized whitefly contained parasitoid pupae, which was determined by counting pupae on ten 2.54 cm^2 leaf discs from mid-plant leaves with a dissecting microscope or, if the inspector was sufficiently skilled, a hand lens. The plants were monitored every 2 days until all parasitoids were fully developed; they were typically ready for harvest about 10–12 days after parasitoid introductions in the summer and 14–20 days in winter. At harvest time, all leaves were stripped from the plants, laid singly on paper towels, placed in large brown paper bags and transported to the laboratory in a large ice chest kept at about 13–16°C. These bags were kept in the laboratory in an incubator and maintained at 16°C until processing.

10.3.3.4 Parasitoid Processing

Parasitoid pupae were removed from eggplant leaves with a fine spray of pressurized water from a spray gun having an adjustable cone shaped nozzle (Triggerjet 22650 Spray Gun™, conejet spray tip #5500-ppb; TeeJet, Spraying Systems Co., Wheaton, Illinois) and separated from dead whitefly nymphs and exuvia using a water and air separation process. During washing each leaf was placed over a piece of nylon organdy stretched over a 30.5 cm diameter metal kitchen colander and held tight with rubber bands. The pupae that collected on the organdy were transferred to a beaker containing 500 ml water and 1.5 ml of liquid detergent (concentrated detergent, low phosphate, no perfume) to reduce the surface tension, mixed well, and allowed to settle for 5 min. Fully developed parasitoid pupae float because of air between the pupa and whitefly cuticle, while whitefly pupae sink. Floating parasitoid pupae were removed off with a small 6.4 cm diameter steel kitchen strainer, transferred to a dry sheet of nylon organdy stretched over a steel frame and allowed to dry. This stirring and straining process was repeated 3–4 times until no more parasitoid pupae were found floating on the surface of water.

Pupae were air dried at room temperatures (20–24°C) for ½ to 1 h using a table fan set at medium speed. After drying, the parasitoid pupae were transferred into a 0.5 l paper container to a depth of 3–4 cm. Dead whitefly nymphs and exuviae are much lighter than parasitoid pupae and could be removed by shaking the container, allowing the exuviae to float up and be removed by suction or blowing air. After processing, the parasitoid pupae were stored until ready for use. These techniques resulted in very clean parasitoid material containing an average of 4.1% unparasitized whitefly pupae (Table 10.2).

A sample of production data collected from 19 generations of *Eretmocerus emiratus* reared between March and July in 1997 is presented in Table 10.2. These data show that parasitoid production averaged 236,000 per cage (range of 65,000–580,000 thousand per cage), which for our production facility translated to between 1,800 and 16,000 parasitoids per ft^2 in 10–28 days, depending on the season. A greenhouse with better temperature control would allow more uniform production at the shorter end of this range. Our average rate of increase was 6.5 times inputs, with a range of 0.8–24.6 (Table 10.2). Lower production was influenced by improper numbers of whiteflies, incorrect timing of parasitoid introduction, poor emergence of adult parasitoids, non optimal greenhouse temperatures and poor

Table 10.2 Production statistics for mass rearing of *Eretmocerus emiratus* on eggplant in 1997.

Harvest date	Generation	Input (×10^3)	No. produced/ cage (×10^3)	Increase factor	% *Bemisia* pupae unparasitized
3/24	4	24	206	8.6	1.6
3/31	5	84	299	3.6	0.5
3/31	6	50	196	3.9	3.0
4/2	7	29	154	5.3	4.0
4/11	9	60	65	1.1	3.6
4/8	10	35	113	3.2	8.0
4/14	11	20	371	18.5	1.6
4/17	12	95	81	0.8	1.5
4/21	13	20	490	24.6	1.2
4/21	14	31	331	10.7	2.3
4/24	15	35	129	3.6	–
4/24	16	44	172	3.9	–
4/29	17	75	114	1.5	2.4
4/29	18	60	133	2.2	2.0
5/1	19	35	66	1.9	1.5
5/5	24	76	499	6.6	1.7
5/6	25	110	580	5.3	3.2
5/7	26	25	347	13.9	6.3
7/7	45	25	186	7.4	11.9
Average	–	36.4	236.4	6.5	4.1

quality plants. Controlling all of these factors was a complex process that required insectary workers to pay attention to many factors so that successful rearing was more an art than science. Refined techniques and worker experience led to an increase in annual production from 5.4 million to nearly 50 million after 4 years, allowing large numbers to be released in the augmentative and classical biological control programs (Chapters 13, 14 and 16).

10.3.3.5 Storage of Parasitoid Pupae

The cleaned and separated parasitoid pupae were stored in paper or plastic containers at 15.5–16°C for 3–5 days before use in the field or shipping to cooperators with little effect on parasitoid emergence (G. Simmons, unpublished data). Longer storage times and lower storage temperatures decreased emergence rates and longevity of adult parasitoids as well as the critical high temperature that adult wasps could survive in the field (J. Gould, unpublished data). Occasionally, commercial shipments of *Eretmocerus* had very poor emergence rates (G. Simmons, personal observation), which suggests that storage time and temperature may have a large effect on parasitoid quality; to negate this problem more research is needed on the effects of cold storage on *Eretmocerus* quality.

10.4 California Department of Food and Agriculture's Greenhouse Rearing System Using Hibiscus

The California Department of Food and Agriculture's (CDFA) Biological Control Program in Sacramento, California provided additional parasitoid production between 1994 and 2000 to supplement the APHIS insectaries in Imperial, California, and Mission, Texas.

10.4.1 Plant Production

Bemisia were reared on *Hibiscus rosa-sinensis* 'Brilliant Red' because this variety had proven to be easy to propagate and grow, was tolerant of low light conditions, and withstood heavy feeding from whiteflies. Plants were propagated using standard practice for taking stem cuttings of evergreen woody shrubs: 10–15 cm long semi-hardwood stem cuttings were taken from older plants maintained in the greenhouse, dipped in rooting hormone with fungicide mix (Rootone-F™) and placed in the propagation bed of crushed granite with bottom heat (24°C) and overhead misting (2 s every 30 min during daylight hours). The greenhouse temperature was maintained at 25–27°C.

Under these conditions it took 6–8 weeks for a cutting to develop into a rooted plant that was then transplanted into a 10.2 × 10.2 × 12.7 cm plastic pot with a soilless mix (Table 10.1) that contained triple phosphate fertilizer (0–45–0) at 1.2 kg/m^3 and crushed oyster shell at 11.7 kg/m^3. To this mix a slow release pelleted fertilizer (22–5–12; Apex) was added at the rate of 8 fl. oz (0.24 l) to 2 ft^3 of soil and reapplied by top dressing every 6–7 months thereafter. Fertilizer was also applied through the drip system (Table 10.1).

Plants were grown with ambient lighting during summer months; during the winter the lighting was supplemented using an equal number of 40 watt Grolux™ and white light fluorescent bulbs for 12 h. Plants were watered regularly as needed to keep the soil in the pots evenly moist; during the warmest months of the year, overhead sprinklers went on three times a day for 3 min to supplement the regular hand watering. In general, hibiscus plants withstood harsh conditions but grew most rapidly if they were supplied plenty of water without reaching the wilting point. The plants in the plant production greenhouse were kept arthropod free using applications of pyrenone (Clean Crop), nicotine sulfate fumigants, Instar2 (growth regulator for immature Homoptera), and Pentac aquiflow (miticide) as needed.

The plants in the 10 cm diameter pots were ready for introduction into the whitefly colony in approximately 8 weeks. Most of these plants were used for whitefly oviposition for introduction into the parasitoid culture. A portion of the plants were transplanted into one gallon pots (16.5 × 17.8 cm) to support continuous production of whitefly adults in the colony.

10.4.2 Whitefly Colony and Host Plant Inoculation

The whitefly culture was initiated using 6-month-old plants in one gallon containers maintained at about 0.6 m in height. Eighteen plants were added to a cage measuring 1.2 × 1.2 × 2.5 m tall with a frame made from 1.9 cm diameter PVC pipe covered with white monofilament polyester fabric with 178 mesh (hole size of less than 90 μm) (Industrial Fabrics Corp, Minneapolis, Minnesota). The cage was placed in a 1.2 × 2.5 × 0.15 m high fiberglass pan to retain water as an aid to irrigation. This also sealed the cage bottom, keeping whitefly in the cages and preventing other pests from entering. The daytime temperature was maintained at 28°C; the nighttime temperature at 26°C. Each plant was watered daily with drip line irrigation containing soluble fertilizer (Table 10.1).

Host plants were inoculated with approximately 2,000 adult whiteflies collected from the leaves of hibiscus plants in the whitefly colony using an insect aspirator connected to an electric vacuum pump (described in Whitefly Colony section above). After adults laid eggs, the plants with second generation 4th instar nymphs were transferred to three whitefly colony cages (six plants per cage) to be used for oviposition onto new plants to be used for parasitoid rearing.

Plants were washed weekly with a soft rain nozzle to remove honeydew and mold. Beneficial insects were released to control two-spotted spider mites and aphids (mainly green peach aphid, *Myzus persicae* (Sulzer) and cotton aphid *Aphis gossypii* Glover), which commonly infested the parasitoid culture (Table 10.1). Day and night temperatures were set at 28°C and 23°C, respectively.

Sixteen fresh 10.2 cm potted plants were added to the cages with the six whitefly host plants and were left for 10 days before being transferred to the parasitoid rearing greenhouse. The one gallon whitefly infested plants used to inoculate new plants in the colony were rotated out of the cages and discarded every 6–9 weeks, depending on their condition, and replaced with new plants. Plants in the whitefly greenhouse were grown only with ambient lighting. The greenhouse's ventilation system was augmented by a 43.2 cm oscillating fan behind each cage, operating 24 h/day to increase air movement and discourage mold formation on the plant leaves.

10.4.3 Parasitoid Rearing

Parasitoids were reared in a 7.5 by 30 m greenhouse. Parasitoid stock cultures were started by placing six host plants (in 16.5 cm diameter pots) infested with first and second instar whitefly inside a rearing cage as described above. Two thousand adult and pupal *Eretmocerus* were introduced into the cage and maintained for one generation. Once parasitoid pupae were mature, four plants were taken from the colony cage and placed on an open bench with other *Hibiscus* in 10.2 cm pots containing 1st and 2nd instar whitefly hosts at the ratio of 16 whitefly host plants to one parasitoid colony plant. Each week 32 new plants with whitefly nymphs were added to the colony. The original plants containing parasitoids were spaced equidistant among the whitefly infested plants to help disperse the parasitoids. The parasitoid colony cages were replenished with 4 new one gallon plants (16.5 m diameter × 17.8 cm) with 1st and 2nd instar whitefly.

Plants exposed to parasitoids were removed after 3–4 weeks when inspection with a × 10 hand lens showed that most of the fourth instar whitefly were visibly parasitized as described previously. Old plants containing parasitoids and any plants in 10.2 cm pots not removed for field use were cut up and the leaves were left in cages to allow parasitoids to emerge and supplement the culture.

Parasitoids were harvested and delivered to the field as pupae on excised leaves, on whole plants (for transplanting as banker plants) or as washed pupae described previously. When this system was operating at peak capacity, production levels as high as 500,000 parasitoids per week (for production of *Eretmocerus emiratus* in August–October in 1997 (Pickett et al. 1997)) were reached with a range of 2.3–5.3 thousand parasitoids per plant per 10.2 cm wide pot. The highest level of production attained was 5,022 parastioids/m^2 of rearing space per week.

Eretmocerus pupae are perishable and have a short shelf life. Storage of pupae at 15–16°C allowed storage times to be extended for up to 5 days; longer storage periods or cooler storage temperatures impacted quality. For progress to be made in the commercialization of exotic *Eretmocerus* spp. more research is required to learn how to extend the shelf life of parasitoid pupae.

Acknowledgements Thanks to Enrique Lozano, Raul Torres, Jose Parra, and Jon Luke Henry for learning how to run the insectary and for their keen observation skills.

USDA Disclaimer Mention of specific pesticides is for informational purposes only and does not constitute a pest control recommendation by USDA. Information provided regarding commercial products does not constitute a recommendation or endorsement by USDA.

References

Goolsby, J.A., J.C. Legaspi, and B.C. Legaspi, Jr. 1996. Quarantine evaluation of exotic parasitoids of the sweetpotato whitefly, *Bemisia tabaci* (Gennadius). Southwestern Entomologist 21: 13–21.

Goolsby, J.A., M.A. Ciomperlik, B.C. Legaspi, Jr., J.C. Legaspi, and L.E. Wendel, 1998. Laboratory and field evaluation of exotic parasitoids of *Bemisia tabaci* (Gennadius) (Biotype "B") (Homoptera: Aleyrodidae) in the Lower Rio Grande Valley of Texas. Biological Control 12: 127–135.

Headrick, D.H., T.S. Bellows, Jr., and T.M. Perring. 1996. Behaviors of female *Eretmocerus* sp. nr. *californicus* (Hymenoptera: Aphelinidae) attacking *Bemisia argentifolii* (Homoptera: Aleyrodidae) on cotton, *Gossypium hirsutum* (Malvaceae) and melon, *Cucumis melo* (Cucurbitaceae). Biological Control 6: 64–75.

Pickett, C.H., J. Brown, G.S. Simmons, L. Van Mantgem, and H. Kumar. 1997. Release of parasites into citrus for control of silverleaf whitefly infesting cotton in the San Joaquin Valley, p. 6. Annual Report, Biological Control Program, California Department of Food and Agriculture, Sacramento, CA.

von Arx, R., J. Baumgärtner, and V. Delucchi. 1983. Developmental biology of *Bemisia tabaci* (Gennadius) (Sternorrhyncha: Aleyrodidae) on cotton at constant temperatures. Mitteilungen der Schweizerischen Entomologischen Gesellschaft 56: 389–399.

Chapter 11
Release and Recovery of Exotic Parasitoids of *Bemisia tabaci* in the Lower Rio Grande Valley of Texas

John A. Goolsby[1] and Matthew Ciomperlik[2]

Abstract An intensive field program was conducted in the subtropical Lower Rio Grande Valley of Texas (LRGV) to evaluate the establishment of the imported parasitoids of *B. tabaci*. Thirty populations/species of *Eretmocerus* and *Encarsia* parasitoids were mass reared for field release in multiple agricultural crops and garden plantings. Small releases of the *Eretmocerus* spp. were also made in the Texas Wintergarden to the northwest of the LRGV. Field tests were conducted on the major crops to determine which species showed the best potential. Banker plants (parasitoid inoculated seedling transplants) were used to facilitate early season augmentative releases of parasitoids in commercial imidacloprid treated cantaloupe melons and watermelons. Following the initial releases, several species showed initial establishment, but only *Eret. hayati* overwintered and dispersed. This result was confirmed by using RAPD-PCR which was used to distinguish the morphologically similar *Eretmocerus* spp. By 1997, *Eret. hayati* was established across the entire LRGV and northeastern Mexico. *Eretmocerus hayati* was predicted to do well in the LRGV because of the high attack rate it exhibited in pre-release quarantine studies and its climatic adaptation to the subtropical LRGV, which is similar to its native range in Pakistan.

11.1 Introduction

The Lower Rio Grande Valley of Texas is situated at the extreme southern tip of Texas and comprises four counties, Cameron, Willacy, Hidalgo and Starr. It adjoins the state of Tamaulipas, Mexico across the Rio Grande River, which forms the southern boundary. The Lower Rio Grande Valley is an area of highly intensive specialized farming covering more than 800,000 ha, of which approximately half

[1]USDA-ARS, Kika de la Garza Subtropical Agricultural Research Center, Beneficial Insects Research Unit, 2413 East Highway 83, Weslaco, TX 78596

[2]USDA-APHIS-PPQ Center for Plant Health Science and Technology Laboratory, Moore Air Base, Bldg. 6414, 22675 N. Moorefield Rd., Edinburg, TX

J. Gould et al. (eds.), *Classical Biological Control of* Bemisia tabaci *in the United States.*

are irrigated. Cotton, grain sorghum, sugarcane, corn, citrus and winter vegetables are the main crops. The climate is warm, subtropical characterized by dry winters and hot humid summers (Jacobs 1981). Tropical maritime air masses are dominant throughout the spring, summer and fall, but are frequently modified by polar air masses in the winter. The average rainfall is 660 mm. Temperatures range from average lows of 10°C in the winter to average highs of 34°C in the summer. Temperatures can go as low as −7°C in the winter and as high as 40°C in the summer. The relative humidity is high in the Lower Rio Grande Valley, with frequent morning dews. The average relative humidity at 3:00 p.m. peaks in December at 68% and reaches a low point in June at 52%.

Cultivated crops that serve as hosts to *Bemisia tabaci* (Gennadius) are available year round in the Lower Rio Grande Valley (Riley and Ciomperlik 1997; Goolsby et al. 1998). This situation allows *B. tabaci* to maintain high population densities throughout much of the year. In the spring, *B. tabaci* populations begin to increase on the spring cucurbits, e.g., melons (*Cucumis melo* L.) and cucumbers (*Cucumis sativus* L.). Once these crops are terminated, whiteflies migrate to cotton for the summer, and then to cole crops (*Brassica* spp.) and weeds such as sowthistle (*Sonchus olereae* L.) during the winter. Populations of *B. tabaci* peak in the late summer following defoliation of cotton. Fall cucurbit crops were once common, but damage by *B. tabaci* caused abandonment of this crop until the late 1990s when the pest was brought under control (M.J. Lukefahr, personal communication). Other crops such as peppers, soybeans, sunflowers, tomatoes and numerous ornamentals also support populations of *B. tabaci*. On the Mexican side of the Rio Grande River, the crop mix consists of sorghum, cotton, corn and several thousand acres of okra, planted in a thin band along the river. Okra is grown year round and often sprayed with broad-spectrum insecticides to minimize damage from sucking bugs (Hemiptera). Populations of *B. tabaci* reach high levels by mid-summer, dispersing with prevailing winds to the north and impacting crops on the US side of the Rio Grande (Vargas Camplis and Goolsby 1997).

The Texas Wintergarden agricultural region lies 300 km to the north of the Lower Rio Grande Valley and is comprised of Webb, Dimmit, Zavala and Uvalde counties. This area was also affected by *B. tabaci*. The main crops in the region are cotton, grain sorghum, corn, winter vegetables, sweet potatoes, peanuts, watermelons, and sesame. The crops are irrigated by ground water and are widely dispersed. The climate is colder than the Lower Rio Grande Valley and the area often experiences frosts and temperatures below freezing for several hours in the winter.

Both the Lower Rio Grande Valley and the Wintergarden agricultural areas were severely affected by *B. tabaci* beginning in the early 1990s. By 1993 a limited number of exotic whitefly parasitoids had been permitted and released in the Lower Rio Grande Valley of Texas, which included *Eretmocerus mundus* Mercet and *Encarsia sophia* (Girault and Dodd). In late 1993, permitting for additional importations was halted pending environmental assessments for the genera *Eretmocerus* and *Encarsia* (USDA 1995a, 1995b). Following the issuance of a FONSI (Finding of No Significant Impact) from USDA-APHIS, primary and heteronomous *Encarsia* and *Eretmocerus* species reared from hosts in the *Bemisia tabaci* complex were permitted for release. Over the next year approximately 30 species/populations of *Encarsia* and *Eretmocerus* were released from quarantine, transferred to mass rearing operations and made available for field colonization programs.

We report on 4 separate studies involving the field release and evaluation of the establishment of these imported parasitoids. Three of these involve releases of exotic parasitoids in the Lower Rio Grande Valley, and one section describes the parasitoid release program in the Texas Wintergarden area.

11.2 Release and Recovery

11.2.1 Releases in the Lower Rio Grande Valley: 1995–1996

These initial releases provided an opportunity for all permitted parasitoid species including 'poor' performing agents to demonstrate their potential in a non-agricultural setting. Commercial crops infested with *B. tabaci* were often treated with broad-spectrum insecticides which made establishment of parasitoids difficult and hindered long term evaluation. We used 'garden plots' for releases and long-term evaluation of exotic parasitoids. These sites were primarily permanent, irrigated vegetable gardens associated with schools, suburban homes, farmsteads, and vegetable variety demonstration plots. Each had a diversity of *B. tabaci* host plants, grown and irrigated in a continuous year round rotation, and free from insecticide use or drift from agricultural areas. This release protocol was particularly important for evaluating species that had not performed well under laboratory and controlled field settings, which allowed us to validate the results of our predictive studies (Goolsby et al. 1996, 1998, 2005).

Each garden plot site was sampled twice before releases were made to characterize the native parasitoid complex. Twenty leaves containing 4th-instar *B. tabaci* were removed from each host plant available at each release site. The leaves were held in 0.5 l paper cans with the tops streaked with honey and placed inside a humiditron (DeBach and Rose 1985) at 26°C and 70% RH for 30 days. Adult parasitoids were collected daily from the top of the container using quarter dram glass vials. After 1 month, the containers were emptied and sorted to recover any additional parasitoids. Identifications of native *Eretmocerus* spp. were made following the keys developed by Rose and Zolnerowich (1997). *Encarsia* spp. were identified using the key of Polaszek et al. (1992). Most exotic specimens were also later analysed with RAPD-PCR to confirm their identity by comparison with established DNA profiles previously developed to characterize the quarantine cultures of exotic parasitoids.

The pre-release evaluations of the field sites revealed that the dominant native parasitoid was *Encarsia pergandiella* Howard which caused 94% of the recorded parasitism. The remaining parasitism (6%) was caused by the native *Eretmocerus tejanus* Rose and Zolnerowich. Both of these native parasitoids were recovered year round. Riley and Ciomperlik (1997) found that *Eret. tejanus* was most common during spring, whereas *Enc. pergandiella* predominated in the summer and fall. During the fall, small numbers (<1%) of *Encarsia luteola* Howard and *Enc.* sp. nr. *meritoria* Gahan were recovered.

All of the exotic *Eretmocerus* and *Encarsia* spp. that were permitted by APHIS for field release were released in the garden plots study. Each site received a combination of populations which could be morphologically and/or genetically distinguished from each other and from the native species by DNA patterns using RAPD-PCR techniques (Black et al. 1992; Legaspi et al. 1996, Chapter 6). Release rates of the various parasitoid populations varied depending on their availability from laboratory colonies and ranged from 500 to 157,000 per site (Table 11.1). Those that were easier to rear and consequently were available in greater numbers were released at higher rates. However, every attempt was made to rear and release sufficient numbers of all the permitted species to allow the possibility of establishment. Hand releases of parasitised whitefly on host plant leaves were made from

Table 11.1 Recovery of exotic parasitoids from garden plots from December 1995 to July 1996 in the Lower Rio Grande Valley, Texas.

Population/species and origin	MBCL accession #	Release site	Release date	Release nos.	No. specimens recovered
Eret. mundus Spain	M92014	Donna	5–6/95	157,000	35
		McAllen	5–6/95	40,000	2
		Mission	8/95	500	6
		La Joya	6/94	50,000	32
Eret. hayati Pakistan	M95012	Mission	8/95	2,000	2
Eret. mundus India	M92019	Monte Alto	n/a[a]	0	1
Enc. sophia Spain	M93003	McAllen	5–6/95	60,000	1
Enc. nr. *hispida* Brazil	M94056	Monte Alto	8/95	2,400	2
Eret. mundus Italy	M94085	Monte Alto	7/95	200	1
Eret. mundus Israel	M94120	Edinburg	6/95	100	1
Enc. lutea Cyprus	M93064	McAllen	5–6/95	320	1
Eret. hayati India	M93005	McAllen	5–6/95	50,000	4
Enc. sophia Malaysia	M94047	Mission	7/95	36,000	1
Eret. staufferi unknown	M94002	La Joya	6/94	50,000	6

[a] *Eret. mundus* from India was not released at Monte Alto but it was recovered there.

May to August 1995. Twenty sites were sampled every 2 weeks from June 1995 to November 1995, and monthly thereafter from December 1995 to July 1996 using the same techniques as described above in the pre-release evaluations.

Recoveries of exotic parasitoids from December 1995 to July 1996 are shown in Table 11.1. Those parasitoids released but not recovered are shown in Table 11.2. These data should be interpreted qualitatively rather than as precise quantitative measures of establishment because those which were available in higher numbers were released at higher rates. Furthermore, many parasitoids were released only once and sometimes only one parasitoid was recovered for some exotic populations. The findings assess establishment only in the short term during 1995–1996. With these limitations in mind, 11 of 29 populations of exotic parasitoids released in the fields were recovered from the release sites within the first year after release.

Identification of exotic parasitoids by integrating the use of morphological characters and RAPD-PCR proved to be a very efficient and accurate method of evaluating field establishment (Chapter 6). The pedicels of the male antennae of the introduced *Eretmocerus* species are uniformly fuscous (darker) as compared to an amber coloration in the native *Eret. tejanus* males. The pattern of setae on the pronotum of females also differs between native and exotic species. Distinguishing morphological characters allowed for separation of exotic and native *Eretmocerus* spp. and RAPD-PCR was used to determine which exotic species were in the recovery samples. The combination of the two techniques allowed us to reduce the cost of genetic characterization by processing only the exotic specimens. *Eretmocerus mundus* (Spain, M92014) and *Eretmocerus hayati* (Pakistan, M95012) were consistently recovered from 1996 onwards. This result was consistent with predictive pre-release studies that found these two species had the highest attack rates for *B. tabaci* on cotton, melons, and broccoli (Chapter 7).

Table 11.2 Exotic parasitoid species not recovered from garden plots from June 1995 to July 1996, Lower Rio Grande Valley of Texas. All parasitoids were released in May–June 1995.

Population/species specimens and origin	MBCL accession #	Release site	Release nos.	No. recovered
Enc. sophia Thailand	M94004	Monte Alto	1,600	0
Enc. sophia Taiwan	M94004	Mission	2,000	0
Eret. melanoscutus Thailand	M94023	Donna	5,400	0
Eret. melanoscutus Thailand	M94036	Monte Alto	20,000	0
Eret. melanoscutus Thailand	M94040	Alamo	20,000	0
Eret. melanoscutus Taiwan	M93058	Monte Alto	5,000	0
Enc. bimaculata India	M92018	Monte Alto	1,600	0
Enc. formosa Thailand	M94040	Monte Alto	14,000	0

11.2.2 Sentinel Plant Survey: 1997

A survey program in the Lower Rio Grande Valley to monitor establishment of exotic parasitoids was implemented using 'sentinel' plants. These were cotton and melon plants pre-infested with immature whitefly and placed in field locations to sample for parasitoid species composition. Plants were placed in the field for 2 days, after which they were returned to the laboratory for rearing and identification of the parasitoids. This technique had two major advantages over conventional leaf sampling methods: (1) sentinel plants provided a standardized test unit across a broad range of locations and crop types; (2) sentinels gave a true measure of primary parasitism without the masking effects of hyperparasitism by the autoparasitoid, *Enc. pergandiella*. However, it is conceivable that some of the parasitoids may have not been attracted to the cotton/melon sentinels.

Ten sentinel locations were selected that represented a varied mix of agricultural and urban sites across the Lower Rio Grande Valley, covering a distance of approximately 80 km. Cantaloupe melons (*Cucumis melo* L. 'Perlita'), followed by cotton in the summer months, were used as sentinel plants. Plants were grown in the greenhouse to the three-leaf stage and were then infested with whitefly and allowed to develop until they contained a mix of 2nd and 3rd instar *B. tabaci* at a density below 5 nymphs per cm^2. The root balls of the plants were placed in sealed plastic containers to retain moisture. Ten sentinel plants were placed monthly throughout the year in each location for a period of 2 days. Plants were then returned to the laboratory for removal of any other plant pests and live parasitoid adults. After recovery each plant was placed in a 150-mm diameter ventilated Petri dish to prevent any additional parasitism and held in an environmental growth chamber at 27°C for development of the parasitoids. *Encarsia pergandiella* pupae, identified by the presence of meconia, were counted and removed prior to emergence to avoid hyperparasitism of the *Eretmocerus* pupae. *Eretmocerus* males and females were removed daily upon emergence and separated for identification. Males were slide mounted to determine if they were exotic or native. This information was used to determine the overall percentage of native versus exotic *Eretmocerus* spp. The percentage assumes a standard female to male sex ratio of 60:40 across *Eretmocerus* spp. (Goolsby, unpublished data). Female *Eretmocerus* were slide mounted and species determinations were made using keys to the imported species of *Eretmocerus* (Zolnerowich and Rose 1998) and RAPD-PCR.

Results from sentinel plant surveys showed that exotics clearly became established, and represented a high percentage of the parasitoids recovered (Table 11.3). At the beginning of the survey in June 1997, native *Eret. tejanus* represented greater than 95% of *Eretmocerus* species. Three months later, during the fall of the 1997 (August–October), exotic populations began to increase, representing 85% of 480 *Eretmocerus* spp. recovered. Exotic parasitoids were more prevalent than natives in six of the eleven samples. Analysis of the female *Eretmocerus* using both morphological techniques and molecular genetics identified three species, *Eret. tejanus, Eret. mundus, and Eret. hayati* (Table 11.4). None of the other exotic *Encarsia* or

Table 11.3 Percentages and total numbers of exotic and native male *Eretmocerus* spp. recovered from sentinel plants in the Lower Rio Grande Valley.

	Jun	Aug	Sep	Oct	Jan	Feb	Mar	May	Jun	Jul	Aug
Date	1997	1997	1997	1997	1997	1998	1998	1998	1998	1998	1998
% Exotic	2	64	91	86	40	70	20	50	25	90	100
% Native	98	36	90	14	60	30	80	50	75	10	0
N	225	141	130	209	201	90	151	8	24	56	100

Table 11.4 Identification of female *Eretmocerus* from sentinel plant recovery surveys in the Lower Rio Grande Valley.

Technique	Date	N	*Eret. mundus*	*Eret. hayati*	*Eret. tejanus*
RAPD-PCR	November 1997	56	0%	64%	36%
Morphology	July 1998	18	38%	62%	0%
Satellite DNA	August 1998	67	34%	38%	28%

Eretmocerus spp. released in the inoculative establishment evaluation was recovered. By the end of 1997, the two exotic species (*Eret. mundus* and *Eret. hayati*) were widely established across the Rio Grande River flood plain.

11.2.3 Releases at Texas Wintergarden: 1995–1996

In 1995, *B. tabaci* biotype B was an emerging pest in the Texas Wintergarden area and outbreaks were common. Releases of exotic parasitoids were made in fall–winter of 1995–1996 (Table 11.5). Recovery sampling was conducted during the same time period on and around the fields where the inoculative releases were made. *Eretmocerus* species were identified as exotic or native based on morphological features, as above. We did not determine which exotic species were recovered from Wintergarden releases, but it was likely to have been *Eret. mundus* and *Eret. hayati* following the trend in the Lower Rio Grande Valley.

11.2.4 Evaluation of Establishment: August 2002

Collections of parasitized whitefly were made in the Lower Rio Grande Valley of Texas in order to supply co-operators with the division of Entomology at the Commonwealth Science and Industry Research Organisation (CSIRO) in Indooroopilly (Brisbane), Queensland, Australia with *Eretmocerus hayati*. This provided an ideal opportunity to assess establishment of exotic parasitoids released in the late 1990s against *B. tabaci*. Three collections were made in Hidalgo Co.

Table 11.5 Releases of exotic parasitoids in the Texas Wintergarden in 1995–1996[a].

County	City	Release site	Plant host	*Eret. mundus* (M92014)	*Eret. hayati* (M95012)	*Eret. emiratus* (M95104)
Webb	San Ygnacio	McDaniel	Cantaloupe	1,000	2,000	1,000
Dimmitt	Carrizo Springs	White	Watermelon	1,000	1,000	1,500
Dimmitt	Carrizo Springs	Castelow	Watermelon	600	1,500	500
Dimmitt	Carrizo Springs	Ward	Sesame	5,500	7,700	5,500
Zavala	Crystal City	Kruze	Cabbage	6,000	8,000	0
Zavala	Batesville	LaFeere	Cotton	0	200	0
Zavala	Batesville	LaFeere	Cabbage	7,000	7,000	5,100
Uvalde	Uvalde	Cargill	Cabbage	7,000	7,000	5,100
Uvalde	Uvalde	Cargill	Cantaloupe	1,500	1,500	1,500
Uvalde	Uvalde	TAES	Cabbage	4,000	7,500	0
Uvalde	Uvalde	Val Tex	Ornamental	4,000	9,500	0
Uvalde	Knippa	Kellings	Sweet potato	2,000	9,400	0
Uvalde	Knippa	Sanderlin	Sesame	9,000	300	1,400
Uvalde	Knippa	Bishop	Sesame	1,400	7,500	200
Uvalde	Knippa	Gilland	Sesame	3,500	10,500	400

[a]In addition to the parasitoids indicated in the table we released the following: *Encarsia sophia* Spain, M93003 (200 per site); *Eretmocerus* nr. *furahashi*, Taiwan (200 per site); *Encarsia sophia*, Pakistan M95017 (400 per site), and *Encarsia* sp., Dominican Republic (250 per site).

from August to October of 2002. This time of year coincides with cotton defoliation, planting of fall cucurbits and germination of weed hosts such as common sowthistle, *Sonchus oleraceus* L.. Approximately 1,000 parasitized whitefly were received in quarantine facilities in Australia. More than 95% of the parasitoids that emerged were *Eretmocerus hayati* (Paul DeBarro, CSIRO Entomology, personal communication.). A subsample of 24 exotic *Eretmocerus* was confirmed as containing only *Eret. hayati* by morphology (M. Rose, Bozeman, MT) or molecular genetic testing. The remaining parasitoids were a mix of native species including *Eret. tejanus* and *Encarsia* spp.

11.2.5 Banker Plants

A novel method, termed 'banker plants', was developed in the *B. tabaci* biological control program for releasing parasitoids in annual vegetable crops (Goolsby and Ciomperlik 1999; Pickett et al. 2004). This method was used for early season augmentation of the exotic silverleaf whitefly parasitoids and aided their establishment in Texas and California. Using this method, seedling transplants were inoculated with *B. tabaci* and held until the cohort reached the 2nd instar and then followed by release of *Eretmocerus* spp. into small shroud cages which covered the transplant

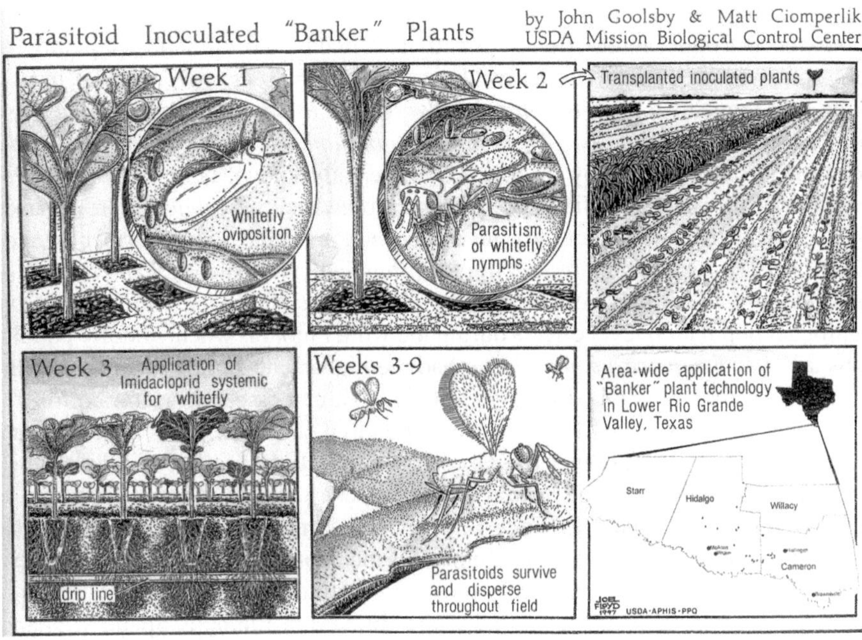

Fig. 11.1 Cartoon illustration of the banker plant method that was used to transfer this technology to commercial growers.

flats. Inoculated transplants were then mixed with standard 'Speedling' flats for use with commercial tractor-mounted transplanters. Use of banker plants in commercial agriculture required evaluating the impact of the insecticide imidacloprid on developing immature parasitoids. Goolsby and Ciomperlik (1999) found that imidacloprid was compatible with the silverleaf whitefly parasitoids if timed correctly (Fig. 11.1). Large-scale field tests incorporating imidacloprid and banker plants were then conducted with the Texas Cooperative Extension service and CDFA with commercial growers over a 2-year period. Banker plant methods were documented to increase control of silverleaf whitefly and increase parasitoid numbers for migration to surrounding crops. However, the establishment of the exotic *B. tabaci* parasitoids precluded the need for early season inoculation of specialist parasitoids and the banker plant research was terminated.

11.3 Summary

In the Lower Rio Grande Valley of Texas it appears that *Eret. hayati*, originally imported from Pakistan, became widely established by 1997 and subsequently became the dominant parasitoid of *B. tabaci*. The successful establishment of this species rather than the others that were introduced may be due to its preference for *B. tabaci*, tritrophic interactions with its host plants and climatic adaptation to subtropical south Texas. The native *Eret. tejanus* and *Enc. pergandiella* have a wider host range than *Eret. hayati* parasitizing both *B. tabaci* and their native host, the banded wing whitefly, *Trialeurodes abutiloneus* (Haldeman) (Rose and Zolnerowich 1997; Goolsby et al. 2005). The only known host of *Eretmocerus hayati* is *B. tabaci*. In the field, both *T. abutiloneus* and *B. tabaci* often occurred together on soybeans and there was no evidence of parasitism of the former by *Eret. hayati*. If *Eret. hayati* is more specific to *B. tabaci* than the native *Eret. tejanus* or *Enc. pergandiella* it may search more effectively for *B. tabaci*, which may have significant population level effects during the winter months when whitefly populations are low. *Eretmocerus hayati* also had the highest attack rate compared to the 15 other *Eretmocerus* and *Encarsia* species evaluated searching on cotton for *B. tabaci* (Goolsby et al. 1998). Several hundred thousand acres of cotton are planted each year in the Lower Rio Grande Valley and the highest *B. tabaci* densities occur on this host. A parasitoid which searches effectively on cotton should have a competitive advantage over other species. Finally, climate may have also played a role in the selective establishment of the exotic *Eretmocerus* spp. The climate of Lower Rio Grande Valley is most similar to the hot subtropical climate of the Indus River of Pakistan compared to the climates of Ethiopia, Spain, Thailand, and the United Arab Emirates where the other *Eretmocerus* spp. were collected. It is likely that each species of *Eretmocerus* from the Palearctic is adapted to its local climate. This climatic adaptation may influence survival of individuals at high summer temperatures and activity during cool temperatures of winter, but this hypothesis has yet to be tested. Goolsby et al. (2005) propose that *Eret. hayati* from Pakistan possessed both the heat and cold tolerance necessary to survive in the Lower Rio Grande Valley of Texas. In summary, *Eret. hayati* is now the dominant parasitoid of *B. tabaci* in the Lower Rio Grande Valley. Several biological attributes such as host specificity, searching ability, and climatic adaptation may account for its apparent success. Further field studies are needed to document the ecological and economic impact of *Eret. hayati* on *B. tabaci*.

References

Black, W.C., N.M. DuTeau, G.J. Puterka, J.R. Nechols, and J.N. Pettorini. 1992. Use of the random amplified polymorphic DNA polymerase chain reaction (WD-PCR) to detect DNA polymorphisms in aphids (Homoptera: Aphididae). Bulletin of Entmological Research 82: 151–159.

DeBach, P. and M. Rose. 1985. Humidity control during shipment and rearing of parasitic Hymenoptera. Chalcid Forum 4: 11–13.

Goolsby, J.A., J.C. Legaspi, and B.C. Legaspi, Jr. 1996. Quarantine evaluation of exotic parasitoids of the sweetpotato whitefly, *Bemisia tabaci* (Gennadius). Southwestern Entomologist 21: 13–21.

Goolsby, J.A., M.A. Ciomperlik, B.C. Legaspi, Jr., J.C. Legaspi, and L.E. Wendel, 1998. Laboratory and field evaluation of parasitoids of exotic parasitoids of *Bemisia tabaci*

(Gennadius) (Biotype "B") (Homoptera: Aleyrodidae) in the Lower Rio Grande Valley of Texas. Biological Control 12: 127–135.

Goolsby, J.A. and M.A. Ciomperlik. 1999. Development of parasitoid inoculated seedling transplants for augmentative biological control of silver leaf whitefly (Homoptera: Aleyrodidae). Florida Entomologist 82: 1–14.

Goolsby, J.A., P.J. De Barro, A.A Kirk, R. Sutherst, L. Canas, M. Ciomperlik, P. Ellsworth, J. Gould, K.A. Hoelmer, S.J. Naranjo, M. Rose, W. Roltsch, R. Ruiz, C. Pickett, and D. Vacek. 2005. Post-release evaluation of the biological control of *Bemisia tabaci* biotype "B" in the USA and the development of predictive tools to guide introductions for other countries. Biological Control 32: 70–77.

Jacobs, J.L. 1981. Soil Survey of Hidalgo County, Texas. US Department of Agriculture, Soil Conservation Service and Texas Agricultural Experiment Station.

Legaspi, J.C., B.C. Legaspi, Jr., R.I. Carruthers, J.A. Goolsby, W.A. Jones, A.A. Kirk, C. Moomaw, T.J. Poprawski, R.A. Ruiz, N.S. Talekar, and D. Vacek. 1996. Foreign exploration for natural enemies of *Bemisia tabaci* from Southeast Asia. Subtropical Plant Science 48: 43–48.

Polaszek, A., G.A. Evans, and F.D. Bennett. 1992. *Encarsia* parasitoids of *Bemisia tabaci* (Hymenoptera: Aphelinidae, Homoptera: Aleyrodidae): a preliminary guide to identification. Bulletin of Entomological Research 82: 375–392.

Riley, D.G. and M. A. Ciomperlik. 1997. Regional population dynamics of whitefly (Homoptera: Aleyrodidae) and associated parasitoids (Hymenoptera: Aphelinidae). Environmental Entomology 26: 1049–1055.

Rose, M. and Zolnerowich, G. 1997. *Eretmocerus* Haldeman (Hymenoptera: Aphelinidae) in the United States, with descriptions of new species attacking *Bemisia* (*tabaci* complex) (Homoptera: Aleyrodidae). Proceedings of the Entomological Society of Washington 99: 1–27.

Vargas Camplis, J. and J.A. Goolsby. 1997. Results from the biological control of boll weevil and the introduction and establishment of new parasitoids of the whitefly for the integrated management of cotton pests. In R. Roberson, ed., Proceedings: International Cotton Pest Work Committee, pp. 75–76. November 19–21, 1997, Mazatlan, Mexico.

US Department of Agriculture, Animal and Plant Health Inspection Service. 1995a. Field releases of nonindigenous parasitic wasps in the genus *Eretmocerus* (Hymenoptera: Aphelinidae) for biological control of whitefly pests (Homoptera: Aleyrodidae). Environmental Assessment. Riverdale, MD.

US Department of Agriculture, Animal and Plant Health Inspection Service. 1995b. Field releases of nonindigenous parasitic wasps in the genus *Encarsia* (Hymenoptera: Aphelinidae) for biological control of whitefly pests (Homoptera: Aleyrodidae). Environmental Assessment. Riverdale, MD.

Zolnerowich, G. and M. Rose. 1998. *Eretmocerus* Haldeman (Hymenoptera: Aphelinidae) imported and released in the United States for control of *Bemisia* (*tabaci* complex) (Homoptera: Aleyrodidae). Proceedings of the Entomological Society of Washington 100: 310–323.

Chapter 12
Release and Recovery of Four Species of *Eretmocerus* against *Bemisia tabaci* Biotype B in Arizona

Juli Gould[1], Diane Waldner[2], Nick Colletto[2], and Paul Merten[3]

Abstract Over 23 million parasitoids in the genus *Eretmocerus* were released in the greater Phoenix metropolitan area and near Yuma, Arizona. The species released were *Eret. hayati*, *Eret. mundus*, *Eret. emiratus*, and *Eret.* nr. *emiratus* from Ethiopia. Two release methods were used with the goal of colonizing and spreading the parasitoids throughout Arizona. Homeowners who agreed to forfeit the use of insecticides were asked to grow whitefly host plants throughout the year. The parasitoids were also released into commercial nurseries with the expectation that they would be distributed regionally through plant sales. All four *Eretmocerus* species became established, although 3 years after release *Eret. emiratus* and *Eret.* nr. *emiratus* could be found in greater numbers. *Eretmocerus* nr. *emiratus* was only released at commercial nurseries, yet it was recovered at 44 sites being monitored through the homeowner release program. Since plants purchased from the nurseries contained *Eret.* nr. *emiratus* and were moved throughout the city as they were sold, releasing parasitoids on nursery plants is a promising method to ensure a wide distribution. Dispersal studies indicated that the exotic parasitoids had dispersed up to 1.5 km within 6 months. Three methods for recovering exotic parasitoids were used: leaf samples, sticky traps, and sentinel plants. Of these, the leaf samples and the sentinel plants were the most likely to recover exotic parasitoids. There was no strong evidence that exotic parasitoids were more likely than the native species to attack whiteflies on certain host plants. Several exotic parasitoid species were able to reproduce on *B. tabaci* regardless of host plant family attacked.

[1]USDA-APHIS-PPQ Center for Plant Health Science and Technology Laboratory, Building 1398, Otis ANGB, MA 02542

[2]USDA-APHIS-PPQ Center for Plant Health Science and Technology Laboratory, 3645 E. Wier Avenue, Phoenix, AZ 85040

[3]USDA-ARS Arid Land Agricultural Research Center, 21881 North Cardon Lane, Maricopa, AZ 85238

J. Gould et al. (eds.), *Classical Biological Control of* Bemisia tabaci *in the United States.*

12.1 Introduction

Agriculture is a major component of the Arizona economy. Although much of the agricultural industry is located in the Sonoran desert of south-central Arizona, ample water for irrigation allows this industry to flourish. In the greater Phoenix metropolitan area in central Arizona, commercial fields are interspersed with urban areas in a mosaic pattern. Many of the commercial crops grown near Phoenix support populations of *Bemisia tabaci* (Gennadius) biotype B (=*B. argentifolii* Perring and Bellows) (Hemiptera: Aleyrodidae). In the spring, populations of *B. tabaci* begin to increase on the spring melon crop. Once that crop is disked under, whiteflies migrate to cotton for the summer, and then to cole crops (*Brassica* spp.) during the winter (Arizona Agricultural Statistics Service 2002). During the years of heaviest impact, Arizona farmers stopped growing melons in the fall because of feeding damage by *B. tabaci*. Other vegetable crops such as lettuce, peppers, and tomatoes also support populations of *B. tabaci*.

The Sonoran desert receives more rainfall than other desert regions in the southwestern USA due to storms from the Pacific coast during winter and monsoon rains from the Gulf of Mexico in summer, and it averages 193 mm of rainfall annually (Arizona Office of Climatology 2004). Average monthly temperatures (over 30 years) in our release areas in central Arizona ranged from a low of 5°C in the winter to a high of 41°C in the summer. Temperatures can go as low as –9°C in the winter and as high as 48°C in the summer. The relative humidity is also low in Arizona as in other desert regions of the world. The average relative humidity peaks in December at 40% and reaches a low point in June at 14%.

At the inception of the interagency *B. tabaci* biological control project, several key crops that supported the growth of whitefly populations, but did not harbor a large fauna of indigenous parasitoids, were identified. These included cole crops (broccoli, cabbage, cauliflower – all winter crops) and melons (spring crop) (Hoelmer 1996). Populations of *B. tabaci* overwinter in cole crops and migrate to spring melons, where large populations build as the spring progresses. Whiteflies then migrate into cotton in the summer in such large numbers that indigenous natural enemies are not able to effect control (Hoelmer 1996). The parasitoid fauna in southwestern USA was less diverse than in the southeastern USA or in other countries (Hoelmer 1996; Chapter 18). Foreign explorations to collect natural enemies to augment parasitoid diversity were conducted from 1992 to 1996 in an attempt to increase parasitoid diversity. Collections were made in a systematic fashion that took into account the climates, cropping systems, host plants and seasonality of *Bemisia* activity in areas of the USA where whitefly outbreaks occur (Kirk and Lacey 1996; Chapter 2). As a result of these collections, 55 different populations of *Encarsia* and *Eretmocerus* were reared at the quarantine facility at the APHIS Mission Biological Control Laboratory in Mission, Texas (Goolsby et al. 2000).

Laboratory and field studies evaluated the performance of 38 exotic and 2 native aphelinid parasitoids on cantaloupe, cotton, and broccoli/kale (Goolsby et al. 1998; see Chapters 7 and 8). Based on the results of these preliminary tests, *Eret. mundus* Mercet, *Eret. hayati* Zolnerowich and Rose, *and Eret. emiratus* Zolnerowich and

Rose were chosen for widespread release. *Eretmocerus emiratus* was not tested in preliminary quarantine laboratory trials because it had not yet been collected; however its native range (United Arab Emirates) was better matched with Arizona in terms of climate than that of other aphelinid species, and it exhibited the highest attack rate in field cage studies in the desert southwest (Hoelmer 2007, Chapter 8). *Eretmocerus* sp. nr. *emiratus* from Ethiopia had not yet been collected at the time of the quarantine studies, but was released in Arizona starting in 1999 based on climate matching and field cage results. This chapter reports on the release and establishment of these parasitoids in central Arizona.

Releases of introduced *B. tabaci* parasitoids were conducted in urban areas of metropolitan Phoenix. Agricultural settings, while they are an ultimate target of biological control, are harsh environments for releasing newly imported parasitoids. The natural desert provides few host plants and agricultural lands are highly unstable habitats for parasitoids. They are frequently disked, eliminating whitefly hosts, and routinely sprayed with insecticides (Ellsworth and Jones 2000). Noncrop hosts, i.e., weedy plants, are also controlled chemically or mechanically, which reduces the availability of *B. tabaci* host plants. It was thought that the probability of successful establishment was higher for parasitoids released in urban areas with a diversity of maintained and watered ornamental host plants harboring *B. tabaci*. In addition, because agriculture is interspersed with housing developments, dispersal distances from release sites to agricultural fields would be minimized. Urban areas may also represent a major reservoir for overwintering whiteflies, which could be reduced by the newly imported parasitoids.

12.2 Mass Rearing Parasitoids for Release

Parasitoids were reared at the Beneficial Insect Rearing Laboratory at the University of Arizona in Tucson Arizona; at the USDA-APHIS Mission Biological Control Laboratory in Mission Texas; and at the USDA-APHIS insectary in Imperial, California. The parasitoid species were reared on *B. tabaci* biotype B, which were in turn reared on eggplant, *Solanum melongena* L. Parasitoids were shipped from insectaries to Arizona by express mail and were released in groups of 2,500. Four species or geographic populations of parasitoids were released in Arizona: *Eret. mundus* (MBCL culture numbers M92014 from Spain and M94120 from Israel), *Eret. hayati* (M95012 from Pakistan), *Eret. emiratus* (M95104 from the United Arab Emirates), and *Eret.* sp. nr. *emiratus* (M96076 from Ethiopia).

12.3 Release of Parasitoids

Two protocols were employed for releasing parasitoids. The goal of each was to colonize and spread the parasitoids throughout central and western Arizona. Selected homeowners who agreed to forfeit use of insecticides were asked to grow host plants that would maintain whiteflies in their yards throughout the year. Secondly, parasitoids were released into commercial nurseries with the expectation that they would be distributed regionally through plant sales. If successful, these methods would promote the movement of colonized parasitoids from urban centers into surrounding agricultural fields.

12.3.1 Release in Home Gardens

In 1997, releases of *Eret. hayati* and *Eret. mundus* were made into home gardens representing seven urban centers in Arizona (Table 12.1; Fig. 12.1). In the winter of 1996, advertisements were placed in local newspapers and newsletters requesting assistance from local residents who had home gardens adjacent to commercial agriculture. Release sites were selected based on size of the garden (approximately 1,000 ft^2), competency of the gardener, and proximity of year-round whitefly host plants such as species of *Hibiscus*, *Lantana*, *Rosa,* and *Justicia,* and needed to be outside the area of insecticide drift. Twenty five release sites throughout Arizona were chosen to release parasitoids; of these 15 were chosen for intensive monitoring (Table 12.1). USDA-APHIS provided to homeowners seeds of plants that can support populations of whiteflies throughout the year: cantaloupe (*Cucumis melo* L.), zucchini (*Cucurbita pepo* L.), sunflower (*Helianthus* sp. L), okra (*Abelmoschus esculentus* (L.) Moench), eggplant (*Solanum melongena* L.), collards (*Brassica*

Table 12.1 Approximate number of parasitoids released into 25 home gardens in Arizona in 1997. The country of origin of the parasitoid and the Mission Biological Control Laboratory culture numbers are given in parentheses.

Release area	No. sites	*Eret. hayati* (Pakistan M95012)	*Eret. mundus* (Israel M94120)	*Eret. emiratus* (United Arab Emirates M95104)
Northeast Phoenix[a]	4	216,000	1,000,000	239,000
Southwest Phoenix[b]	3	155,000	409,000	161,000
Coolidge/Casa Grande	3	119,000	458,000	107,000
Southeast Phoenix[c]	4	308,000	939,000	148,000
Gila Bend	2	68,000	115,000	97,000
Colorado River[d]	7	200,000	543,000	244,000
Northwest Phoenix[e]	2	222,000	550,000	139,000
Total	25	1,288,000	4,014,000	1,135,000

[a] Includes Scottsdale and Lehi
[b] Includes Goodyear, Tolleson, and Buckeye
[c] Includes Gilbert, Tempe, Chandler, and Mesa
[d] Includes Tacna, Wellton, Yuma, and Parker
[e] Includes Sun City and Goodyear

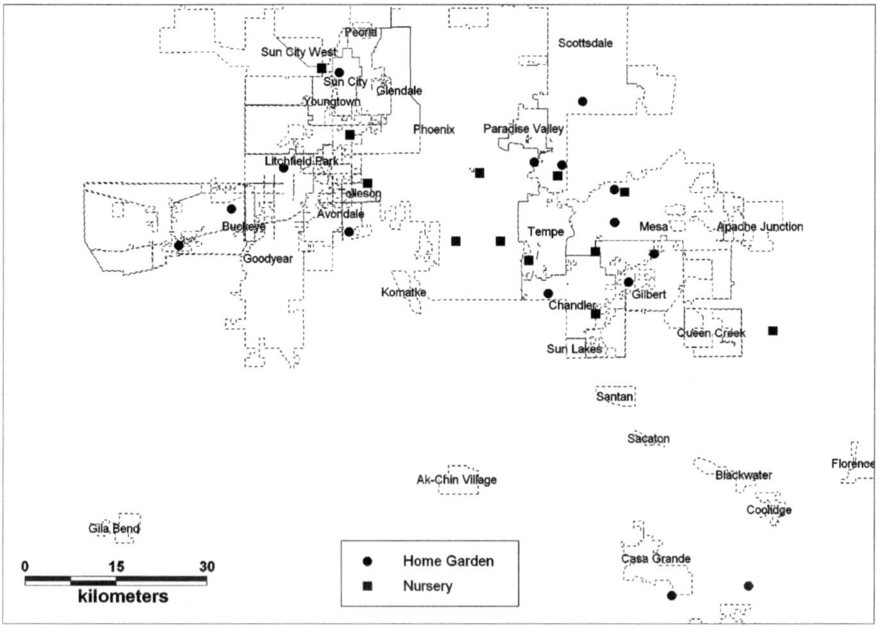

Fig. 12.1 Map of the greater Phoenix, Arizona metropolitan area showing the location of home garden and retail nursery release sites.

oleracea L. var. *acephala* group L.), broccoli (*Brassica oleracea* L. var. *botrytis* L.), and basil (*Ocimum basilicum* L.). Parasitoids were released once every 3 weeks after whitefly nymphs became plentiful around mid-May. Parasitoid pupae were released from 40dram vials placed on the soil under the shade of plants. A metal screen covered the vial to exclude predators that might eat parasitoid pupae but allowed emerging parasitoid adults to escape.

To monitor the establishment and dispersal of exotic parasitoids plants were intensively sampled at 15 home garden release sites in the Phoenix metropolitan area and Coolidge/Casa Grande. At each release site permission was obtained to sample plants at up to 27 other homes within a 3.2km radius. In total, 15 home garden release sites and 233 sites surrounding the release sites (248 sites total) were sampled for establishment of exotic parasitoids. Leaf samples were collected at release sites monthly (June–December, 1997–1999; October–November 2000) and from surrounding gardens only once, at the end of November, each year. It was assumed based on the authors' observations that densities of whiteflies and parasitoids would be highest at the end of the growing season, increasing the probability of parasitoid detection. Twice a year (July or August and December) two sentinel hibiscus plants were placed at the release site, and at 1.6km and 3.2km away. In 1999 and 2000, yellow sticky traps were also placed at each sample site.

12.3.2 Releases at Commercial Nurseries

From 1998 to 2000 most of the releases were conducted in commercial nurseries in Yuma, Arizona and the Phoenix Metropolitan area to increase the number of release sites beyond those at private home gardens. Fourteen nurseries were selected to receive parasitoids based on (1) abundance of whitefly host plants, (2) high whitefly densities, (3) number of commercial clients, and (4) extent of sales throughout the Phoenix and Yuma metropolitan areas. The nurseries were located in the following areas: Tolleson, Phoenix, Mesa, Scottsdale, Sun City, Glendale, Chandler, and Yuma (Fig. 12.1). Parasitoids were released as pupae in 40 dram vials that were placed inside of horizontal paper cups (473 ml) that protected them from rain and irrigation. The parasitoids were released on the following *B. tabaci* infested host plants: *Datura innoxia* Miller (angel's trumpet), *Tecoma stans* (L.) (Arizona yellow bells), *Tecoma capensis* (Thunb.) Lindl. (cape honeysuckle), *Ficus* spp. (fig), various herbs, *Hibiscus rosa-sinensis* Willd. (hibiscus), *Alcea* (hollyhock), *Jacaranda* (jacaranda), *Justicia spicigera* Schlect. (justicia), *Lantana* spp. (lantana), *Passiflora cincinnata* Masters (passion vine), *Euphorbia pulcherrima* Willd. Ex Klotzsch (poinsettia), *Rosa* spp. (rose), *Ruellia brittoniana* Leonard (ruellia), *Cistus salvifolius* L. (salvia), *Vigna caracalla* (L.) Verdc. (snail vine), various vegetables, and *Verbena rigida* Spreng. (verbena). The number of parasitoids released on a given date depended on the availability of reared parasitoids and the density of whiteflies at each nursery (Table 12.2).

Leaves containing mature whiteflies were collected from each commercial nursery prior to every release. In 1999, leaf collections at nurseries were made monthly from July to November; in 2000, samples were taken in April, May, and July. Beginning in 1999, in consultation with the commercial nurseries, sites were located where 14 landscape companies had placed plants purchased from nurseries where exotic parasitoids had been released. Leaves from these sites were collected throughout the year; in 2000, samples were taken in September only.

Table 12.2 Approximate number of exotic parasitoids released at 12 nurseries in the greater metropolitan Phoenix area between 1998 and 2000. The country of origin of the parasitoid and the Mission Biological Control Laboratory culture numbers are given in parentheses.

City	Number of nurseries	*Eret. mundus* (Spain 92014)	*Eret. hayati* (Pakistan M95012)	*Eret. emiratus* (United Arab Emirates M95104)	*Eret.* nr. *emiratus* (Ethiopia 96076)
Chandler	1	452,663	120,600	103,550	355,800
Glendale	1	597,030	185,472	147,614	1,544,630
Mesa	2	2,445,068	21,630	204,502	1,060,388
Phoenix	4	860,517	694,992	506,318	1,839,052
Scottsdale	1	218,344	93,870	62,216	104,524
Sun City	1	107,523	53,932	36,044	53,600
Tempe	1	105,470	201,000	45,108	509,000
Queen Creek	1	1,920,000	0	0	885,000
Total	12	6,706,615	1,371,496	1,105,352	6,351,994

12.4 Sampling Whitefly Parasitoids

Post-release monitoring was conducted to determine establishment and spread of released parasitoids. Three methods were used to determine if parasitoids had established at a given site: (1) leaf samples, (2) sentinel plants, and (3) yellow sticky traps. The strengths and weaknesses of each are described below.

12.4.1 Leaf Samples

Leaves containing mature whitefly nymphs were collected from host plants and placed in emergence canisters, 30 cm by 12 cm diameter paper tubes with a funnel and a glass collection jar on top. Parasitoids were sorted under a dissecting microscope (×160), and the numbers of native and exotic parasitoids that emerged were recorded. Exotic *Eretmocerus* females were slide-mounted and identified to species (Zolnerowich and Rose 1998). The number of parasitoids recovered depended on the number of leaves collected and the density of whiteflies on those leaves, not necessarily the number of parasitoids at that site. At sites where whiteflies were scarce, leaf samples could not be taken to verify the presence or absence of released parasitoids.

12.4.2 Sentinel Hibiscus Plants

Low whitefly densities at a monitoring site could preclude detection of established parasitoids. Sentinel potted hibiscus plants infested with whitefly nymphs were used to attract exotic parasitoids. Sentinel plants were placed at each release site in 1997, and beginning in 1998, at one site each 1.6 and 3.2 km distant. Two hibiscus plants were placed beneath or close by plants containing whiteflies, and left at the site for 1 week. Ten hibiscus plants from the same lots as those placed outdoors were kept in the laboratory and checked for exotic parasitoids to ensure that sentinel plants were parasitoid free at the time of deployment. Sentinel plants were placed in the field in the summer and the fall each year from 1998 to 2000.

12.4.3 Sticky Traps

Another method of detecting exotic parasitoids at low densities was to capture them on yellow sticky traps (Hoelmer et al. 1998). In 1999 and 2000, two 15 × 30 cm sticky traps were hung from the branches of plants containing whitefly nymphs. The traps were collected 1 week later and the numbers of male and female exotic and native parasitoids were recorded. Since it was difficult and time-consuming to remove the parasitoids from the traps, species identifications were not made.

12.5 Determining Species of Parasitoids Collected

The species of parasitoid was determined using the morphology and DNA of adult specimens. Exotic and native males could be separated by the color of the antennal pedicel. Adult females were cleared and placed on glass slides for species determination using morphological characters (Zolnerowich and Rose 1998). *Eretmocerus mundus, Eret. hayati,* and *Eret. emiratus* were identified by antennal characteristics. It was not possible to distinguish *Eret. emiratus* from the United Arab Emirates from *Eret.* sp. nr. *emiratus* from Ethiopia based on known morphological characters. Beginning in 1999, male *Eretmocerus* were preserved in 95% ethanol and sent to the USDA-APHIS Mission Biological Control Laboratory for species identification using RAPD-PCR (see Chapter 6) to distinguish between *Eret. emiratus* and *Eret.* sp. nr. *emiratus* from Ethiopia.

12.6 Results and Discussion

All three exotic parasitoid species (*Eret. mundus, Eret. hayati, and Eret. emiratus*) were recovered at all 15 of the intensively monitored home garden release sites in 1997. That year, 49 of the 233 (21%) sites surrounding home garden releases showed evidence of reproduction by exotic *Eretmocerus* species (Table 12.3). Because of the timing of the collections, these parasitoids were recovered more than one generation after release. In 1998, exotic *Eretmocerus* were recovered from 19 of 248 sample sites (7.7%) and at only 9 of the 15 release sites. The number of sites with exotic *Eretmocerus* increased in 1999 to 45 of 248 sites (18%), and by 2000 reproduction by exotic *Eretmocerus* was evident at 98 sites (40%) at and surrounding the home garden release sites. By 1998, *Eret. mundus* was only found at three sites and *Eret. hayati* at only five sites. In contrast, recoveries suggested that *Eret. emiratus* (*ex* UAE) and *Eret.* sp. nr. *emiratus* (*ex* Ethiopia) had established

Table 12.3 Number of homeowner release sites and surrounding locations where exotic *Eretmocerus* were recovered (total 248 sites).

Year	*Eret. mundus*	*Eret. hayati*	*Eret. emiratus* females	*Eret. emiratus* males	*Eret.* nr. *emiratus* males	Total
1997	17	16	26	n/a	n/a	64 (26%)
1998	3	5	7	n/a	n/a	19 (8%)
1999	1	2	28	12	21	45 (18%)
2000	4	2	80	11	44	98 (40%)

Table 12.4 Number of commercial nursery release sites where exotic *Eretmocerus* were recovered.

Year	*Eret. mundus*	*Eret. hayati*	*Eret.* nr. *emiratus females*	*Total sites with exotic Eretmocerus*
1998	5	5	6	7
1999	2	0	3	8
2000	1	0	4	4

locally and were increasing in abundance. In 2000, *Eret.* sp. nr. *emiratus* was recovered from half of the original release sites and from nearby properties of 13 out of 15 release sites. Because we did not release *Eret.* sp. nr. *emiratus* into any of these sites, this species must have become established through the nursery release program.

There is less evidence of establishment in the Colorado River area than in Phoenix; reproduction of exotic *Eretmocerus* was seen only at three of the seven release sites. Because establishment was low, release of some species continued through 1999. It could not be determined whether the parasitoids that were recovered had survived at the sites for more than a year. Establishment of *Eret. hayati* at two locations and *Eret. emiratus* at one site a year after release was documented.

Recovery data showed that parasitoids were reproducing on whiteflies on plants at commercial nurseries and at properties to which nursery plants were moved. Parasitoid reproduction was documented at all 12 nurseries in the Phoenix metropolitan area (Yuma was not sampled), but not in all years (Table 12.4). *Eret. hayati* was only recovered during the first year of the 1998–2000 release program. Exotic parasitoids were recovered at 6 of the 14 sites where landscape companies had planted plants purchased from commercial nurseries where releases had been made over the 3-year program. This, coupled with the discovery of *Eret.* nr. *emiratus* (ex. Ethiopia) at homeowner release sites where this species had never been released, indicates that releasing parasitoids on nursery plants is a promising method to ensure a wide distribution.

12.6.1 Dispersal from Release Site

Global Positioning System (GPS) coordinates were taken for each release site, as well as for all surrounding sample sites, analyzed by MapInfo Professional Version 6.5 (MapInfo Corporation Troy, N.Y.) and the distance between sites was determined. Releases were first conducted in June 1997, and samples at the surrounding sites were taken in December 1997 giving the parasitoids up to 6 months to reproduce and disperse to the sites where they were recovered. An exotic male *Eretmocerus* was recovered 1.5 km from one of the release sites, but at the time the techniques to distinguish species among male specimens were not available. *Eret.*

mundus females were found an average of 0.37 km from the release site, with a maximum of 0.67 km (N = 2). Only one female *Eret. hayati* was found at a surrounding sample site, 0.77 km from the release site. *Eret. emiratus* was found at 11 sites and it dispersed the greatest distance of the three exotic species (mean = 0.45 km, maximum = 1.03 km). The exotic parasitoids could have dispersed these distances in one event, perhaps carried by wind, or in several stages over the course of several generations. Studies conducted on the short-range dispersal of *Eretmocerus* (Hagler et al. 2002; Bellamy and Byrne 2001) concluded that members of this genus do not usually disperse great distances. However, Byrne (personal communication) found three *Eret. eremicus* marked with Day-glo dust in passive fan traps 1.0 km from the melon field in which they had been released.

12.6.2 Comparison of Recovery Methods

At most of the sample sites, especially in 1997 and 1998, only a single method of sampling, leaf collection, was used to recover exotic parasitoids. In 1997 we began placing sentinel plants at the homeowner release sites and in 1998 we began placing plants at two sites surrounding each homeowner site as well. After 1999 we added yellow sticky traps at all sites being monitored. From 1998 to 2000 there were 111 instances recorded where two or three methods were used at the same site in the same year (not necessarily at the same time) to recover exotic parasitoids. To compare the efficacy of leaf collection, sentinel plants, and sticky traps in detecting exotic parasitoids, t-tests were used and only sites where exotic parasitoids were recovered using at least one method during a given year and where more than one method was employed were considered.

There was no significant difference in efficacy between leaf collection and sentinel plants (Table 12.5). Both sentinel plants and leaf collections were significantly more likely to detect the presence of exotic parasitoids than were sticky traps. Figure 12.2 shows graphically the percentage of the time parasitoids were captured by each of the three combinations of sampling methods when two methods were used in combination. When any two sampling methods were used in combination, they detected exotic parasitoids more than 95% of the time when exotic parasitoids were known to be present at a site. In other words, although each method was successful in detecting exotic parasitoids at a high percentage of sites, that percentage approached 100% when two methods were used. The sticky trap method was less useful in enhancing discovery of exotic parasitoids as it added only 9–15% detection to the sentinel plant and leaf collection methods (Fig. 12.2).

Although the sentinel plant and leaf collection methods were equally effective, they differed greatly in the amount of effort required for deployment. Rearing plants, infesting them with whiteflies, and subsequently keeping them free of parasitoids are labor-intensive activities. When whiteflies are abundant at a site leaf collection should be the method of choice. When whiteflies are at low density, finding enough leaves with whiteflies at the correct stage becomes time consuming

Table 12.5 Two-way comparisons of the efficacy of three pairs of methods used to detect exotic parasitoids, as measured by the percent of instances where exotic parasitoids were recovered when the methods were used at the same site during the same year.

Leaf	Sentinel plant	Sticky trap	d.f.	p-value
80.9	72.3	–	45	0.3326
–	87.0	52.2	22	0.0115[a]
81.3	–	58.6	73	0.0025[a]

[a]significant at or above $\alpha = 0.05$

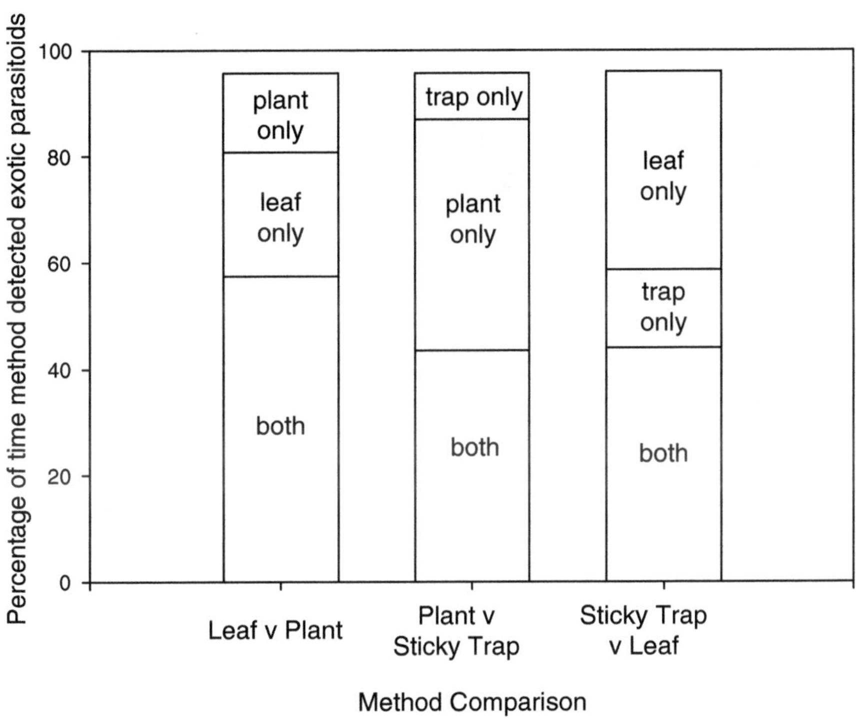

Fig. 12.2 Comparison of the utility of three methods for detecting exotic parasitoids at sites where exotic parasitoids were known to occur.

also. In these situations, sentinel plants can play an important role. Sticky traps are not only less effective, but aphelinid parasitoids cannot be identified to species without removing the individuals from the sticky material. Sticky traps do collect exotic parasitoids, however, and they are much simpler and less labor intensive than sentinel plants. Depending on the research goals, they may play a role in detecting parasitoids when whiteflies are at low density.

Table 12.6 The percentage and fraction of sites where an *Eretmocerus* species was found on a sampled plant family. The category "all exotic *Eretmocerus*" includes *Eret. hayati* and those individuals we could not identify to species[a].

Plant family	Plant species	Plant common names	All exotic *Eretmocerus*	*Eret. mundus*	*Eret. emiratus*	Native *Eret. eremicus*
Acanthaceae	*Justicia*; *Ruellia*	Mexican honeysuckle, ruellia	42% (5/12)	100% (1/1)	25% (3/12)	54% (13/24)
Asteraceae	*Cynara scolymus, Lactuca, Chrysanthemum, Helianthus,*	Artichoke, Jerusalem artichoke, Lettuce, chrysanthemum, sunflower	67% (10/15)	33% (3/9)	45% (5/11)	50% (8/16)
	Zinnia	Zinnia				
Bignoniaceae	*Tecoma stans, Jacaranda*	Cape honeysuckle, jacaranda	72% (26/36)	0% (0/1)	56% (18/32)	24% (14/59)
Brassicaceae	*Brassica oleracea, Raphanus*	Broccoli, brussel sprouts, cabbage, cauliflower, collards	57% (19/33)	59% (10/17)	30% (9/30)	56% (25/45)
Caprifoliaceae	*Lonicera*	Honeysuckle	33% (2/6)	no info	25% (1/4)	36% (4/11)
Cucurbitaceae	*Cucumis, Cucurbita,*	Melon, squash, cucumber	56% (80/143)	59% (23/29)	41% (52/126)	61% (103/169)
	Citrullus lanatus	Pumpkin, zucchini				
Fabaceae	*Vigna*	Beans, snail vine	71% (5/7)	50% (1/2)	50% (3/6)	43% (6/14)
Lamiaceae	*Ocimum basilicum, Mentha*	Basil, mint, hyssop spurge	25% (4/16)	17% (1/5)	8% (1/13)	37% (10/27)
Malvaceae	*Hibiscus lunariifolius, Gossypium hirsutus, Alcea rosea, Abelmoschus esculentus*	Hibiscus, cotton, hollyhock, Okra	64% (145/225)	38% (27/72)	44% (83/190)	49% (173/351)
Moraceae	*Ficus carica, Morus*	Fig, mulberry	27% (6/22)	0% (0/3)	25% (4/16)	37% (14/38)
Rosaceae	*Rosa*	Rose	50% (64/128)	8% (1/13)	35% (38/109)	42% (86/206)
Solanaceae	*Capsicum, Lycianthes, Datura, Solanum melongena, Solanum* spp.	Jimson weed, eggplant, potato, tomato	50% (62/123)	54% (20/37)	39% (37/94)	69% (96/139)
Verbenaceae	*Petrea volubilis, Lantana camara, Verbena polystachya*	Lantana, verbena	78% (132/168)	14% (1/7)	52% (72/137)	51% (158/309)
Vitaceae	*Vitis*	Grape	40% (4/10)	0% (0/1)	60% (3/5)	43% (3/7)

[a]A sufficient number of *Eret. hayati* were not collected and this species was not included separately in the analysis.

12.6.3 Comparison of Recovery on Different Plant Types

Eretmocerus species have been shown to prefer certain plant species (Headrick et al. 1996; Goolsby et al. 1998; Gruenhagen and Perring 2001; Simmons et al. 2002). The hypothesis that some species have a higher attack rate on one infested plant over another although the plants are equally infested with *B. tabaci* was addressed by examining sites/year/host plant combinations where exotic parasitoids were found. Those sites at which any exotic parasitoids were found were called "positive" sites. Overall, at the "positive" sites exotic parasitoids were found most often on Verbenaceae (78% of the time), Bignoniaceae (72%), Fabaceae (71%), Asteraceae (67%) and Malvaceae (64%) when these plant families were present (Table 12.6). At the other extreme, only 25% of plants in the Lamiaceae produced exotic parasitoids at positive sites.

We also examined the range of plant families attacked by specific parasitoid species. In general, the two exotic species and the native *Eret. eremicus* were recovered from all plant families. For the majority of plant family/parasitoid species combinations, the percentage of sites at which the parasitoid was found on that plant family ranged from 25% to 75%. If one discounts instances where sample size was less than five sites, there were no instances where one was either sure to find or sure to not find a particular parasitoid species reproducing on a specific plant family. In other words, the exotic parasitoids, like the native species, can be found attacking *B. tabaci* on a wide variety of plant families, although certain plants appear to be more favored than others.

Unfortunately, our data cannot shed light on the impact that the different parasitoid species are having on whitefly populations on different plants. The sampling methods permitted recording only the presence or absence of exotic parasitoids, not their density or impact. One plant family of interest at the beginning of this project was Brassicaceae (broccoli, kale, cauliflower). The native *Eret. eremicus* was not causing sufficient whitefly mortality on these crops and it was hoped one of the new exotic parasitoid species would do better. There were no differences found in the presence of native and exotic *Eretmocerus* species on Brassicaceae in these surveys, however. Exotic and native *Eretmocerus* were found on Brassicaceae 56% and 59% of the time, respectively.

In conclusion, several of the exotic parasitoid species released in Arizona are able to reproduce on *B. tabaci* regardless of host plant family attacked. *Eretmocerus mundus* and *Eret. hayati* initially became established and were recovered at a few sites. *Eretmocerus emiratus* (UAE M95104) and *Eret.* sp. nr. *emiratus* (Ethiopia M96076) became widely established in the Phoenix metropolitan area, while *Eret. mundus* and *Eret. hayati* became rare 4 years after releases began.

References

Arizona Agricultural Statistics Service. 2002. 2002 Annual Statistics Bulletin. URL: http://www.nass.usda.gov/az/02bul/main.htm.

Arizona Office of Climatology. 2004. Phoenix, Arizona: Normals, Means, and Extremes. URL: http://www.wrcc.dri.edu/cgi-bin/clilcd.pl?az23183.

Bellamy, D.E. and D.N. Byrne. 2001. Effects of gender and mating status on self-directed dispersal by the whitefly parasitoid *Eretmocerus eremicus*. Ecological Entomology 26: 571–577.

Ellsworth, P.C. and J. Jones. 2000. Arizona Cotton Insect Losses. The University of Arizona, Cooperative Extension. Web Publication #AZ1183. Tucson, AZ. URL: http://ag.arizona.edu/crops/cotton/insects/cil/cil.html

Goolsby, J.A., M.A. Ciomperlik, B.C. Legaspi, Jr., J.C. Legaspi, and L.E. Wendel. 1998. Laboratory and field evaluation of exotic parasitoids of *Bemisia tabaci* (Gennadius) (Biotype "B") (Homoptera: Aleyrodidae) in the Lower Rio Grande Valley of Texas. Biological Control 12: 127–135.

Goolsby, J.A., M.A. Ciomperlik, A.A. Kirk, W.A. Jones, B.C. Legaspi, J.C. Legaspi, R.A. Ruiz, D.C. Vacek, and L.E. Wendel. 2000. Predictive and empirical evaluation for parasitoids of *Bemisia tabaci* (Biotype "B"), based on morphological and molecular systematics. In A. Austin, and M. Dowton, eds., Hymenoptera: Evolution, Biodiversity, and Biological Control. 4th International Hymenopterists Conference; 6–11 January 1999, Canberra, ACT., Australia, *pp*. 347–358. CSIRO, Collingwood, Victoria.

Gruenhagen, N.M. and T.M. Perring. 2001. Impact of leaf trichomes on parasitoid behavior and parasitism of silverleaf whiteflies (Homoptera: Aleyrodidae). Southwestern Entomologist 26: 279–290.

Hagler, J.R., C.G. Jackson, T.J. Henneberry, and J.R. Gould. 2002. Parasitoid mark-release-recapture techniques: II. Development and application of a protein marking technique for *Eretmocerus* spp., parasitoids of *Bemisia argentifolii*. Biocontrol Science and Technology 12: 661–675.

Headrick, D.H.., T.S. Bellows, and T.M. Perring. 1996. Host-plant effects on the behavior of *Eretmocerus* sp. nr. *californicus* females raised from melon. Entomophaga 41: 15–26.

Hoelmer, K.A. 1996. Whitefly parasitoids: Can they control field populations of *Bemisia*? In D. Gerling and R.T. Mayer, eds., *Bemisia* 1995: Taxonomy, Biology, Damage, Control and Management, pp. 451–476. Intercept, Andover, Hants, UK.

Hoelmer, K.A. 2007. Field cage evaluation of introduced *Eretmocerus* species (Hymenoptera: Aphelinidae) against *Bemisia tabaci* strain B (Homoptera: Aleyrodidae) on cantaloupe. Biological Control 43: 156–162.

Hoelmer, K.A., W.J. Roltsch, C.C. Chu, and T.J. Henneberry. 1998. Selectivity of whitefly traps in cotton for *Eretmocerus eremicus* (Hymenoptera: Aphelinidae), a native parasitoid of *Bemisia argentifolii* (Homoptera: Aleyrodidae). Environmental Entomology 27: 1039–1044.

Kirk, A.A. and L.A. Lacey. 1996. A systematic approach to foreign exploration for natural enemies of *Bemisia* and some current results. In D. Gerling and R.T. Mayer, eds., *Bemisia*: 1995 Taxonomy, Biology, Damage, Control and Management, pp. 531–533. Intercept, Andover, Hants, UK.

Simmons, A.M., S. Abd-Rabou, and G.S. McCutcheon. 2002. Incidence of parasitoids and parasitism of *Bemisia tabaci* (Homoptera: Aleyrodidae) in numerous crops. Environmental Entomology 31: 1030–1036.

Zolnerowich, G. and M. Rose. 1998. *Eretmocerus* Haldeman (Hymenoptera: Aphelinidae) imported and released in the United States for control of *Bemisia* (*tabaci* complex) (Homoptera: Aleyrodidae). Proceedings of the Entomological Society of Washington 100: 310–323.

Chapter 13
Release and Recovery of Exotic Natural Enemies of *Bemisia tabaci* (Biotype "B") (Hemiptera: Aleyrodidae) in Imperial Valley, California

William J. Roltsch[1], Kim A. Hoelmer[2], Gregory S. Simmons[3], and Earl Andress[4]

Abstract Numerous methods of release were used to establish parasitoids of *Bemisia tabaci* in the deserts of southern California. Despite the release of a large number of species, only a few species became well established. Because of the intensive effort put forth toward establishing new species it is probable that the failure of some species to establish was due to their being poorly adapted to this desert agricultural region. The results presented here are convincing evidence that establishment by *Eretmocerus emiratus* and *Eret.* sp. nr. *emiratus* (M95104 and M96076) has occurred at numerous sites within the Imperial Valley and that population growth and dispersal are continuing. The climate in the countries of origin of these populations in the United Arab Emirates and Ethiopia closely matches the climate in southwestern US desert valleys. It is likely, however, that their populations have yet to reach equilibrium in Imperial Valley. Data suggest that the establishment process of those exotic *Eretmocerus* that successfully established has been dynamic, involving a consistent increase in exotic parasitoid populations through 2000 on numerous host plants of *Bemisia*, and a steep decline in 2001 followed by a population rebound in 2002. Data collected as recently as 2004 indicate that exotic *Eretmocerus* populations and *Encarsia sophia* from Multan, Pakistan, continue to represent a significant proportion of the *Bemisia* parasitoid community.

[1]California Department of Food & Agriculture, Biological Control Program, 3288 Meadowview Rd., Sacramento, CA 95832

[2]USDA-ARS, Beneficial Insect Introduction Research Unit, 501 South Chapel Street, Newark, DE 19713-3814

[3]USDA-APHIS-PPQ, Center for Plant Health Science and Technology Laboratory, 3645 East Wier Avenue, Phoenix, AZ 85040

[4]USDA-APHIS-PPQ Pink Bollworm Rearing Facility, 3645 E Chipman Road, Phoenix, AZ 85040

J. Gould et al. (eds.), *Classical Biological Control of* Bemisia tabaci *in the United States.*

13.1 Introduction

The Imperial Valley represents a region of year-round crop production consisting of a variety of field and vegetable crops grown on approximately 200,000 ha. Alfalfa (*Medicago sativa* L.), is the dominant crop, grown for hay, pasture grazing and seed on nearly 81,000 ha throughout the year (Birdsall 1992, 1999). In addition to alfalfa, prominent crops grown in the summer include upland cotton (*Gossypium hirsutum* L.), sugar beets (*Beta vulgaris* L.), Sudan grass (*Sorghum bicolor* L.), and Bermuda grass (*Cynodon dactylon* L.). During the cooler periods of the year, a variety of vegetable and cucurbit crops are grown, including cole crops (*Brassica oleracea* L.), lettuce (*Lactuca sativa* L.), carrots (*Daucus carota L.*) and numerous melon varieties (*Cucumis melo* L.). Many of the crops grown in Imperial Valley are hosts of whitefly, *Bemisia tabaci* (Genn.), and cropping seasons of several host plants overlap. For example, spring melon production extends well into June, and cotton is grown from March to mid-fall (Perring et al.1991; Gonzalez et al. 1992). In turn, a fall crop of cantaloupe is planted in August. Although alfalfa and annual crops are the dominant production crops, citrus acreage increased throughout the 1990s, with more than 2,400 ha present by 2002 (Birdsall 2002). In addition to crops, a variety of ornamental plants are present in urban areas and at rural home sites, and various weedy and native desert host plant species also occur in the region. As a result, attractive whitefly host plants are available throughout the year. Very high summer temperatures occur in the lower desert climate of Imperial Valley, with temperatures exceeding 37°C on more than 140 days during most years. Over a 1-year period in 1999–2000, temperatures exceeded 43°C on 24 days, while daily low temperatures on 14 days were less than 4.4°C, with freezing temperatures occurring on several days (University of California's IPM Worldwide Web site [http://www.ipm.ucdavis.edu]).

Reproducing predominantly on cotton in the desert Southwest, *Bemisia tabaci* biotype A has been a significant pest sporadically since the early 1900s, serving as a vector of plant viruses to vegetables and cucurbits, encouraging sooty mold on cotton, and causing limited direct damage. During the late 1980s several introductions of parasitoids of *B. tabaci* from Pakistan were made in the Imperial Valley against the original 'A' strain of *B. tabaci* (Chapter 18) without any apparent establishment. In the early 1990s a new whitefly pest problem began to unfold in Imperial Valley and elsewhere in the region. Extensive outbreaks of *Bemisia tabaci* (B biotype) (=*Bemisia argentifolii* Bellows and Perring) occurred on a broader range of host plants, including melon, cole crops (*Brassica* sp.) and numerous ornamental species (Perring et al. 1991; Parrella et al. 1992; Bellows et al. 1994). During a 1-year period from May 1991 to April 1992, crop losses due to *Bemisia* were estimated at $129.7 million (Gonzalez et al. 1992).

Native natural enemies of *B. tabaci* in the Imperial Valley of California consist of several predator and parasitoid species, including *Geocoris* spp. (Hemiptera: Lygaeidae), *Orius* spp. (Hemiptera: Anthocoridae), *Chrysoperla* sp. (Neuroptera:

Chrysopidae), *Semidalis* sp. (Neuroptera: Coniopterygidae), and the parasitic hymenopteran species of *Eretmocerus eremicus* Rose and Zolnerowich, *Encarsia luteola* Howard and *Encarsia meritoria* Gahan (Aphelinidae). *Eretmocerus eremicus* was the most important parasitoid of *B. tabaci* biotype A, especially in cotton (Gerling 1967; Bellows and Arakawa 1988) and continues to account for considerable parasitism of *B. tabaci* biotype B. Data have shown that parasitism of *B. tabaci* biotype A by *Enc. luteola* and *Enc. meritoria* was considerably less than parasitism by *Eret. eremicus* in cotton (Gerling 1967). Little historical data are available that identify parasitoid species composition associated with other host plant species of *B. tabaci* in the region (see Chapter 18). To increase the impact of natural enemies on *B. tabaci* biotype B, a suite of non-native parasitoid wasp species and two predatory beetle species were introduced into Imperial Valley to complement the existing resident natural enemy fauna.

Exotic parasitoid species released in Imperial Valley were reared by USDA-APHIS facilities in Imperial, California (1994–2000) and Mission, Texas (1992–1999), the University of Arizona, Tucson, Arizona (1995–1999), and the California Department of Food and Agriculture (CDFA) insectary in Sacramento (1994–2000). The University of California, Riverside reared and released *B. tabaci* parasitoids from 1993 to 1997 (D. H. Headrick, personal communication). Table 13.1

Table 13.1 Parasitoids released in Imperial Valley and the adjoining desert areas of Palo Verde Valley, Winterhaven, California, Yuma, Arizona, and Mexicali Valley, Mexico from 1992 to 1999.

Species	Accession code	Source institution(s)[a]	Years released	No. released
Amitus bennetti[b]	–	UCR	1994	382,800
Encarsia bimaculata	M92018	MBCL, IV	1994–1997	65,644
Encarsia formosa	M92017	USDA-ARS Stoneville, Mississippi	1992	160
Encarsia formosa	M92030	MBCL	1992–1994	273,500
Encarsia formosa	M94051	MBCL	1995	1,000
Encarsia nr. *hispida*	M94056	MBCL, IV, CDFA	1995–1996	1,287,085
Enc. nr. *pergandiella*	M94055	MBCL, IV, CDFA	1995–1996	23,500
Encarsia sophia	M93003	MBCL, IV, CDFA	1994–1996	492,300
Encarsia sophia	M94041	MBCL	1995	3,635
Encarsia sophia	M94047	MBCL	1995	3,204
Encarsia sophia	M95107	MBCL, IV	1998–1999	698,580[c]
Eretmocerus emiratus	M95104	IV, AZ, CDFA	1996–1998	56,839,066[d]
Eret. sp. nr. *emiratus*	M96076	IV, MBCL, AZ	1997–1999	14,191,200[e]
Eretmocerus hayati	M95012	MBCL, IV	1995–1997	7,735,058
Eretmocerus hayati	M95105	IV	1995–1997	20,000
Eret. melanoscutus	M94023	IV, CDFA	1995–1996	1,673,398
Eret. melanoscutus	M94036	CDFA	1995	2,000
Eret. melanoscutus	M94040	MBCL	1995	600

(continued)

Table 13.1 (continued)

Species	Accession code	Source institution(s)[a]	Years released	No. released
Eretmocerus mundus	M92014	MBCL, IV, AZ	1992, 1994–1996, 1998–1999	7,498,324[f]
Eretmocerus mundus	M92019	MBCL, IV	1993–1996	7,554,605
Eretmocerus mundus	M92027	MBCL, IV	1994–1995	121,000
Eretmocerus mundus	M93005	MBCL, IV	1994–1995	84,050
Eretmocerus mundus	M93058	MBCL	1995	1,000
Eretmocerus mundus	M94092	MBCL	1995	1,000
Eretmocerus mundus	M94120	MBCL, IV, AZ	1995–1998	4,355,111
Eretmocerus mundus	M94124	MBCL	1995	700
Eretmocerus mundus	M94125	MBCL	1995	1,000
Eretmocerus staufferi	M94002	CDFA, IV	1994	2,625,550
Eretmocerus tejanus	M94003	CDFA	1994–1995	2,709,650
Eretmocerus sp.[g]	–	UCR	1997	26,180

[a]UCR = University of California, Riverside, California; IV = APHIS greenhouse facilities in Imperial Valley, California; AZ= University of Arizona facilities in Tucson, Arizona; MBCL = APHIS Mission Biological Control Laboratory in Mission, Texas; CDFA = California State facilities in Sacramento, California.
[b]Population origin: Puerto Rico.
[c]500,000 of total released in Mexicali Valley, Mexico.
[d]360,000 of total released in Mexicali Valley, Mexico.
[e]276,000 of total released in Mexicali Valley, Mexico.
[f]198,000 of total released in Mexicali Valley, Mexico.
Similar numbers were released in Mexico during 2000.
[g]Population origin: Hong Kong.

presents the overall release numbers by species and APHIS Mission Biological Control Laboratory quarantine culture number over successive years. In addition two predatory coccinellid species were mass-produced and released. In 1992–1993 *Delphastus catalinae* (Horn) was reared at the CDFA insectary in Sacramento, California from a population that was originally collected in Florida in 1992. Twenty-three thousand beetles were released by CDFA at 23 rural and urban home sites in Imperial Valley during 1993, and over 110,000 beetles (reared at the CDFA insectary) were released by University of California scientists during the evaluation of augmentative biological control (Pickett et al. 1999; Heinz et al. 1999; Hoelmer and Pickett 2003). The second coccinellid species, *Serangium parcesetosum* Sicard, was reared by USDA-APHIS from material collected in 1993 in Podumbu, India. Approximately 10,000 insects were released during the fall of 1994 in four urban communities in Imperial County. A second population (origin uncertain) of *Serangium parcesetosum* was reared and released in the Imperial Valley by University of California Riverside researchers in 1993 (R. Staten, personal communication).

13.2 Release Site Characteristics

13.2.1 Long-Term Insectary Gardens

Long-term insectary gardens were set up in 1995 to provide continual availability of whitefly host plants across successive years for the production and maintenance of introduced parasitoid species. This followed a study in 1993–1995 that utilized sunflower (*Helianthus annuus* L.) and collard (*Brassica oleracea* L. Acephala group) to provide year-round habitat in the form of refuges for conserving native parasitoid species. For several years, the insectary gardens contained a number of whitefly host plants (Chapter 15) which was reduced over time. Beginning in 1998 the plots consisted of two or more of the following plant types at any one time: cole crops (collards, broccoli), sunflower, okra (*Abelmoschus esculentus* (L.)), sweet basil (*Ocimum basilicum* L.), cotton and cantaloupe. Two garden plots (0.25–0.5 ha each) were maintained at the Imperial Valley Research Center, near Brawley, California (north end of Imperial County) through 2002, and two garden plots of similar size were maintained at an organic farm at the south end of Imperial County until July of 2000. Plot details are presented in Chapter 15.

13.2.2 Short-Term Insectary Gardens

Short-term insectary gardens were set up in 1998 and 1999 at several locations in the Imperial Valley and Palo Verde Valley in Imperial County. Arrangements were made with local farmers to grow a large garden (approx. 0.5 ha.) consisting of four beds each of okra and basil on one edge of a commercial field from April to November. Two gardens, each composed of okra and basil, were grown each year within Imperial Valley. No winter crop was grown to create a bridge across years. Each garden was inoculated with exotic parasitoid species in early summer coincident with increasing densities of *Bemisia* with the intent of producing large numbers of parasitoids throughout the summer and early fall to migrate locally and facilitate their regional establishment (details in Chapter 15).

13.2.3 Urban Releases

Urban releases were made in more than 50 home yards and vegetable gardens containing several whitefly host plant species to maximize the probability of establishing exotic parasitoids. Several releases were made at many of the sites each year, with most releases taking place during 1994–1997. The most common plant species included hibiscus (*Hibiscus rosa-sinensis* L.) and mulberry (*Morus alba* L.), although a wide range of other whitefly host plants were also available including orchid tree (*Bauhinia variegata* L.), roses (*Rosa* sp.), fig (*Ficus carica* L.), cape honeysuckle (*Tecomara capensis* Thunb.), lantana (*Lantana camara* L.) and snail vine (*Vigna caracalla* L.). In addition to utilizing existing whitefly host plants in yards, efforts were made to grow several annual host plant species in a number of home gardens (details in Chapter 15).

13.2.4 Commercial Organic Cantaloupe Crops

Commercial organic cantaloupe crops in large multi-hectare plots were utilized to experimentally test augmentation techniques to control *Bemisia* on spring melons (*Cucumis melo* L.) (details in Chapter 16). Crops were inoculated with several exotic parasitoid species. In addition to providing information on the potential of augmentative biological control in cantaloupe, these multiyear studies are believed to have facilitated the effort to permanently establish exotic parasitoids.

13.2.5 Commercial Crops

Commercial crops receiving few or no pesticide applications were utilized for releases of greenhouse-reared parasitoids. In cooperation with the Imperial County Agricultural Commissioner's office, melon, cotton, alfalfa and cole crop fields that were grown organically or were expected to receive few pesticide applications were identified.

13.2.6 Wildlife Refuges

Wildlife refuges provided a unique opportunity for making releases of insectary-reared parasitoids. A Memorandum of Understanding was developed with the US Fish & Wildlife Service for parasitoid releases and surveys for establishment on the substantial acreage of alfalfa located on the Salton Sea National Wildlife Refuge. This acreage was managed by the Service to provide food for migrating waterfowl, and was not treated with insecticides except for treatment of the windrows for alfalfa weevil, *Hypera postica* (Gyllenhal), after cutting and raking. Alfalfa cutting practices on the refuge (as elsewhere in Imperial Valley) typically left substantial amounts of uncut foliage that served as a potential reservoir for whitefly and parasitoids. The refuge was surrounded on three sides by commercial agriculture and provided a good site for establishing populations of natural enemies that could then migrate outwards. During 1996 and 1997, 700,000 non-indigenous *Eretmocerus* and *Encarsia* were released in alfalfa fields on the refuge.

13.2.7 Commercial Plant Nurseries

Parasitoids were released in commercial nurseries on whitefly-infested ornamental plants such as hibiscus, orchid tree, roses, cape honeysuckle, and lantana, which were purchased by business and home owners and incorporated throughout the area into gardens and landscaping. As many as seven separate releases were made at each of eight plant nurseries (one wholesale, seven retail) in four Imperial Valley communities from February to May 1998. In all, a total of 2.3 million (estimated from mean pupal weight) *Eret. mundus* Mercet (MBCL culture M94120), *Eret. emiratus* Zolnerowich and Rose (M95104), and *Eret.* nr. *emiratus* (M96076) produced by the APHIS insectary in Imperial, California, were distributed in nurseries as pupae.

13.3 Release Procedures

Several release methods were employed to maximize the opportunity for establishment of *B. tabaci* parasitoids in the often harsh climate of this desert region. During the early releases through 1994 parasitoids were placed directly into agricultural crops as parasitized whitefly nymphs on hibiscus and eggplant leaves shipped by air courier from the APHIS rearing facility in Mission, Texas, or as adults contained in 9 dram vials with a honey soaked adsorbent paper strip from CDFA in Sacramento. Leaves with pupae or adult wasps in vials were hand-distributed under the crop canopy in maximal shade to delay leaf desiccation and promote the highest possible levels of parasitoid emergence and survival. Nevertheless, the mortality of parasitoids was relatively high on these leaves in the hot, dry climate of Imperial Valley especially during summer months, and emergence was often less than 50%.

From 1995 to 1998, releases of pupae on clipped leaves were continued because collecting adult parasitoids was not practical when parasitoid production increased dramatically. Eventually, however, the Imperial, Mission and Tucson rearing facilities switched to a more efficient parasitoid harvesting system in which parasitoids were washed from leaves as pupae and easily strained from water (Chapter 10). Subsequent releases used loose parasitoid pupae contained in small paper or plastic containers (20 cm^3) placed under the shade of the crop canopy or attached to plant stems. This method was used extensively from 1998 to 2000. Parasitoids were also distributed as larvae or pupae within parasitized whiteflies on live potted plants that were transplanted into insectary field plots or among ornamentals or home garden plants in certain instances. This method was frequently used from 1996 to 1998. As a result, nearly all parasitoid species were released using several methods.

Assessments of parasitoid emergence were made in several ways. During July 1995, leaves with parasitoid pupae of *Enc.* sp. nr. *hispida* (M94056) were attached to six 13 × 18 cm white paper cards using white glue and placed in the canopy of an insectary planting of cotton using paper clips. The cards were removed from the field after 12 days and counts of emerged and dead parasitoids were made. On average, 33% of the parasitoids emerged. A similar study was conducted in mid-November of

1995, using pupae of *Enc.* sp. nr. *hispida* and *Eret. melanoscutus* Zolnerowich and Rose (M94023). After 8 days of exposure, it was found that 42% of the *Encarsia* and 44% of the *Eretmocerus* had successfully emerged. It was estimated that an additional 2% were still viable and likely to emerge and the remainder was dead. During November 1996, 553 pupae of *Eret. hayati* Zolnerowich and Rose (M95012) on 20 hibiscus leaves were attached to cotton leaves in a field insectary planting; the leaves were returned to the lab after 14 days and inspected. Emergence was only 8%, 58% were dead and 35% were missing. In early October 1997, the fate of 100 *Eret. emiratus* on 17 hibiscus leaves attached directly to leaves of cotton insectary plants using paper clips was examined over four consecutive days. Individual pupae were circled with a permanent marking pen. During this time 20% emerged, 63% were missing, 12% died and 5% were viable pupae. It was assumed that the majority of the missing pupae were removed by predators, especially ants that were commonly observed. Emergence of parasitoids from loose pupae distributed in containers was also examined; adult emergence using this method sometimes exceeded 90%. These results show that field releases of whitefly parasitoid pupae vary extensively in terms of successful emergence, which is likely to be a function of species and overall fragility of these whitefly parasitoids when subject to handling, shipping and storage.

13.4 Monitoring

In most instances, monitoring involved the collection of leaves from multiple plants at a site, whether an urban yard, insectary field plot, or commercial field site. Leaves were selected that had late stage larval and pupal parasitoids. If parasitism was lacking, leaves with approximately 50% whitefly exuviae were selected. It had been determined that selecting leaves older than those containing predominantly late-stage whitefly nymphs is necessary to determine whether a specimen is parasitized, when viewing specimens using a microscope (Gould and Naranjo 1999; Naranjo 2001). Within a population of unparasitized and parasitized whitefly, the exuviae from recently emerged healthy whiteflies coincide phenologically with the late larval and pupal stage of parasitoids. These parasitoid life stages are visible through the cuticle of a parasitized whitefly nymph without dissection. Therefore, selecting leaves with the population age structure described above provided a consistent basis for characterizing and comparing whitefly and parasitoid population patterns. For identification purposes, this approach assured sampling of parasitized whiteflies containing parasitoid life stages capable of completing development and emerging when removed from plant leaves. Life stage counts were typically conducted on two 1 cm^2 portions of the lower surface of each leaf. Small portions of each sample leaf were used to obtain population counts because densities were typically very high, making whole leaf counts impractical.

Parasitoid collections were made by removing pupae from a leaf sample and placing up to 100 *Eretmocerus* and *Encarsia* pupae sorted by genus into each of two 0.5 dram glass shell vials with a cotton plug. Vials with pupae were held for emergence in a 0.2–0.5 m^2 plastic box 16 cm deep, containing a dish (0.1–0.5 l) with rock salt saturated with water to provide approximately 60% humidity. Parasitoid specimens were sorted after emergence to separate exotic species from native species; female specimens of exotic species were slide mounted and identified by a systematist, and some male and female specimens were identified using RAPD-PCR DNA analysis at USDA-APHIS, Mission, Texas.

The separation of native male *Eret. eremicus* from exotic Palearctic *Eretmocerus* species (including nearly all species released in the region) was easily and very reliably done by comparing the color of the antennal pedicel; i.e., light for *Eret. eremicus* and dark for all Palearctic species. Of particular importance during the early phases of establishment when densities of a newly introduced species were low, this technique provided an easy means of rapidly sorting large numbers of field-collected *Eretmocerus* specimens. This resulted in a simple relative estimate of how well one or more exotic *Eretmocerus* species were performing compared to *Eret. eremicus*. The commonly occurring native *Eret. eremicus* was used as a benchmark to facilitate expressing the recovery of exotic *Eretmocerus* species as a percentage of all *Eretmocerus* collected. As previously noted, *Eret. eremicus* was common (>10% parasitism) at most times of the year, attacking whitefly on a variety of plant species. From mid-summer through fall, attack rates peaked on plants such as cotton and mulberry.

Distinguishing exotic female *Eretmocerus* from native female *Eret. eremicus* using a stereo microscrope was considerably more time consuming than separating male specimens. As a result, to achieve an overview of exotic *Eretmocerus* establishment from field collections, fewer female than male *Eretmocerus* specimens were obtained. For this reason, *Eretmocerus* data reported in the following tables consists either of male specimens only or it includes a comparatively smaller number of female counts with male data. To validate using mostly male data that could be biased because of differences in sex ratios among species, sex ratios were compared by sorting *Eretmocerus* samples taken from 15 cotton fields during September and October 1999. There were 773 native *Eret. eremicus* (85%) and 133 specimens (15%) of exotic *Eretmocerus* spp. consisting primarily of *Eret. emiratus* and nr. *emiratus*. Examination of this group of samples showed that the sex ratios were equal in both native and exotic *Eretmocerus* populations, supporting the use of male *Eretmocerus* field data to reflect both male and female parasitoid population patterns.

13.4.1 Long-Term Insectary Gardens

Long-term insectary gardens were monitored monthly from May to September, and approximately every 2 months during the remainder of the year by the collection of 15 leaves from each plant type. Based on sample data analysis during the first year, graphing the standard error (as a percent of the sample mean) versus population density demonstrated that a sample consisting of a count of whiteflies on two leaf discs per leaf for15 leaves kept the SE below 30% when whitefly densities exceeded three per disc (unpublished data, WJR).

13.4.2 Short-Term Nursery Gardens

Short-term nursery gardens were sampled intermittently in order to confirm that the introduced parasitoid species were reproducing. Further information is presented in Chapter 15 (see: Summer–Fall Insectary Habitats, 1998 and 1999).

13.4.3 Urban Releases

Samples from urban sites were collected periodically from various locations in several towns in Imperial Valley starting in fall 1995 and continuing through 2000. Urban sites were thought to be more stable environments than agricultural crops for whitefly, and thus parasitoid, populations. The most commonly sampled host plant was hibiscus, but samples included other perennial species such as roses, lantana, orchid tree, mulberry, fig, and various annual vegetables such as tomatoes, beans and melons. Samples consisting of whitefly-infested leaves were collected throughout spring, summer and fall and held in the laboratory for parasitoid emergence.

13.4.4 Sentinel Plants

A survey using sentinel plants was conducted to document the establishment and spread of introduced parasitoids. This survey began in May 1998 and was conducted twice a year in the late summer and in the late fall–early winter from 1998 to July 2000. Sentinels were greenhouse-grown *Hibiscus rosa-sinensis* in four liter pots that were exposed to adult whiteflies for oviposition for several days inside mesh screened 1 m^3 cages in the greenhouse and taken into the field when third instars were first observed. Plants used as sentinels typically contained a minimum of several hundred to a thousand nymphs when taken to the field. Samples of leaves were collected from plants as checks against contamination by parasitoids before field exposure; plants were transported to and from the field in cages or in paper bags to exclude contaminating parasitoids. Sentinel plants were placed at or near 15 sites throughout the irrigated portion of the Imperial Valley where introduced *Eretmocerus* had been found reproducing at release sites prior to 1998. For each of the 15 sites, two plants were placed at each recovery site, two were placed 1.6–3.2 km (1–2 miles) from the site, and two at 6.5–8.1 km (4–5 miles) from the site, the distant sites being isolated from present or previous releases. Sentinel plants were placed in proximity to other whitefly hosts that included crops, ornamental plants, or garden vegetables to expose them to local parasitoid populations. Plants were inspected regularly in the field and watered when necessary. After 1 week of field exposure, the plants were taken back to the laboratory and held in screened cages until the first whiteflies or parasitoid emerged; the leaves were harvested and held until emergence was complete. Parasitism rates were recorded and emerged parasitoids were identified as native or introduced.

13.4.5 Commercial Cotton Fields

Commercial cotton fields were sampled from late August through mid-October of 1998 to 2000 to determine the extent of establishment of exotic *Eretmocerus* and *Encarsia*. Leaf samples consisting of 30–50 leaves containing appropriate stages of parasitized whiteflies were taken from two borders of each commercial cotton field, primarily the east and south sides. Leaves were placed in paper bags and held in an ice chest maintained at 17°C while in the field and were stored overnight in a walk-in chamber at 13°C. Leaves were classed by host whitefly density as follows: 1: <25, 2: 26–75, 3: 76–150, 4: >150 per leaf. Individual parasitized whitefly pupae were placed in shell vials for identification of emerged parasitoids.

Because whiteflies parasitized by *Enc. sophia* (Girault and Dodd) are particularly noticeable (its pupal case is an opaque glossy black color) and visually distinct from *Eretmocerus* and native *Encarsia luteola* and *Enc. meritoria* (pupal cases either non-pigmented or partially pigmented), a separate method was used to detect its presence. Each leaf was visually scanned under a dissecting microscope and the number of *Enc. sophia* pupae were recorded. As a result, the probability of detecting *Enc. sophia* was greater than for detecting exotic *Eretmocerus* species that are indistinguishable from native *Eretmocerus* spp. upon general inspection of pupae.

13.4.6 Desert Survey

A survey was conducted from May 1993 to April 2001 in non-irrigated desert regions adjoining the Imperial Valley. The principal purposes of this survey were to identify wild host plants of *B. tabaci,* document native parasitoids attacking *Bemisia* on these hosts, and to identify potential non-target whitefly species and their indigenous parasitoids (see Chapter 18). A secondary goal included the documentation of exotic parasitoids that eventually were found in desert regions. Samples were collected several times throughout the year at 55–60 desert sites along a 160 km route around the perimeter of the Imperial Valley (Hoelmer and Culver 1997). Many species of desert vegetation including introduced invasive weeds and native plants were examined for the presence of whiteflies and parasitoids. Leaf samples containing whiteflies were held in the laboratory for emergence and identification of parasitoids.

13.5 Field Recovery Results

Large numbers of parasitoids were released for most species, and several species were released over successive years (Table 13.1). Despite the large numbers of parasitoids released, very few exotic *Eretmocerus* and *Encarsia* species were recovered during the first 2 years (1994 and 1995). In 1994 the uniparental species *Eretmocerus staufferi* Rose and Zolnerowich was recovered on several occasions during the summer, but only at home sites where it had been released. In 1995, CDFA made 40 collections from four insectary field sites and one home site. In total, 637 male *Eretmocerus* were identified to species. Of the three specimens that were exotic species, one was identified as probably *Eret. hayati* based on PCR analysis and the other two specimens were not identified to species due to poor specimen quality. Similarly, *Encarsia* sp. nr. *hispida* was rarely found during 1995 and never found thereafter. This species was easily identifiable in the field due to its opaque black pupal case and meconia located laterally along the sides of the pupa. This latter characteristic distinguished it from *Enc. sophia* released in subsequent years.

By 1997 many recoveries were made in agricultural crops or other rural sites relatively distant from urban and agricultural release sites. Exotic *Eretmocerus* were found at seven rural sites including one nearly 2 miles from the nearest release location. All other rural sites were located at greater distances (as many as 15–20 miles) from any of the current year's release sites. Rural recoveries were made from spring melons, mulberry and edible fig trees located at rural residences or schools. The number of sites, the proportion of sites with exotics present, and the proportion of exotic parasitoids among the total reared were all considerably higher during 1997 than in 1996.

13.5.1 Long-Term Nursery Gardens

Long-term nursery gardens provided considerable information on the establishment of new species, especially *Eretmocerus* spp. (Table 13.2). Data are presented according to two periods of the year in accordance with the seasonality of certain crop plants and parasitoid occurrence. For example, the cropping period for spring cantaloupe is relatively short (i.e., March–June) compared to cotton, okra and basil, which grow over half of the year (i.e., March–October); however, these latter crops are most heavily attacked by whitefly from July onward. The predominant native *Encarsia* species (*Enc. luteola*) is most common during spring, whereas native *Eret. eremicus* is particularly abundant from mid-summer through fall. In 1996 and 1997 several species of exotic *Eretmocerus* were collected in the insectary gardens more frequently than in the previous several years. During this time the gardens were being inoculated with exotic parasitoid species each year. From 1998 onward, no additional releases were made in or in close proximity to the gardens. From 1998

Table 13.2 Percentage of exotic male *Eretmocerus* in long-term insectary gardens that received inoculative releases through 1998.

Plant host by year	No. of samples[a]		No. of male specimens		% exotic[b]	
Sample period	Jan.–Jun.	Jul.–Dec.	Jan.–Jun.	Jul.–Dec.	Jan.–Jun.	Jul.–Dec.
Cantaloupe						
1995	1	–	23	–	0	–
1997	4	–	81	–	10	–
1999	1	–	19	–	11	–
2000	5	–	112	–	73	–
2002	3	–	27	–	62	–
Cole crops						
1995	2	1	10	15	0	0
1996	13	2	325	49	29	26
1997	5	2	68	35	44	63
1998	6	3	104	93	45	74
1999	12	4	259	141	75	99
2000	2	1	20	17	100	100
2002	–	1	–	12	–	83
Cotton						
1995	–	7	–	228	–	6
1996	–	2	–	9	–	10
1997	–	3	–	99	–	57
1998	–	10	–	286	–	45
1999	1	3	5	51	0	67
2000	3	10	35	105	45	84
2001	–	2	–	26	–	3
2002	–	2	–	39	–	32
2003	–	6	–	40	–	90
Okra						
1996	4	8	140	246	28	14
1997	3	12	81	293	7	30
1998	2	17	21	385	5	55
1999	2	11	65	278	23	80
2000	1	5	4	84	75	96
2001	–	4	–	84	–	13
2002	3	3	56	51	39	72
2003	–	6	–	56	–	100
Basil						
1996	3	5	88	162	45	3
1997	2	5	28	98	6	18
1998	1	13	5	143	0	20
1999	2	7	17	85	4	55
2000	1	1	9	5	100	100
2003	–	5	–	14	–	100

[a]Samples came from one or more fields in each period.

[b]Using the commonly occurring native *Eret. eremicus* as a benchmark, the recovery of exotic *Eretmocerus* species is expressed as a percentage of all *Eretmocerus* collected.

onward, despite the lack of parasitoid releases, exotic *Eretmocerus* became considerably more prevalent, often representing well over 50% of a large *Eretmocerus* population with the predominant species being *Eret.* sp. nr. *emiratus* (M96076, Ethiopia) and *Eret. emiratus* (M95104, United Arab Emirates). DNA analysis indicated that the former population was dominant. During 2001 the two gardens on the organic farm were discontinued, limiting sampling to the two gardens at the Imperial Valley Research Center. Although sampling was appreciably reduced in 2001, data obtained from okra and cotton strongly suggested a significant reduction in exotic parasitoid activity for that year (Table 13.2); exotic species comprised only 13% of the *Eretmocerus* found on okra and 3% on cotton. The exotic *Eretmocerus* spp. rebounded in 2002. Fourteen samples taken from cotton, okra, basil, cantaloupe and collard from June to August 2002 showed the *Eretmocerus* population (193 male specimens) consisted of an average of 48% exotic species. Morphological identification (M. Rose) determined the presence of a mixed population consisting of predominantly *Eret.* sp. nr. *emiratus* and *Eret. emiratus* with two specimens identified as *Eret.* sp. nr. *mundus.*

From 1996 through the spring of 1997, *Enc. sophia* (M93003, Spain) was collected repeatedly at two of the four field plots and one home site where it had been released. During this time period it was common at the home site, whereas in the insectary garden field plots it was rarely collected. This species was easily identified in the field by having a black pupal case in combination with female meconia located posteriorly. *Encarsia sophia* (M95107, Multan, Pakistan) was released following field cage evaluations in 1997. Approximately 200,000 *Enc. sophia* (M95107) individuals were collected from experimental field cages and moved to an adjacent nursery garden (data not reported in Table 13.1). Based on a 1 day survey that involved the inspection of urban plants along a transect north of the field site in September of 1997, this parasitoid was found in neighboring residential sites on roses and mulberry trees at least 3 km away approximately 3 months following its release. This was the only exotic *Encarsia* species that clearly became established in Imperial Valley. Specimens of this population were differentiated from those of M93003 through RAPD-PCR DNA analysis. Its presence in the insectary gardens relative to all *Eretmocerus* and other *Encarsia* species was as high as 50% in late summer. Up to the final year samples were taken (2004), *Enc. sophia* (M95107) was frequently collected in moderate to high numbers in urban settings, commonly causing in excess of 25% relative parasitism of fourth instar whitefly.

13.5.2 Urban Releases

Urban site samples showed a seasonal increase in populations of parasitoids on host plants, with the greatest numbers in late fall and lowest numbers in winter (Table 13.3). This was a consequence of a seasonal increase of whitefly on urban host plants driven by migration from adjacent croplands in late spring and summer. Of the parasitoids identified across all sample dates, native *Eret. eremicus* comprised

Table 13.3 Number and proportion of native and exotic parasitoids recovered per site in urban surveys from 1997 to 2000[a].

Date of survey	No. sites	*Eretmocerus* Native	*Eretmocerus* Exotic	*Encarsia* females Native	*Encarsia* females Exotic	*Encarsia* males Unidentified
Feb. 1997	2	84 (0.47)	87 (0.48)	0 (0)	0 (0)	9 (0.05)
Aug. 1997	12	301 (0.44)	306 (0.44)	30 (0.04)	1 (0)	51 (0.07)
Oct. 1997	22	1311 (0.70)	367 (0.20)	20 (0.01)	0 (0)	174 (0.09)
Sept. 1998	38	811 (0.76)	136 (0.13)	60 (0.06)	2 (0)	62 (0.06)
Nov. 1998	40	1664 (0.77)	240 (0.11)	97 (0.04)	13 (0.01)	144 (0.07)
Jan. 1999	20	333 (0.83)	14 (0.04)	8 (0.02)	0 (0)	45 (0.11)
Aug. 1999	41	444 (0.74)	96 (0.16)	39 (0.06)	1 (0)	22 (0.04)
Oct. 1999	42	899 (0.66)	210 (0.15)	29 (0.02)	107 (0.08)	121 (0.09)
Feb. 2000	28	245 (0.68)	27 (0.07)	10 (0.03)	22 (0.06)	57 (0.16)
Sept. 2000	29	405 (0.55)	242 (0.33)	35 (0.05)	13 (0.02)	42 (0.06)

[a]Both sexes of *Eretmocerus* are pooled; female *Encarsia* are shown separately, as males could not be identified reliably.

about 66% and exotic *Eretmocerus* about 21%. There was a reduction in new releases in 1998 and no releases were made in Imperial Valley during 1999 except for several inoculative releases in the city of Mexicali, Mexico, located adjacent to Imperial County's southern border. The presence of *Enc. sophia* began to be noticeable in urban samples in 1999 and had become widespread by 2000.

In urban samples collected in 1996, only two sites out of more than fifty had a large proportion of exotic *Eretmocerus* (up to 95%). Most of the urban sites sampled during 1997 included some introduced *Eretmocerus*. At two thirds of the sites where exotics were recovered in 1997, the proportion of introduced parasitoids among the total reared from the sample typically ranged from 10% to 40% while the remaining third had exotics in lower or higher proportions.

Since no releases were made early in the year in urban areas, the source of early season recoveries was most likely overwintering populations or dispersal from agricultural releases outside of town. Late season recoveries during the fall months showed that whitefly populations became very large on mulberry trees, which are widely planted throughout the valley. Introduced *Eretmocerus* were found in many of these samples in 1997. However, a later sample of *B. tabaci* on mulberry trees at eleven urban sites taken during summer 1999 contained only two percent exotic *Eretmocerus* of a total of 749 *Eretmocerus* reared. Parasitism by *Encarsia* was also low in this sample.

13.5.3 Sentinel Plants

Sentinel plant samples consistently indicated the presence of exotic *Eretmocerus* species within the Imperial Valley by late 1998 and through the survey conclusion at the end of 2000 (Table 13.4). By late 1999 there were large populations of native

Table 13.4 Number and proportion of native and exotic parasitoids recovered per site in sentinel surveys from 1998 to 2000[a].

Sample	*Eretmocerus*		*Encarsia* females		*Encarsia* males
Distance (miles)	Native	Exotic	Native	Exotic	Unidentified
May 1998					
0	221 (0.59)	0 (0)	140 (0.38)	0 (0)	12 (0.03)
1–2	85 (0.58)	0 (0)	59 (0.40)	0 (0)	3 (0.02)
4–5	36 (0.57)	0 (0)	22 (0.35)	0 (0)	5 (0.08)
Aug. 1998					
0	117 (0.76)	14 (0.09)	5 (0.03)	0 (0)	18 (0.12)
1–2	126 (0.78)	20 (0.12)	3 (0.02)	0 (0)	13 (0.08)
4–5	88 (0.90)	0 (0)	4 (0.04)	1 (0.01)	5 (0.05)
July 1999					
0	251 (0.62)	52 (0.13)	46 (0.11)	17 (0.04)	37 (0.09)
1–2	108 (0.52)	73 (0.35)	5 (.002)	0 (0)	21 (0.10)
4–5	119 (0.60)	53 (0.27)	1 (0.01)	0 (0)	26 (0.13)
Nov. 1999					
0	1292 (0.58)	558 (0.25)	224 (0.10)	116 (0.05)	45 (0.02)
1–2	934 (0.70)	96 (0.07)	162 (0.12)	104 (0.08)	34 (0.03)
4–5	1094 (0.61)	445 (0.25)	107 (0.06)	96 (0.05)	40 (0.02)
July 2000					
0	1420 (0.72)	223 (0.11)	137 (0.07)	48 (0.02)	140 (0.07)
1–2	896 (0.70)	204 (0.16)	94 (0.07)	5 (0)	83 (0.06)
4–5	308 (0.65)	111 (0.23)	40 (0.08)	0 (0)	15 (0.03)
Dec. 2000					
0	34 (0.20)	74 (0.43)	31 (0.18)	23 (0.13)	10 (0.06)
1–2	3 (0.02)	18 (0.15)	48 (0.39)	49 (0.40)	4 (0.03)
4–5	10 (0.05)	28 (0.15)	93 (0.51)	38 (0.21)	13 (0.07)

[a]Both sexes of *Eretmocerus* are pooled; female *Encarsia* are shown separately as males could not be identified reliably.

and introduced parasitoids, but a year later, in 2000, a general decline in parasitoid populations occurred for unknown reasons in both native and introduced species. Significant parasitism by exotic *Enc. sophia* was not apparent until the latter part of 2000. Contrary to our expectation that exotic species would be detected first in sentinels placed closest to pre-sentinel release sites rather than in those most distant from these sites, there was no clear trend in parasitoid numbers captured by sentinel hibiscus plants at increasing distances from original release points. This suggests that exotic *Eretmocerus* had dispersed more quickly than anticipated and therefore were already present throughout most areas chosen for the sentinel surveys, even where we had not found them in earlier surveys. At the final sampling in December 2000 the number of sample sites located in urban areas was reduced due to insufficient numbers of survey plants; the sites dropped had been sampled in the urban survey in September 2000. Across all sample dates, native *Eret. eremicus* comprised about 63% of parasitoids reared, and exotic *Eretmocerus* about 18%, which was similar to the relative proportions found in urban surveys, and showed no significant change in this proportion over the 2-year sampling period. Native *Eret. eremicus* consistently had a 1:1 sex ratio, whereas exotic *Eretmocerus* exhibited a slightly biased ratio of 0.85:1.0 female:male throughout the study. This may reflect difficulty experienced by exotic females in locating exotic males among a larger population of native parasitoids, because an unmated female will lay eggs that produce only male progeny.

A subset of 92 exotic female *Eretmocerus* specimens from the July 2000 samples was identified to species; at seven sites only *Eret. emiratus* was found, at five sites only *Eret. mundus*, and a mixture of these species and possibly *Eret. hayati* at seven other sites. At the latter seven sites 68% of the specimens were *Eret. emiratus* or nr. *emiratus* whose origin (UAE or Ethiopia) was uncertain because they had been slide mounted and were not available for PCR assay. Two thirds of the remainder were identified as *Eret. mundus,* and one third possessed intermediate gradations of morphological characters used in species identification. Slide-mounted specimens with intermediate characters could not be conclusively identified; these specimens may have been *Eret. hayati, Eret. mundus, Eret. emiratus,* or possibly hybrid crosses. Thirty-seven exotic specimens were identified from December 2000 samples; at four sites only *Eret. emiratus* was found, at two sites only *Eret. mundus*, and a mixture of *Eretmocerus* species at two other sites. These latter two sites yielded 35% *Eret. emiratus* and 57% *Eret. mundus,* the remainder were intermediate in form.

13.5.4 Commercial Cotton Fields

Sampling of commercial cotton fields provided valuable information on the extent of exotic parasitoid establishment. The percentage of cotton fields with detectable levels of exotic *Eretmocerus* species increased from 43% in 1998 to over 95% by the fall of 2000 (Table 13.5 and Fig. 13.1) Exotic species represented less than 5% of the *Eretmocerus* present in 1998 and increased to nearly 50% in each field by 2000. During this time, nearly all exotic specimens were identified as *Eretmocerus* sp. nr. *emiratus* (M96076, Ethiopia) or *Eret. emiratus* (M95104) based on PCR assays. By 1999 the majority of parasitoid releases were conducted in conjunction with augmentation studies in cantaloupe, which grew predominantly in the southern half of Imperial Valley while the majority of the cotton crop was grown in the north half of the valley.

Native *Encarsia* species and exotic *Enc. sophia* were found in low numbers, approximately one individual per infested leaf. However, the survey revealed that *Enc. sophia* became considerably more prominent over the 3-year time period so that by 2000 it was present in 75% of the commercial cotton fields in Imperial Valley (Table 13.5). PCR assays revealed that the majority was *Enc. sophia* (M95107, Multan, Pakistan), and only two out of nearly 100 specimens tested were M93003 from Spain. This was the only evidence that the Spanish strain of *Enc. sophia* (M93003) still remained in Imperial Valley since it was last detected in 1996.

Table 13.5 Survey of commercial cotton fields for exotic *Eretmocerus* spp. and *Encarsia sophia* in the Imperial Valley.

	Fall 1998	Fall 1999	Fall 2000
Eretmocerus spp.			
Number of fields sampled	23	42	24
Fields with exotic *Eretmocerus*	43%	74%	96%
Mean exotics in fields where detected	4%	28%	48%
Encarsia sophia			
Fields sampled	23	42	32
Fields with *Encarsia sophia*	4%	64%	75%

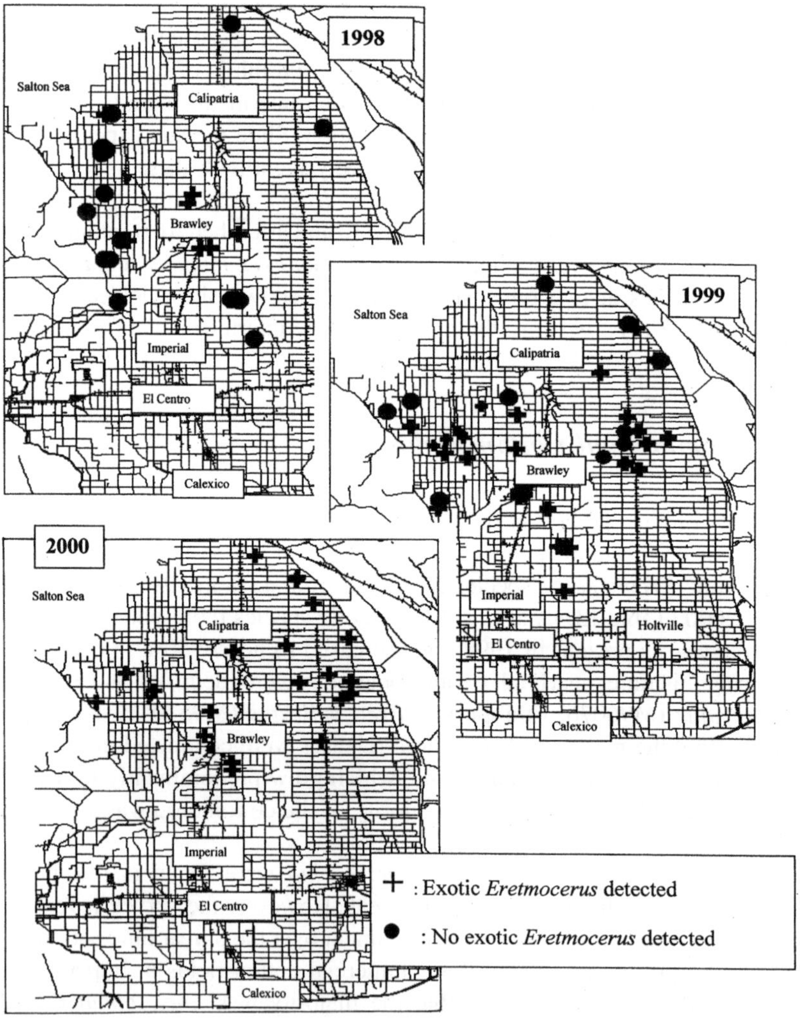

Fig. 13.1 Fall surveys illustrating the presence or absence of exotic *Eretmocerus* parasitizing *Bemisia tabaci* in commercial cotton fields in Imperial Valley, CA, from 1998 to 2000. The distance from the bottom to top of each map is 65 km (40 miles).

13.5.5 Desert Surveys

Desert survey samples revealed that *Eret. eremicus, Enc. luteola, Enc. meritoria* and *Enc. coquilletti* Howard were the only indigenous parasitoids reared from *B. tabaci* on desert vegetation. The first exotic recoveries of *Eret. mundus* and *Eret. emiratus* were made in 1997 at three widely separated desert sites located west of the cultivated portion of the valley. During 1999 and 2000, exotic *Eretmocerus* were recovered from *B. tabaci* at eight additional desert sample sites both east and west of the irrigated valley. The majority (28 of 48 specimens) were *Eret. emiratus,* two were *Eret. mundus,* and the remainder were intermediate in morphology and so could not be conclusively identified to species. By 2000 introduced *Eretmocerus* species had been recovered from *B. tabaci* on different occasions from numerous widely separated desert sites both east and west of the Imperial Valley, providing additional evidence that establishment and dispersal of introduced species had occurred. Some of these sites were up to 50 km from the nearest release location, suggesting that significant dispersal had taken place. It is likely that prevailing winds provided the means for such dispersal, given the low density of hosts in the desert compared with urban and agricultural areas. No specimens of any exotic *Encarsia* had been recovered in samples from desert surveys as of 2001.

References

Bellows, T.S., Jr., T.M. Perring, R.J. Gill, and D.H. Headrick. 1994. Description of a species of *Bemisia* (Homoptera: Aleyrodidae). Annals of the Entomological Society of America 87: 195–206.

Bellows, T.S., Jr. and K. Arakawa. 1988. Dynamics of preimaginal populations of *Bemisia tabaci* (Homoptera: Aleyrodidae) and *Eretmocerus* sp. (Hymenoptera: Aphelinidae) in Southern California cotton. Environmental Entomology 17: 483–487.

Birdsall, S.L. 1992. Imperial County Agriculture Crop and Livestock Report. Imperial County, CA. 9 pp.

Birdsall, S.L. 1999. Imperial County Agriculture Crop and Livestock Report. Imperial County, CA. 9 pp.

Birdsall, S.L. 2002. Imperial County Agriculture Crop and Livestock Report. Imperial County, CA. 9 pp.

Gerling, D. 1967. Bionomics of the whitefly-parasite complex associated with cotton in southern California (Homoptera: Aleurodidae; Hymenoptera: Aphelinidae). Annals of the Entomological Society of America 60: 1306–1321.

Gonzalez, R.A., G.E. Goldman, E.T. Natwick, H.R. Rosenberg., J.I. Grieshop, S.R. Sutter, T. Funakoshi, and S. Davila-Garcia. 1992. Whitefly invasion in Imperial Valley costs growers, workers millions in losses. California Agriculture 46 (Sept.–Oct.): 7–8.

Gould, J.R. and S.E. Naranjo. 1999. Distribution and sampling of *Bemisia argentifolii* (Homoptera: Aleyrodidae) and *Eretmocerus eremicus* (Hymenoptera: Aphelinidae) on cantaloupe vines. Journal of Economic Entomology 92: 402–408.

Heinz, K.M., J.R. Brazzle, M.P. Parrella, and C.H. Pickett. 1999. Field evaluations of augmentative releases of *Delphastus catalinae* (Horn) (Coleoptera: Coccinellidae) for suppression of *Bemisia argentifolii* Bellows & Perring (Homoptera: Aleyrodidae) infesting cotton. Biological Control 16: 241–251.

Hoelmer, K.A. and G. Culver. 1997. Survey of desert host plants for whiteflies and parasitoids. In T.J. Henneberry, N.C. Toscano, T.M. Perring, and R.M. Faust, eds., Silverleaf Whitefly: 1997 Supplement, 5-Year National Research and Action Plan – 5th Annual Review, p. 149. San Diego, CA. January 28–30, 1997. US Department of Agriculture, ARS 1997–02, Washington, DC.

Hoelmer, K.A. and C.H. Pickett. 2003. Geographic origin and taxonomic history of *Delphastus* spp. (Coleoptera: Coccinellidae) in commercial culture. Biocontrol Science and Technology 13: 529–535.

Naranjo, S.E. 2001. Conservation and evaluation of natural enemies in IPM sytems for *Bemisia tabaci*. Crop Protection 20: 835–852.

Parrella, M.P., T.S. Bellows, R.J. Gill, J.K. Brown, and K.M. Heinz. 1992. Sweetpotato whitefly: prospects for biological control. California Agriculture 46 (Jan.–Feb.): 25–26.

Perring, T.M., A. Cooper, D.J. Kazmer, C. Shields, and J. Shields. 1991. New strain of sweetpotato whitefly invades California vegetables. California Agriculture 45 (Nov.–Dec.): 10–12.

Pickett, C.H., K.A. Casanave, S.E. Schoenig, and K.M. Heinz. 1999. Rearing *Delphastus catalinae* (Coleoptera: Coccinellidae): practical experience and a model analysis. Canadian Entomologist 131: 115–129.

Chapter 14
Releases of Exotic Parasitoids of *Bemisia tabaci* in San Joaquin Valley, California

Charles H. Pickett[1], Gregory S. Simmons[2], and John A. Goolsby[3]

Abstract In 1991 *Bemisia tabaci* was reported in the southern San Joaquin Valley infesting crops outside of greenhouses for the first time. From 1994 to 1996, 24 species/strains of imported aphelinids, primarily species of *Eretmocerus*, were released in urban and agricultural settings for control of this whitefly. Based on results of colonization and from other field and laboratory studies, several million individuals of the five highest ranked candidates were released from 1997 to 2000 into four citrus orchards in the southern San Joaquin Valley: *Eretmocerus mundus*, *Eret. hayati*, *Eret. emiratus*, *Eret.* sp. nr. *emiratus*, and *Encarsia sophia*. Citrus, in combination with surrounding weeds, aided in the initial establishment of several imported parasitoids. Although low numbers of parasitoids were found in citrus during this study, the high proportion of parasitism and the large number of trees showing parasitism would suggest that several hundred thousand parasitoids could be produced in each of our release sites during the fall. During surveys on weeds and cotton in 2002, all exotic parasitoids emerging from *B. tabaci* were *Eret. mundus*.

14.1 Introduction

The San Joaquin Valley of central California is the major agricultural production region of the state. Many of the approximately 250 commodities grown in California are represented in this valley. Cotton (*Gossypium hirsutum* L.) and alfalfa (*Medicago sativa* L.) constitute much of the acreage in the southern part

[1] California Department of Food & Agriculture, Biological Control Program, 3288 Meadowview Rd., Sacramento, CA 95832

[2] USDA-APHIS-PPQ, Center for Plant Health Science & Technology Laboratory, 3645 E. Wier Avenue, Phoenix, AZ 85040

[3] USDA-ARS, Kika de la Garza Subtropical Agricultural Research Center, Beneficial Insects Research Unit, 2413 E. US Hwy 83, Weslaco, TX 78596-8344

J. Gould et al. (eds.), *Classical Biological Control of* Bemisia tabaci *in the United States.*

of the valley where releases of parasitoids were conducted. Nevertheless there is a rich diversity of field and orchard crops along the eastern floor near the foothills of the Sierra Nevada Mountains. The roughly 400 × 100 km valley has a Mediterranean climate (30 year averages; Western Regional Climate Center, http://www.wrcc.dri.edu/). Virtually no rain falls from late spring until winter. The winter-spring rainy season brings an annual precipitation varying from 159 mm in the southern part of the valley to over 317 mm in the north. Irrigation supplements rainfall for most crops. Temperatures vary from an average January low of 3.1–3.6°C (southern to northern end of valley) to a summer high of 34.5–36.0°C.

In 1991 the biotype-B strain of *Bemisia tabaci* was first reported in the San Joaquin Valley (Gill 1992) and within 5 years this strain of *B. tabaci* was found throughout central California (Keaveny 2000). Prior to this discovery, *B. tabaci* was confined to greenhouses in this valley, where they could easily survive winter conditions. Surveys by the California Department of Food & Agriculture's (CDFA) Biological Control Program and the University of California (Ball and Casanave 1994; Godfrey et al. 1995; Rose and Zolnerowich 1997) found limited populations of three different species of aphelinids associated with *B. tabaci* infestations in central California: *Eretmocerus eremicus* Rose and Zolnerowich, *Encarsia meritoria* Gahan, and *Encarsia pergandiella* Howard. *Eretmocerus eremicus* and *Enc. meritoria* were also present in Imperial Valley, a desert agricultural growing region in southeastern California, where they and other native parasitoids failed to prevent outbreaks of *B. tabaci* (Chapters 13 and 18).

From 1994 to 1996, several species/populations of parasitoids being reared at the USDA APHIS Mission Biological Control Laboratory in Mission, Texas, were released into residential backyards in and near Bakersfield, California. Based on parasitoid overwintering survivorship from these releases and field and laboratory studies (Goolsby et al. 1996, Chapter 8), large numbers of individuals of selected species/strains were reared and released into four agricultural sites over a 4-year period, from 1997 to 2000.

14.2 Releases in Bakersfield and Surrounding Regions, 1994–1996

14.2.1 Methods

The first releases of imported parasitoids into the San Joaquin Valley were conducted in Bakersfield and surrounding communities within Kern, Tulare, and Fresno counties (Fig. 14.1). The goal of these releases was to conduct field screenings

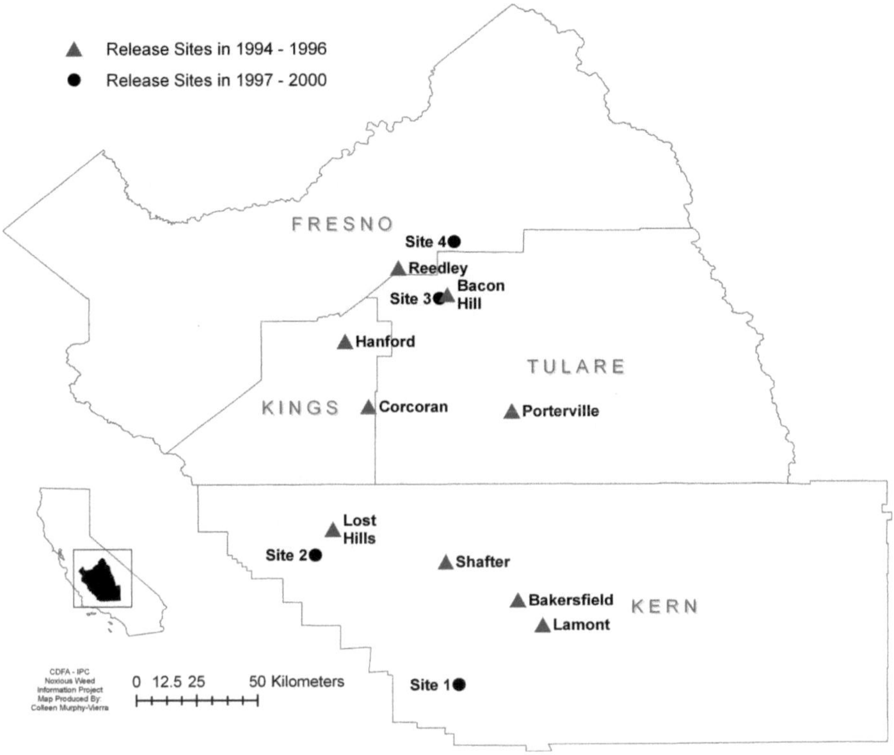

Fig. 14.1 Map of parasitoid release sites, southern San Joaquin Valley, California.

of a large number of exotic aphelinids (*Eretmocerus* spp. and *Encarsia* spp.) and compare field results with laboratory studies of attack rates (Goolsby et al. 1998). Parasitoids used in these releases were reared at the USDA-APHIS PPQ Mission Biological Control Laboratory and at the CDFA Biological Control Program's mass rearing facility in Sacramento. Releases were made into yards of private homeowners who agreed not to spray insecticides and allowed access to their property. The sites had a combination of woody perennial plants (e.g., *Hibiscus* spp., *Lantana* spp.) and vegetables [e.g., okra (*Abelmoschus esculentus* (L.), melons (Cucurbitaceae), broccoli (*Brassica oleracea* L.)] susceptible to *B. tabaci*, which provided a year-round reservoir of whiteflies. Some homeowners planted perennials known to harbor *B. tabaci* parasitoids (Roltsch 1999). Releases were made such that each site received combinations of parasitoid species that could be separated morphologically or through genetically unique DNA patterns (Legaspi et al. 1996).

14.2.2 Results and Discussion

From 1994 to 1995, a total of 283,214 parasitoids, representing 27 species/strains of aphelinids were released into 20 homeowner's yards (Table 14.1). Releases were conducted in September and October when whitefly infestations could easily be found on homeowner's plants. Parasitoids were released either as pupae or emerging

Table 14.1 Species and geographic populations of parasitoids released in the southern San Joaquin Valley, California, 1994–1996, in private home gardens and some agricultural crops.

MBCL Culture number	Species	MBCL RAPD pattern	Country of origin	Number released
M92014	*Eretmocerus mundus*	ERET-1	Murcia, Spain	128,497
M92017	*Encarsia formosa*	ENC-2	Angelohori, Greece	2,000
M92018	*Encarsia bimaculata*	ENC-1	Parbhani, India	600
M92019	*Eretmocerus mundus*	ERET-1	Padappai, India	39,400
M92027	*Eretmocerus mundus*	ERET-1	Cairo, Egypt	1,400
M93003	*Encarsia sophia*	ENC-7	Murcia, Spain	13,900
M93005	*Eretmocerus hayati*	ERET-2	Thirumala, India	17,400
M93058	*Eretmocerus mundus*	ERET-1	Tainan, Taiwan	6,000
M93064	*Encarsia lutea*	ENC-10	Mazotos, Cyprus	1,750
M94002	*Eretmocerus staufferi*	ERET-5	College Station, Texas	14,500
M94003	*Eretmocerus tejanus*	ERET-6	Mission, Texas	22,000
M94016	*Encarsia sophia*	ENC-11	Shan-Hua, Taiwan	200
M94023	*Eretmocerus melanoscutus*	ERET-3&8	Sai Noi Klong Ha Roi, Thailand	34,500
M94040	*Eretmocerus melanoscutus*	ERET-3	Kampang Saen, Thailand	6,100
M94041	*Encarsia sophia*	ENC-5	Chiang Mai, Thailand	2,400
M94047	*Encarsia sophia*	ENC-5	Kuala Lumpur, Malaysia	8,700
M94055	*Encarsia* sp. nr. *pergandiella*	ENC-15	Sete Lagoas, Brazil	4,998
M94056	*Encarsia* sp. nr. *hispida*	ENC-16	Sete Lagoas, Brazil	24,500
M94085	*Eretmocerus mundus*	ERET-1	Frascati, Italy	250
M94092	*Eretmocerus mundus*	ERET-1	Italy, mix of sites	5,700
M94103	*Eretmocerus mundus*	ERET-1	Gat, Israel	1,800
M94105	*Eretmocerus mundus*	ERET-1	Gat, Israel	1,200
M94107	*Encarsia lutea*	ENC-10	mixture of Mediterranean sites	1,450
M94120	*Eretmocerus mundus*	ERET-1	Golan, Israel	28,000
M94124	*Eretmocerus mundus*	ERET-1	Negev Desert, Israel	1,600
M94125	*Eretmocerus mundus*	ERET-1	Golan, Israel	2,800
M94129	*Encarsia lutea*	ENC-10	Mazarron, Casas Nueva, Spain	4,200
M95012	*Eretmocerus hayati*	ERET-10	Multan, Pakistan	924,266
M95104	*Eretmocerus* nr. *emiratus*	ERET-11	United Arab Emirates	18,521
Total				1,231,274

adults from 500 ml paper cans with 5 cm diameter screened openings as an exit that were hung in the shade of plants. Parasitoids were also released as pupae sprinkled onto plants. During 1996, two additional cultures of *Eretmocerus* (M95012, M95104) that performed well in laboratory and field tests (Goolsby et al. 1998, Chapters 7 and 8), including M92014, were released at homesites in Bakersfield and into agricultural settings nearby. Slightly over 1 million parasitoids were released in agricultural crops (cotton, citrus, small plantings of okra) or in nearby non-crop weedy vegetation that could serve as overwintering host to *B. tabaci*. Release sites were increased to 32 in 1996.

During September 1997, 14 of the previous 32 release sites were sampled for parasitoids. Exotic male *Eretmocerus* were recovered from yard plants hosting *B. tabaci* at six properties. Based on genetically unique RAPD-PCR patterns (Legaspi et al. 1996; Goolsby et al. 1999), *Eretmocerus mundus* Mercet (M92014) was repeatedly recovered at one site in 1996 and 1997, accounting for 25–56% of the recovered males (n = 10–20 individuals), the remainder being natives. This was a private home in the northeastern area of greater Bakersfield (Kern County) with a high diversity of imported and native vegetation and large vegetable garden. *Eret. mundus* and *Eret. hayati* Zolnerowich and Rose (Pakistan, M95012) were recovered from citrus 5 months after releases and *Eret. mundus* from nearby sowthistle (*Sonchus* sp.) 9 months later.

14.3 Releases in Citrus, San Joaquin Valley, California: 1997–2000

14.3.1 Methods

During the mid-1990s some of the most severe *B. tabaci* infestations were found in cotton grown in the San Joaquin Valley of central California near areas of high citrus acreage (Summers, personal communication). The problem was most serious in eastern Tulare County where there is a patchwork of citrus groves and cotton. The degree of *B. tabaci* infestation in late summer cotton decreased moving east to west, away from the citrus belt of eastern Tulare County. High overwintering populations of this pest in citrus, close to cotton, suggested that the proximity of these two crops allowed for an unusually high buildup of *B. tabaci* in each. *Bemisia tabaci* is not considered a pest in citrus since it appears to migrate out of these trees in spring. Based on degree day accumulation (*c.* 290 DD; Powell and Bellows 1992), *B. tabaci* would rarely produce more than two generations on citrus from mid-October to mid-February. However, *B. tabaci* can be extremely damaging to the quality of cotton lint (Bellows and Arakawa 1988); the honeydew produced by outbreak populations collects on leaves and other plant parts. A sequence of crops susceptible to *B. tabaci* has been cited as a major reason for the large outbreaks in Imperial County and other regions of the USA (Wilhoit et al. 1994; Hoelmer et al.

1994). Cole crops serve as an overwintering bridge for *B. tabaci* in mixed crop agroecosystems of the southern USA at a time when parasitism is low on these plants. It appears citrus plays a similar role in San Joaquin Valley.

One serious difficulty in establishing newly imported parasitoids for control of pests attacking field and vegetable crops is the adverse environment in which they find themselves upon release. These crops are typically sprayed several times a season with insecticides then plowed under after a short period of time, i.e., 4–8 months. Previous attempts at establishing *B. tabaci* parasitoids in California failed, probably due to the use of insecticides in the release sites (Nuessly 1990) and limited resources for conducting large-scale releases. During the current importation effort, outdoor insectaries and field refugia were used to aid in the initial field establishment of parasitoids in Imperial Valley, California and southern Texas (Chapters 9, 13, and 15). Establishing refugia is an expense to growers that requires a long-term commitment, which is often difficult to obtain. It was proposed that releases of parasitoids specific to *B. tabaci* could be made into citrus during fall, when trees are inundated with migrating whiteflies. Subsequent generations of parasitoids could move from citrus onto surrounding weeds, then to cotton the following spring and back into citrus the next fall. The proximity and phenology of *B. tabaci* plant hosts would help bridge the parasitoid populations, preventing local extinction and allowing for their regional dispersal and establishment. Since citrus is typically not treated with insecticides during the winter months, releases into an orchard with large numbers of hosts had the potential for producing large numbers of parasitoids, which could eventually establish permanent populations in these areas, increasing year-round natural control of *B. tabaci* in adjacent cotton.

14.3.2 Site Selection

In summer 1997, two release sites (Sites 3 and 4) were selected from the eastern half of Tulare County where a mixture of citrus and cotton exists and a third site was selected at a large farm near Lost Hills, Kern County (Site 2), approximately 110 km southwest of the first two (Fig. 14.1). A fourth release site was selected in summer 1998 (Site 1), since Site 2 was ceasing cotton production. All citrus orchards were adjacent to cotton and both had a history of *B. tabaci* infestations. Most cotton (70%) grown in this region is the ‘Acala’ variety (CDFA 2004).

14.3.3 Releases

The four populations/species of Aphelinidae that were mass reared from 1997 to 2000 by CDFA and at USDA-APHIS-PPQ facilities in Mission, Texas, and in Imperial Valley, California were: *Eretmocerus emiratus* Zolnerowich and Rose (United Arab Emirates, M95104), *Eret.* sp. nr. *emiratus* (Ethiopia, M96076),

Table 14.2 Species of exotic parasitoids released in citrus in the southern San Joaquin Valley, California. Roughly equal numbers were released into each of four study sites, 1997–2000.

Year	*Eret. emiratus* (M95104)	*Eret.* nr. *emiratus* (M96076)	*Eret. hayati* (M95012)	*Eret. mundus* (M92014)	*Enc. sophia* (M95107)	Total
1997	4,050,000	0	0	0	0	4,050,000
1998	3,200,000	6,900,000	100,000	28,000	0	10,228,000
1999	0	1,200,000	163,000	1,950,000	1,400	3,314,400
2000	0	0	0	0	124,000	124,000
Total	7,250,000	8,100,000	263,000	1,978,000	125,400	17,716,400

Eret. mundus Mercet (Spain, M92014), *Eret. hayati* Zolnerowich and Rose (Pakistan, M95012) and *Encarsia sophia* (Girault and Dodd) (Pakistan, M95107). During field cage trials all candidate species exceeded the native *Eret. eremicus* Rose and Zolnerowich in fecundity rates (Chapter 8). Parasitoids were reared in greenhouses on either hibiscus (*Hibiscus rosa-sinensis* L.) or eggplant (*Solanum melongena* L. 'Ichiban') infested with *B. tabaci*.

Hibiscus plants in 10 cm pots and infested with parasitized whiteflies were shipped by car and placed in water filled plastic tubs (15 × 4 × 2.5 cm) located near or at the base of marked trees. Tubs provided enough water for plants to survive up to 2 weeks, enough time for most of the developing parasitoids to eclose and disperse. Parasitoids reared on eggplant were released as pupae or adults. Parasitized late fourth instar whitefly nymphs were washed from leaves, separated from non-parasitized nymphs, and then dried (see Chapter 10). These were shipped to the field using an overnight carrier and placed in paper cans hanging from the canopy of citrus trees. Parasitoids were released into late season cotton as adults or pupae beginning around August 1, 1998, and 1999 in fields where insecticides were not being used.

Parasitoid releases into citrus were initiated after first and second instar whiteflies were found on citrus leaves, on or about September 1. Each year from 1997 to 1999, each site received approximately 100,000 parasitoids per week for 10 weeks (Table 14.2). Most had been reared on eggplants. It is estimated that over 17 million parasitoids were released during these 3 years. However, the actual number of adults entering the system may have been lower since estimates were based on parasitized whitefly pupae and did not account for adults that did not emerge. Only *Enc. sophia* was reared and released in 2000; approximately 10,000 *Enc. sophia* pupae were released weekly for 10 weeks starting in late August at the southern-most release site, Site 1 (Fig. 14.1). In 1999, a small number of *Eret. mundus, Eret. hayati, and Enc. sophia* (<130,000) were released at a second farm about 5 km north of Site 1 that produced organic cotton.

14.3.4 Sampling

Whitefly densities in citrus were monitored from 1997 to 1999 to optimize release dates of the parasitoids, evaluate specific citrus groves as release sites, and measure parasitoid establishment and potential impact. Individual leaves represented a sample unit for citrus. Whiteflies were monitored from two blocks, each consisting of 20 trees, centrally located in a row, at each release site. Block 1 was located two to three rows from the citrus/cotton interface and was used for releasing parasitoids; Block 2 was located in the center of the orchard and monitored for whitefly densities distant from the orchard's edge. The distance between blocks at each site varied from 30 to 60 m. Forty citrus leaves were collected from each block to determine the age distribution of whiteflies and establishment of exotic parasitoids. Four leaves from areas likely to be infested with whiteflies were removed from every other tree in a block (10 trees per block). A collector, standing at the perimeter of the tree canopy, removed leaves that were within arms reach and at the terminal end of the branch, where eggs and nymphs were most likely to be found. Both sides of citrus leaves were examined for the presence of eggs, young nymphs (first and second instar), and old nymphs (third and fourth instar), and insects were counted using a dissecting microscope. Whitefly densities are reported by leaf area, which for whole leaves was based on the product of leaf length and width measured for each sample. Prior studies have shown that this product closely estimates leaf area for a variety of leaf shapes (Pickett et al. 1996). At high whitefly densities (>50 per leaf), leaves were sub-sampled by randomly selecting up to five 1 cm diameter circles on each leaf.

The same leaves used for monitoring whitefly densities were used to determine number of parasitized whitefly nymphs. Fourth instar whiteflies with asymmetrical mycetomes, which are caused by the presence of young parasitoid larvae, or with visible late instar parasitoid larvae or pupae inside, were recorded as parasitized.

Monitoring of whiteflies and parasitoids in spring 1998 and 1999 included the use of *B. tabaci*-susceptible weeds, and of cotton in late spring. Weeds such as common mallow (*Malva neglecta* Wallr.), sowthistle (*Sonchus* sp.), and spotted spurge (*Euphorbia maculata* L.) were collected monthly to determine if *B. tabaci* had been attacked by exotic parasitoids. Each sample, consisting of weed leaves from one to three locations in or near each citrus orchard release site, was returned to the laboratory and placed in paper cans within 48 h. Three to six weeks later, adult *B. tabaci* and aphelinids that had emerged in the paper cans were counted. Plants were then dried for 24 h in a drying oven and weighed so that insect counts could be adjusted by plant dry weight. The high variation in plant architecture among the sampled weeds precluded the use of single leaves as sample units.

Cotton leaves were collected twice monthly to determine the level of parasitism and presence of exotic parasitoids. Three transects at right angles to the citrus/cotton interface and equidistant along the block of 20 citrus trees were established for sampling silverleaf whitefly and parasitoids in cotton. Ten leaves were collected arbitrarily at 5, 50 and 100 m along these transects. Only leaves with whitefly nymphs present were retained; they were placed in paper bags, labeled with location and date, and stored at 11°C for up to 1 week. The number of *B. tabaci* eggs, nymphs and parasitized whiteflies were recorded from the bottom (abaxial) side of each leaf prior to placement in paper cans. Leaves were retained in paper cans for 3 weeks to allow emergence of adult parasitoids. Adult whiteflies and aphelinids from weeds were identified to genus, and exotic *Eretmocerus* males were separated from natives based on the pigmentation of the antennal pedicel (Zolnerowich and Rose 1998). When a sample contained a large number of exotic males, the female *Eretmocerus* were removed, slide mounted and identified to species following Rose and Zolnerowich (1997) and Zolnerowich and Rose (1998).

14.3.5 Results and Discussion

Eretmocerus mundus was the exotic parasitoid most commonly recovered from 1998 to 2002 (Table 14.3). Although a minor component of species released initially, the proportion of *Eret. mundus* increased each year of sampling. Initially we recovered all three species of *Eretmocerus* released: *Eret. mundus*, *Eret. hayati*, and *Eret. emiratus*. By 2002, 3 years after the last releases of *Eretmocerus* spp., only *Eret. mundus* was recovered from *B. tabaci* samples. We have yet to recover *Enc. sophia* in the San Joaquin Valley.

14.4 Parasitism in Citrus and Weeds

Survival of *B. tabaci* from eggs to pupae in citrus over the winter was low; varying from 0.48% to 5.0% (Pickett unpublished data), resulting in a low number of whitefly nymphs in citrus (Fig. 14.2); parasitized nymphs in citrus were therefore also few in number. Data were averaged over Blocks 1 and 2 because the number of whitefly nymphs was so low on leaves coming from these trees. Densities were usually less than 0.1 per cm^2 leaf area, although leaves used in sampling were picked from a region of the tree likely to be infested with whiteflies. Nevertheless, parasitism at times was relatively high. In over 20% of the samples, 100% of the late instar nymphs were parasitized, and on average, 28% of nymphs on citrus leaves were found to be parasitized throughout the study (Fig. 14.2). An earlier survey in 1993 and 1994 in the same region of California found 1.5% of 20,000

Table 14.3 Species composition of exotic parasitoid recoveries from weeds and cotton, combined, infested with *B. tabaci*. No adult parasitoids were recovered from citrus for identification.

		Percentage			
Year	Number parasitoids identified	*Eret. mundus* (ex. Spain)	*Eret. hayati* (ex. Pakistan)	*Eret. emiratus* (Ethiopia + UAE)	*Enc. sophia* (ex. Pakistan)
1998	18	33	61	5	–
1999	29	58	21	21	–
2000	14	86	14	0	–
2002	19	100[a]	0	0	0

[a] Some *Eret. mundus* appear as hybrids with *Eret. hayati*.

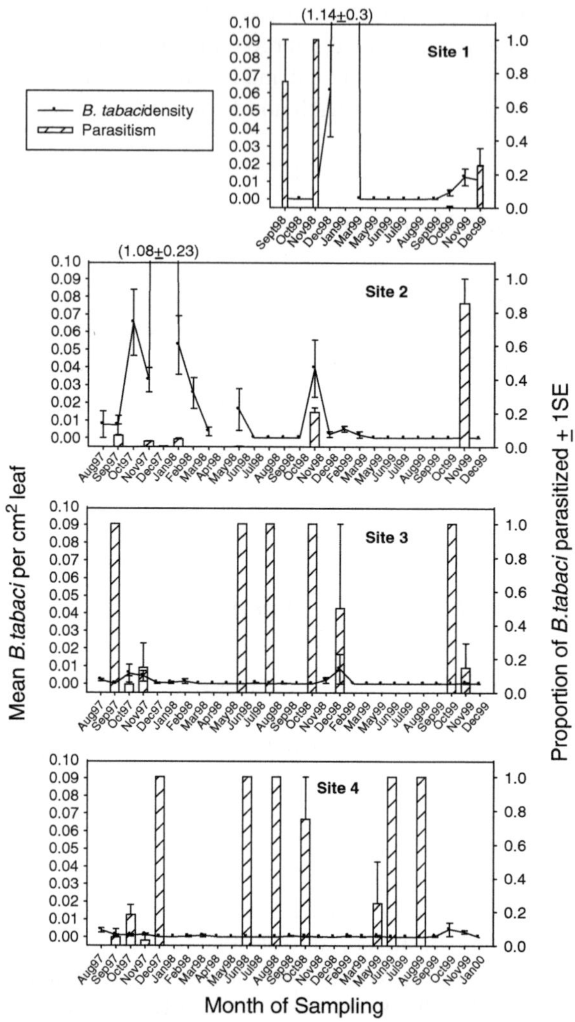

Fig. 14.2 Density of *B. tabaci* nymphs and proportion parasitized on citrus leaves in release sites.

B. tabaci parasitized by native parasitoid species on samples taken from an array of crop species and associated weeds (Godfrey et al. 1995). A small number of parasitoids on a single tree translates to a much larger number produced on a per hectare basis. Assuming 100 leaves on a single tree had at least one parasitized whitefly each, a conservative estimate, a single hectare of citrus (250 trees/ha) would produce 25,000 parasitoids per generation.

The low number of parasitized nymphs in citrus precluded regular sampling for species identification. It was known from greenhouse and field studies using citrus

(Pickett et al. 1999; Chapter 8) and limited recoveries of *Eret. mundus* that imported Palearctic *Eretmocerus* attacked and successfully developed on *B. tabaci* reared on citrus. In fall 1996, three species of exotic *Eretmocerus* (*Eret. mundus*, *Eret. hayati*, and *Eret. emiratus*) were released into a single citrus tree at a farm in Kern County and all three were recovered in February 1997; their identity was confirmed using RAPD-PCR. The same site was sampled August 1997, prior to mass releases, and all five adult parasitoid males collected were *Eret. mundus*; none were native species. Because of the lack of whitefly nymphs in citrus samples, however, weed samples provided the best estimate of the percentage of exotic parasitoids in citrus orchards. Numbers of exotic parasitoids were based on emerged adult male *Eretmocerus*. These could be identified based on the clearly discernable dark pigmentation of the antenna pedicel (Zolnerowich and Rose 1998). The vast majority (81–95%) of adult parasitoids recovered from surrounding weeds between 1997 and 1999 were the exotic *Eretmocerus* species which had been released; the only species of *Encarsia* released, *Enc. sophia*, was never recovered.

Weeds most likely helped support populations of parasitoids released into the citrus–cotton interface. *Bemisia tabaci* and parasitoid recoveries from weeds during the spring were so rare that data were averaged across weed species. The highest numbers of parasitoids recorded on weeds occurred when parasitoids were being released in fall (Fig. 14.3). Parasitoids were also present in very low numbers on weeds in the spring when whiteflies were absent from citrus (Fig. 14.2). Recoveries of parasitoids from *B. tabaci* before September, when releases began, demonstrate the overwintering survivorship of released parasitoids. The movement of parasitoids back and forth between citrus and weedy plants most likely helped maintain their presence in the citrus year round. The recovery of parasitoids from weeds shows that they were able to disperse from paper cans and hibiscus plants and survive and reproduce through multiple generations in their new environment of citrus and surrounding weedy plants. By August of the following year, the released parasitoids would have survived at least 10 generations (based on a developmental requirement of of 295 degree-days for a single generation of the related aphelinid species *Encarsia inaron* (Gould et al. 1995)).

The abundance of whiteflies and parasitoids found on weeds varied by plant and insect taxon (Fig. 14.4). The exotic *Eretmocerus* and *B. tabaci* tended to occur in highest numbers on the same species of plants. The number of exotic *Eretmocerus* on all weeds combined was significantly correlated with *B. tabaci* ($r = 0.43$, $p < 0.0001$, $n = 99$ plant collections); as was the number of native *Encarsia* ($r = 0.25$, $p < 0.01$, $n = 99$). The number of native *Eretmocerus* on the same group of weeds, however, was not correlated with *B. tabaci* ($r = 0.04$, $p < 0.67$, $n = 99$). These results suggest that the released parasitoids either have a greater affinity for *B. tabaci* than do the native *Eretmocerus* or the latter parasitoid is not attracted as strongly to the same plants as both the exotic *Eretmocerus* and *B. tabaci*. Exotic *Eretmocerus* were commonly found on nightshade (*Solanum* sp. probably *elaeagnifolium* Cav.), rarely found on sunflower (*Helianthus annuus* L.), while the opposite was true for native *Eretmocerus*. Although the native *Encarsia* (*Enc. pergandiella* and *Enc. meritoria*) were positively correlated with

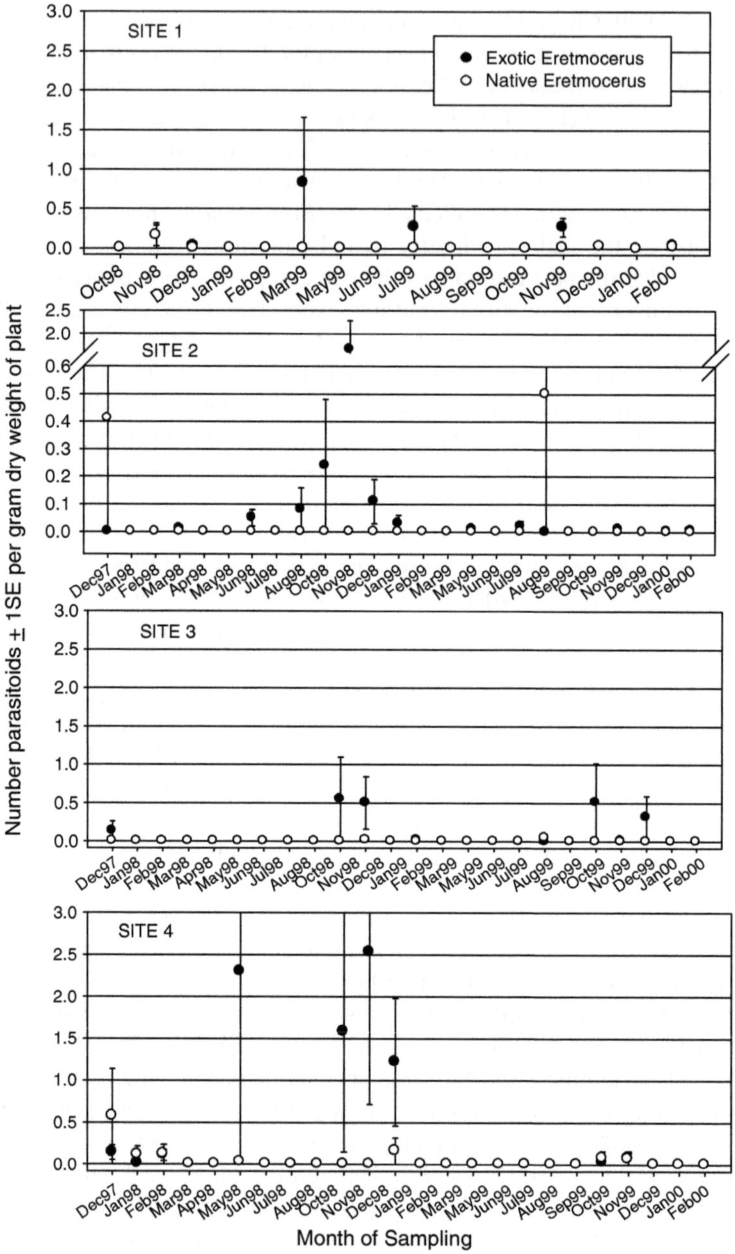

Fig. 14.3 Parasitoid dynamics on weeds associated with citrus release sites. Mean based on sample size of 2–14 collections of weeds per date.

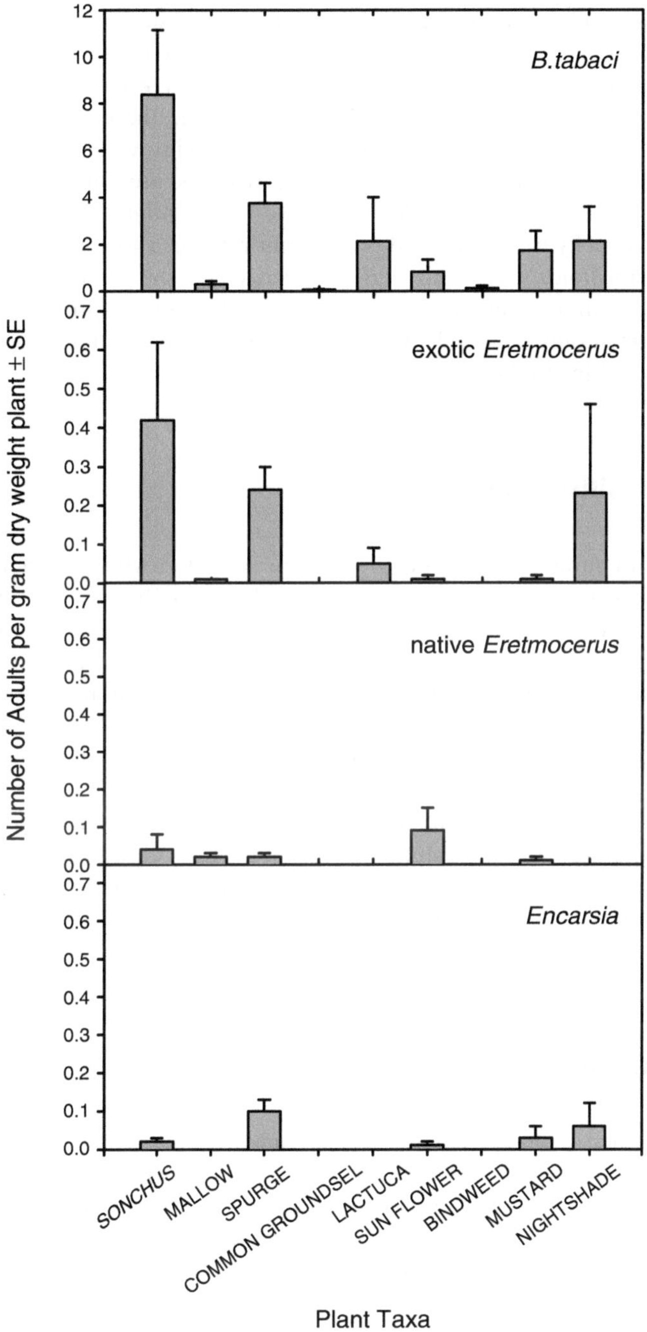

Fig. 14.4 Distribution of *B. tabaci* and its parasitoids on different weeds associated with citrus release sites.

B. tabaci, their lower average numbers on these weeds suggest they do not reproduce as well and are less specialized than the introduced exotic *Eretmocerus* spp. (Fig. 14.4). Most likely native *Eretmocerus* (probably *eremicus*) preferentially attacks other whitefly species in the San Joaquin, e.g., the greenhouse whitefly, *Trialeurodes vaporariorum* (Westwood) or banded wing whitefly *T. abutiloneus* (Haldeman). *Bemisia tabaci* has only been known to be present as a host for these parasitoids for about 8 years or less. A survey of *B. tabaci* on plants grown in homeowner's backyards conducted prior to our releases showed that only 16% of recovered parasitoids were native *Eretmocerus* and the remaining individuals were native *Encarsia* (Ball and Casanave 1994). The present survey showed that the native parasitoids on weeds were 35% *Eretmocerus* and the remainder were *Encarsia* spp. (Fig. 14.4).

14.5 Parasitism in Cotton

Parasitism in cotton was difficult to measure since the whitefly population had dropped to low levels during the summers of 1998 and 1999 as growers began to use growth regulators to control their whitefly infestations. Whitefly-infested cotton leaves could not be collected prior to releases to determine movement of overwintering exotic parasitoids into this crop. Cotton samples from Imperial Valley (Chapters 8 and 13) show that both introduced exotic and native *Eretmocerus* readily attack *B. tabaci*. In general, *Eretmocerus* numbers on cotton leaves were correlated with numbers of *B. tabaci* on the sample leaves (Fig. 14.5), suggesting a density dependent relationship. Although these *Eretmocerus* were not identified to species, the extremely low numbers of parasitoids reported in an earlier survey (Godfrey et al. 1995), and the high proportion of exotic parasitoids found on weeds near citrus and cotton suggest many of these were probably exotic.

14.6 Summary

Citrus, combined with surrounding weeds, aided in the establishment of several imported parasitoids. At two of four release sites sampled in fall 2000 following the final releases, the majority of the parasitoids recovered from weeds in and adjacent to the citrus orchards were exotic. Although low numbers of parasitoids were found in citrus during this study, the high proportional levels of parasitism, and large number of affected trees, would suggest that several hundred thousand parasitoids might be produced in each of our release sites during the fall. Data showed that released parasitoids had a much higher affinity for attacking *B. tabaci* on weeds than did native *Eretmocerus*. In 2002, all exotic parasitoids emerging from B. *tabaci* collected from weeds were *Eret. mundus*. During surveys of cotton in the

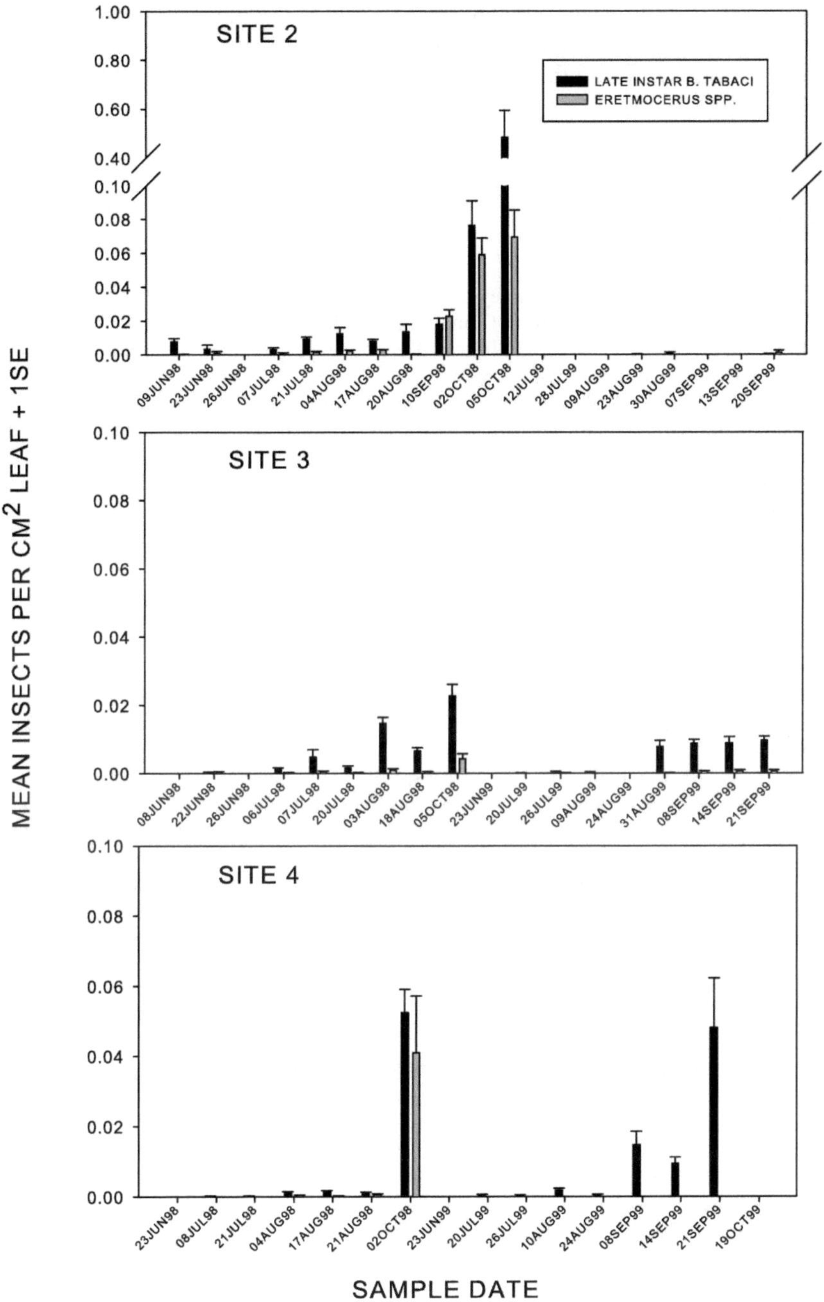

Fig. 14.5 Density of *B. tabaci* and *Eretmocerus* parasitoids on leaves of cotton adjacent to citrus release sites.

southern San Joaquin Valley conducted in the fall of 2004, *Eret. mundus* was found attacking *B. tabaci* on cotton leaves in 6 of 37 fields; no other exotic aphelinids were recovered. Of the adult male exotics recovered from cotton leaves, 83% were *Eret. mundus*.

References

Ball, J. and K.A. Casanave. 1994. Survey for resident parasitoids of the silverleaf whitefly and the release of natural enemies in Kern County. In L.G. Bezark, ed., Biological Control Program Annual Summary, 1993, 32 pp. California Department of Food and Agriculture, Division of Plant Industry, Sacramento, CA.

Bellows, T.S. Jr., and K. Arakawa. 1988. Dynamics of preimaginal populations of *Bemisia tabaci* (Homoptera: Aleyrodidae) and *Eretmocerus* sp. (Hymenoptera: Aphelinidae) in Southern California cotton. Environmental Entomology 17: 483–487.

CDFA, 2004. Pink Bollworm Program's September 2004 report to the Cotton Chronicle, a monthly newsletter for cotton ginners and growers, www.ccgga.org.

Gill, R.J. 1992. A review of the sweetpotato whitefly in southern California. Pan Pacific Entomologist 68: 144–152.

Godfrey, L., C. Goodell, C.G. Summers, W.J. Bentley, and T.M Perring. 1995. Seasonal development of silverleaf whitefly populations in cotton and other crops in the San Joaquin Valley, pp. 831–834. Proceedings of the Beltwide Cotton Conference. San Antonio, TX, 4–7 January 1995 National Cotton Council of America.

Goolsby, J.A., J. Crisostomo Legaspi, and B.C. Legaspi, Jr. 1996. Quarantine evaluation of exotic parasitoids of the sweetpotato whitefly, *Bemisia tabaci* (Gennadius). Southwestern Entomologist 21: 13–21.

Goolsby, J.A., M.A. Ciomperlik, B.C. Legaspi Jr., J.C. Legaspi, and L.E. Wendel. 1998. Laboratory and field evaluation of exotic parasitoids of *Bemisia tabaci* (Gennadius) (Biotype B) (Homoptera: Aleyrodidae) in the Lower Rio Grande Valley of Texas. Biological Control 12: 127–135.

Goolsby, J.A., M.A. Ciomperlik, A. Kirk, W. Jones, B.C. Legaspi, R. Ruiz, D.C. Vacek, and L.E. Wendel. 1999. Importation, evaluation, and colonization of parasitic Hymenoptera attacking *Bemisia tabaci* (Genn.) (Biotype "B"); integration of morphologically-based systematics and molecular methods in support of an action program. In A. Austin and M. Dowton, eds., Hymenoptera: Evolution, Biodiversity, and Biological Control, pp. 347–358. Proceedings of the 4th International Hymenopterists Conference, 6–11 January 1999, Canberra ACT, Australia. CSIRO, Collingwood, Victoria.

Gould, J.R., T.S. Bellows, Jr., and T.D. Paine. 1995. Preimaginal development, adult longevity and fecundity of *Encarsia inaron* (Hym.: Aphelinidae) parasitizing *Siphoninus phillyreae* (Hom.: Aleyrodidae) in California. Entomophaga 40: 55–68.

Hoelmer, K.A., L.S. Osborne, F.D. Bennett, and R.K. Yokomi. 1994. Biological control of sweetpotato whitefly in Florida. In D. Rosen, F.D. Bennett, and J.L. Capinera, eds., Pest Management in the Subtropics: Biological Control- a Florida Perspective, pp. 101–113. Intercept, P.O. Box 716, Andover, Hants, SP10 1 YG, UK.

Keaveny, D. 2000. Silverleaf Whitefly Report for the San Joaquin Valley. California Department of Food & Agriculture, Pink Bollworm Program.

Legaspi, J.C., B.C. Legaspi, Jr., R.I. Carruthers, J. Goolsby, W.A. Jones, A.A. Kirk, C. Moomaw, T.J. Poprawski, R.A. Ruiz, N.S. Talekar, and D. Vacek. 1996. Foreign exploration for natural enemies of silverleaf whitefly *Bemisia tabaci*, from Southeast Asia. Subtropical Plant Science 48: 43–48.

Nuessly, G.S. 1990. Biological control of the sweetpotato whitefly in California. In R. Yokomi, K. Narayanan, and D. Schuster, eds., Sweetpotato Whitefly-Mediated Vegetable Disorders in Florida, pp. 81–82. Proceedings of the University of Florida IFAS Workshop, Homestead, Florida.

Pickett, Charles H., Joe C. Ball, Kathleen C. Casanave, Karen Klonsky, Karen Jetter, Larry Bezark, and Steve E. Schoenig. 1996. Establishment of the ash whitefly parasitoid, *Encarsia inaron* (Walker) and its economic benefit to ornamental street trees in California. Biological Control 6: 260–272.

Pickett, C.H., J.A. Brown, H. Kumar, and D.A. Mayhew. 1999. Host feeding by *Eretmocerus* spp. on citrus, In Dale M. Woods, ed., Biological Control Program Annual Summary, 1998, 89 pp. California Department of Food and Agriculture, Plant Health and Pest Prevention Services, CA.

Powell, D.A. and T.S. Bellows, Jr. 1992. Preimaginal development and survival of *Bemisia tabaci* on cotton and cucumber. Environmental Entomologist 21: 359–363.

Roltsch, W.J. 1999. Perennial plant refuges for silverleaf whitefly natural enemy conservation in Imperial Valley, California. In T.J. Henneberry and R.M. Faust, eds., Silverleaf Whitefly, National Research, Action, and Technology Transfer Plan, 1997–2001. Second Annual Review of the Second 5-year Plan. US Department of Agriculture, Agricultural Research Service 1999–01, 85 p.

Rose, M. and G. Zolnerowich. 1997. *Eretmocerus* Haldeman (Hymenoptera: Aphelinidae) in the United States, with descriptions of new species attacking *Bemisia* (*tabaci* complex) (Homoptera: Aleyrodidae). Proceedings of the Entomological Society of Washington 99: 1–27.

Wilhoit, L., S. Schoenig, D. Supkoff, and B. Johnson. 1994. A regional simulation model for silverleaf whitefly in the Imperial Valley. Pest Management Analysis and Planning Program, State of California Department of Pesticide Regulation. PM 94–01.

Zolnerowich, G. and M. Rose. 1998. *Eretmocerus* Haldeman (Hymenoptera: Aphelinidae) imported and released in the United States for control of *Bemisia* (*tabaci* complex) (Homoptera: Aleyrodidae). Proceedings of the Entomological Society of Washington 100: 310–323.

Chapter 15
Habitat Management for the Establishment of *Bemisia* Natural Enemies

William J. Roltsch[1], Charles H. Pickett[1], Gregory S. Simmons[2], and Kim A. Hoelmer[3]

Abstract Challenges to the establishment of newly introduced parasitoids of *Bemisia tabaci* biotype B in the desert southwestern USA were presented by the severe climate of the region, the short-lived presence of most of its host plants, and the high mobility of the pest. To increase the chances of establishing new parasitoids capable of surviving in this region, three strategies were employed: (1) multiple releases of large numbers of exotic parasitoid species throughout the growing season, (2) creation of short-term and permanent refuges using a mixture of annual plants, and (3) managed releases in home gardens and backyard vegetation in urban settings. Long-term insectary habitat provided the opportunity to observe parasitoids over a wide range of seasonally related changes, such as weather and host plant. Continuous plantings of annual whitefly hosts were grown in field plots over successive years to provide a year-round habitat for newly introduced parasitoid species in the desert environment of Imperial Valley, California. Collard plants served as a bridge host during the winter season. Two introduced species, *Eretmocerus emiratus* and *Eret.* sp. nr. *emiratus,* represented over 50% of the *Eretmocerus* attacking *Bemisia* in the field insectaries in cultivated field insectaries from 1997 to 2000. *Encarsia sophia* from Pakistan also became established after its release in 1997. In contrast, summer–fall field insectaries promoted the mass production of several introduced parasitoids. Short-term plots of okra and basil, selected for high rates of parasitism sustained by *B. tabaci,* were planted along the edges of commercial fields to produce large numbers of parasitoids from April to November. Although this conservation approach only functioned for a portion of the year, millions of parasitoids were produced at several locations in Imperial Valley and Palo Verde Valley in California. Home sites containing ornamental host plants of *B. tabaci* also provided perennial habitats for establishment of exotic parasitoids. Selected commercial crops were also utilized as field insectaries. From 1994 to

[1]California Department of Food & Agriculture, Biological Control Program, 3288 Meadowview Rd., Sacramento, CA 95832

[2]USDA-APHIS-PPQ, Center for Plant Health Science and Technology Laboratory, 3645 E. Wier Avenue, Phoenix, AZ 85040

[3]USDA-ARS Beneficial Insect Introduction Research Unit, 501 S. Chapel St., Newark, DE 19713

J. Gould et al. (eds.), *Classical Biological Control of* Bemisia tabaci *in the United States.*

1999, large-scale releases of exotic parasitoids were made in spring melons. The fields effectively served as insectaries where large populations of exotic parasitoids developed, greatly enhancing the probability of new species establishing regionally. In southwestern desert valleys alfalfa comprises a large proportion of irrigated acreage and represents a significant percentage of host crops impacted by *B. tabaci* and a potentially large reservoir for natural enemies of *B. tabaci*. Frequent cutting removes a large part of whitefly populations but substantial amounts of alfalfa along section borders, in furrows and along field edges escape cutting, and parasitism of *B. tabaci* in this alfalfa was significant. Releases were also made in alfalfa fields located on a National Wildlife Refuge. In the San Joaquin Valley of central California, citrus foliage serves as an overwintering refuge for *B. tabaci*, which colonizes during the fall months; imported parasitoids were colonized in citrus to take advantage of this fall–winter habitat.

15.1 Introduction

Conservation biological control represents the "manipulation of the environment to favor natural enemies; either by removing or mitigating adverse factors or by providing lacking requisites" (DeBach 1974). This includes isolation from insecticide treatments and dust, and/or improved availability of hosts, foods, and overwintering sites (DeBach 1974; Dyer and Landis 1996; Bugg and Pickett 1998; Coll 1998). While such variables are of key importance for the performance of a native or resident species, they also play a key role in the establishment of a newly introduced species.

Considerable challenges exist in the establishment of newly introduced parasitoids against *Bemisia tabaci* biotype B (= *B. argentifolii* Bellows and Perring). Agricultural areas of the southwestern USA are characterized by a desert climate with high summer temperatures, low rainfall, occasional winter frost and frequent low humidity, which could impact establishment and the effectiveness of new species. A further challenge to whitefly control by parasitoids is that many cultivated and wild *Bemisia* host plants in desert areas are short-lived annual species, and whiteflies are highly mobile (Byrne et al. 1996; Birdsall 2002). Therefore, parasitoids must be able to search for hosts and reproduce on many different ephemeral plants. Parasitoid performance may vary markedly among *Bemisia* host plants (Simmons et al. 2002). Such habitat constraints limit the number of generations and number of parasitoids per generation that can be produced, and thus their impact on *B. tabaci*, with the result that establishment and effective regional biological control are impacted. Despite the availability of perennial host plant species such as *Citrus* spp., alfalfa (*Medicago sativa* L.), and ornamental species (largely in urban communities) including mulberry (*Morus alba* L.), hibiscus (*Hibiscus rosa-sinensis* L.), orchid tree (*Bauhinia* sp.), and roses (*Rosa* spp.), the portion of the year these perennial species support high populations of whiteflies is usually a fraction of the entire growing season, as illustrated by data collected from hibiscus (Fig. 15.1., WJR personal observation). This restricts the time available for parasitoids to reproduce at a location before having to disperse in search of other favorable habitats. Furthermore, newly introduced parasitoid species often have to search for plant species with which they have little or no previous historical association. Such circumstances create significant temporal and spatial challenges for *Bemisia* natural enemies, forcing their movement between a variety of plants and habitats each year.

An additional obstacle in the establishment of new species may include competition with native natural enemy species. For instance, very high levels of parasitism by native *Eretmocerus eremicus* Rose and Zolnerowich were recorded repeatedly in the late fall of each year on cotton and several ornamental species (Bellows and Arakawa 1988, authors' unpublished data). This occurs at a time when *Bemisia* populations are declining on most ornamental plants in residential areas, following peak activity in late summer and early fall. These events conceivably provide considerable competition between newly introduced species and native species whose densities are peaking.

To maximize the opportunity to establish parasitoids capable of surviving in the US desert southwest, three strategies were employed: (1) multiple releases of large

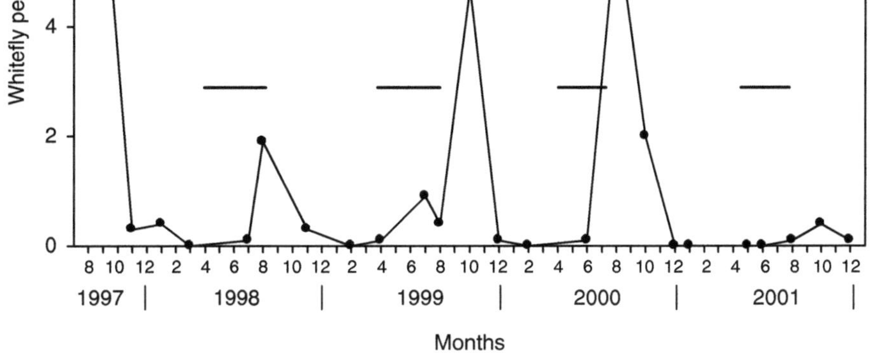

Fig. 15.1 Densities of apparently unparasitized fourth instar whitefly over successive years on hibiscus at a home site in Brawley, California. Whiteflies containing late stage parasitoid larvae or pupae were omitted.

numbers of exotic parasitoid species throughout the growing season, (2) planting a mixture of annual plants to create short-term and permanent refuges, and (3) using home gardens and enhancing backyard vegetation in urban settings. Commercial crops were used to aid in the regional colonization of introduced parasitoids.

15.2 Long-Term Insectary Habitats

Continuous plantings of annual hosts that support whitefly populations were grown in field plots to provide a year-round whitefly habitat over successive years for newly introduced parasitoid species in the desert environment of Imperial Valley. Field sites were located at the Imperial Valley Research Center (IVRC) in Brawley, California (formerly the USDA-ARS Irrigated Desert Research Station) (1993–2002) at the northern end of Imperial Valley, and at an organic farm (1993 to spring of 2000) at the south end of the valley. Each insectary plot was 0.2–0.5 ha in size. In addition to plants listed in Table 15.1, the plot included two to four beds of cantaloupe (*Cucumis melo* L. 'Topnet') grown in the spring, and six to twelve beds of cotton (*Gossypium hirsutum* L. 'Delta Pine 5415') grown from March to September.

Table 15.1 Traits of several *Bemisia tabaci* biotype B host plants relative to their use in insectary plantings in the Imperial Valley of California.

Plant	Growing period (months)	Whitefly (months)	Comments
Collard 'Champion'/Broccoli 'Italian Green Sprouting', *Brassica oleracea* L.	Sept.–Jul.	Oct.–Jul.	Easy to grow, potential for high whitefly densities in late spring, variable parasitization of whitefly
Sunflower, *Helianthus annuus* 'Giant Gray Stripe' and numbered oil seed variety	Sept.–Jan.	Oct.–Jan.	Powdery mildew prone, variable parasitization of whitefly
Okra, *Abelmochus esculentus* (L.) 'Clemson Spineless'	Mar.–Nov.	Jun.–Nov.	Easy to grow, moderate to high whitefly densities, consistently high % parasitization
Sweet basil, *Ocimum basilicum* L. 'Italian Large Leaf'	Mar.–Nov.	Jul.–Nov.	Easy to grow, low densities of whitefly, high % parasitization
Eggplant, *Solanum melongena* L. 'Black Beauty'	Mar.–Nov	Jun.–Oct.	Unreliable growth, variable parasitization, potential for high whitefly densities
Kenaf, *Hibiscus cannabinus* 'Cubano'	Mar.–Nov.	Jun.–Oct.	Easy to grow, high % parasitization, moderate to high whitefly densities, destruction is difficult
Roselle, *Hibiscus sabdariffa* L. 'Sabdariffa'	Mar.–Nov.	Jul.–Oct.	Easy to grow, high % parasitization, low whitefly densities

15.2.1 Plantings

Two beds of collards (*Brassica oleracea* L. Acephala group) and four to eight beds of sunflowers (*Helianthus annuus* L.) were planted in late September to October. The destruction of young stands of these insectary crops by high densities of whitefly in early fall and sporadic damage by birds and rodents often led to replanting. Sunflower survival was variable between years due to powdery mildew (*Erysiphe cichoracearum* DC) and frost. Four to eight beds of okra (*Abelmoschus esculentus* (L.)) and two to eight beds of basil (*Ocimum basilicum* L.) were planted in March; both established relatively easily. Eggplant (*Solanum melongena* L.) was grown for 2 years, but was discontinued because stand establishment was highly inconsistent. The fiber crop kenaf, *Hibiscus cannabinus* L., was easy to grow; however it was grown for only 3 years because the fibrous nature of the plant made crop destruction difficult.

15.2.2 Parasitoid Release/Inoculation Methods

Parasitoids were released into plots using several techniques. Initially, parasitoids were released as adults. Subsequently, production and handling of parasitoids were simplified by releasing pupae in one of two ways. Parasitoids were commonly released by transplanting potted plants (greenhouse grown hibiscus and eggplant) containing several thousand parasitoid pupae per plant into the field insectary plots or released into the permanent insectary plots on excised leaves. For most species, tens of thousands were released into each plot on multiple occasions; these releases ended by the fall of 1997. No additional exotic parasitoids were released in close proximity to the refuge plots so that post-release population dynamics and survivorship could be monitored. Release procedures and number released by species throughout the valley are further explained in Chapter 13.

15.2.3 Monitoring

Plots were sampled throughout the year to monitor *Bemisia* and parasitoid populations (see Chapter 13 for further details). Additional sampling was employed to estimate exotic parasitoid production numbers specifically in refuge plantings, which included estimating of the number of okra plants in a plot, the number of whitefly and parasitoid infested leaves per plant, the number of parasitoid pupae per leaf, and the proportion of total *Eretmocerus* comprised of exotic species. The number of okra plants was estimated in each plot by counting the number of plants in one meter per row in each of three rows and multiplying by the number of row-meters in the plot. The mean number of infested leaves per plant and parasitoids per leaf was estimated from 15 randomly selected plants. The numbers of *Eretmocerus* and *Encarsia sophia* (Girault and Dodd) pupae on these leaves were counted in the laboratory with the aid of a stereomicroscope and the adult native and exotic *Eretmocerus* were separated. The relative proportion of *Eretmocerus* species on a leaf was determined by placing 50 *Eretmocerus* pupae (several from each leaf) into a 0.5-dram glass vial with a cotton plug. Emerged adult *Eretmocerus* were sorted and identified as native *Eret. eremicus* or exotic *Eretmocerus* spp. *Encarsia sophia* could be identified to species in the pupal stage due to a uniquely identifiable black pupal case and meconia placement that differs from other exotic and native *Encarsia*. From these data, the total number of exotic *Eretmocerus* and *Enc. sophia* pupae were estimated for a given day. Estimates of total daily adult parasitoid production were made by assuming that a parasitoid spends an average of 5 days in the pupal stage during late summer and fall temperatures (Greenberg et al. 2000, authors' unpublished data). Therefore, daily production was tabulated by dividing the total number of estimated exotic *Eretmocerus* and *Enc. sophia* pupae by five. Estimates of parasitoid production on basil were not made.

15.2.4 Parasitoid Production

Several species of exotic *Eretmocerus* were common or abundant in the continually maintained field insectaries from 1997 to 2000 (Chapter 13). *Eretmocerus mundus* Mercet was found in low numbers, whereas two other exotic species, *Eret. emiratus* Zolnerowich and Rose (APHIS Mission Biological Control Laboratory culture number M95104, *ex* United Arab Emirates) and *Eret.* sp. nr. *emiratus* (M95076 *ex* Ethiopia), were found in higher numbers. Combined, these latter two species represented well over 50% of the *Eretmocerus* population attacking *Bemisia* in the field insectaries. Of these two very closely related species, DNA analysis indicated that *Eret.* sp. nr. *emiratus* was dominant. DNA analysis could not distinguish between the different geographic populations of *Eret. mundus* that were released in the Imperial Valley (M92014 *ex* Spain, M92019 *ex* India, and M94120 *ex* Israel). *Encarsia sophia* (M95107) from Multan, Pakistan also became firmly established after its initial release in 1997. Through fall 2000, these insectary gardens have annually contributed millions of these new parasitoids to Imperial Valley (Table 15.2). Despite having been released in large numbers (see Chapter 13) on multiple occasions, other exotic *Eretmocerus* and *Encarsia* species were either never recovered, or were present for only brief periods of time, i.e., several months or less. The potential for parasitoid production in such insectary sites is illustrated in Table 15.2. It is estimated that several hundred thousand adult parasitoids emerged each day on okra from late summer well into fall (Table 15.2, Fig. 15.2a). Although it is suggested by Fig. 15.2b that whitefly and parasitoid productions on basil may also be considerable, this was not the case because fewer leaves per basil plant contained the densities that occurred on okra leaves, and basil leaves are much smaller than okra leaves (8 cm^2 vs. 100 cm^2).

Table 15.2 Estimated number of adult exotic parasitoids produced on okra in a 0.04 ha permanent field insectary at IVRC, Brawley, California (field M). There were 2,156 okra plants in each plot and eight leaves per plant and 15 plants were sampled.

	Mean		
Date of sampling	Pupae per leaf	Pupae per plot	Daily production per plot[b]
	Eretmocerus spp.		
4 Sept. 1997	67	1,155,616[a]	205,700
16 Sept. 1997	92	1,586,816	–
24 Sept. 1997	90	1,552,320	–
	Encarsia sophia		
4 Sept. 1997	117	2,018,016	403,603
16 Sept. 1997	94	1,621,312	324,262
24 Sept. 1997	88	1,517,824	303,565

[a]Eighty-nine percent of all *Eretmocerus* collected on this date were exotic species. The proportion of exotic to native species on the following two dates was not determined.
[b]Estimate of exotic adult parasitoid species assuming a 5-day pupal developmental period. Daily *Eretmocerus* production = [*Eret.* pupae per plot]*[% exotic *Er.*/100]/5 day development period. Daily *Enc. sophia* production = [*Enc. sophia* per plot/5 days].

15.3 Summer–Fall Insectary Habitats, 1998 and 1999

Based on the results from the long-term insectary gardens in 1997, okra and basil were used in single season plantings in 1998 and 1999 to further the establishment and dispersal of exotic parasitoids in the region. The best parasitoid production period was from July through early November, with peak production coinciding with regional whitefly peak activity in August and September. A two-plant system was developed to facilitate high parasitoid production levels and minimize risk of *Bemisia* outbreaks. Okra was selected because it was easy to grow and consistently supported large numbers of *Bemisia* with high rates of parasitism (i.e., 50–90% parasitism of 4th instar nymphs) by *Eretmocerus* and *Encarsia* species. Basil supported lower numbers of whitefly in the Imperial Valley, but most were heavily parasitized. Had the okra become heavily infested and contained few parasitized whitefly, it would have been destroyed, whereas basil would have been retained, thereby allowing for the preservation of a portion of a refuge's parasitoid production. The destruction of okra due to excessively high densities of healthy whitefly was never necessary.

15.3.1 Plantings

Arrangements were made in 1998 and 1999 with several local growers to plant and maintain rows of okra and basil along the edge of several commercial fields from April to November. For each year, two plots were grown in Imperial Valley, one

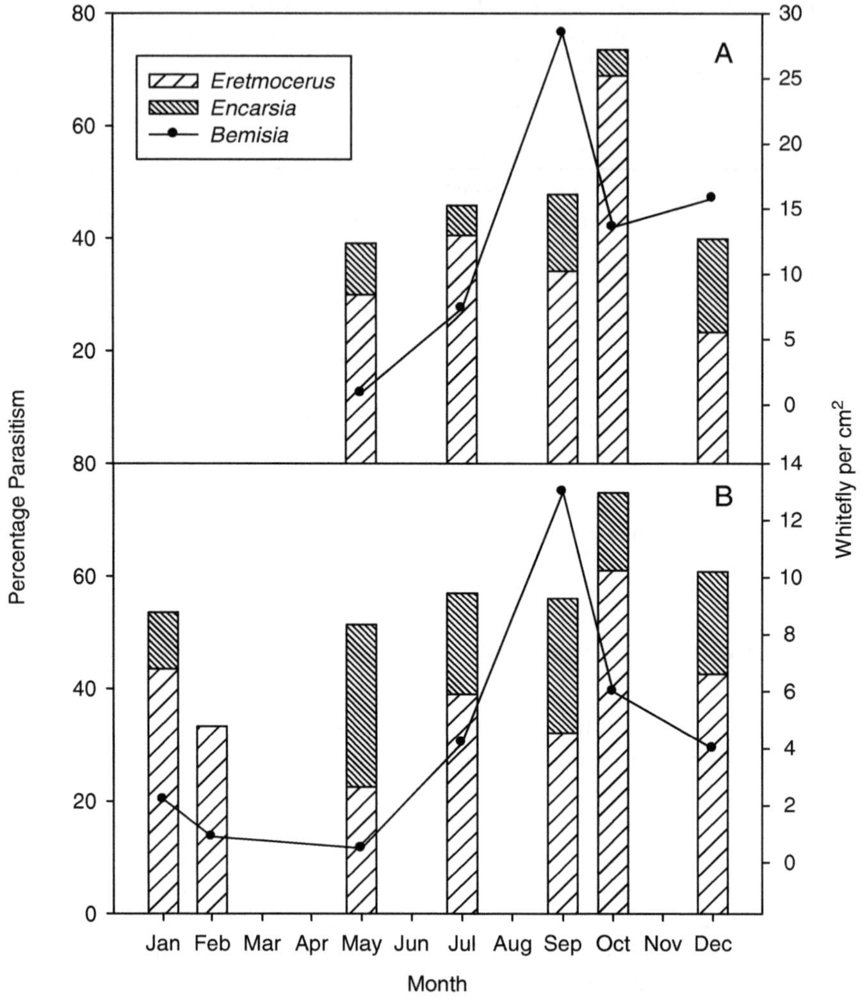

Fig. 15.2 Whitefly density and parasitism patterns on insectary okra (a) and basil (b) in 1997 at the IVRC, Brawley, California.

each in the north and south ends of the valley. Another two plots were grown in Palo Verde Valley, approximately 65 km northeast of Imperial Valley. Plots consisted of four to eight beds (approximately 1 m between-row spacing) each of okra and basil grown from seed, representing a total plot size of 0.3–0.6 ha. Planting took place in late March or April. In some instances, the plantings were on the edge of cotton or alfalfa fields, both host plants of *Bemisia*. Plantings were also made adjacent to fields planted with Sudan grass (*Sorghum bicolor* L.). Although Sudan grass is not a *Bemisia* host, *Bemisia* host plants such as alfalfa and cantaloupe were located within several kilometers. Insectary plants were never treated with insecticides.

15.3.2 Parasitoid Release/Inoculation and Sample Methods

Once the okra and basil became lightly infested with whitefly (typically in late June), the refuge sites were inoculated with selected parasitoid species produced in a USDA or CDFA greenhouse insectary. The goal was to release 100,000 parasitoids per species into each plot. It was assumed that on average, a female parasitoid would produce at least 20 offspring (Chapter 8). Release of large numbers of parasitoids enabled a rapid increase in parasitoid population density, accomplishing the main goal of massive parasitoid production while minimizing the chance of a whitefly outbreak. In practice, over 200,000 parasitoid pupae of each of two to three exotic species were usually released at each site. Species included: *Eret. mundus*, *Eret. emiratus*, *Eret.* sp. nr. *emiratus*, and *Enc. sophia*. Releases commenced when third instar whitefly density was 20 or more per leaf. Studies of *Eretmocerus* and *Encarsia* survivorship and host stage preference indicated the former preferred second and third instars whereas the latter preferred third and early fourth instars (Headrick et al. 1996; Ball 1998; Antony et al. 2003).

15.3.3 Production Results

In 1998, one site in the Imperial Valley (north Imperial Valley site) and one site in Palo Verde Valley represented healthy parasitoid insectaries. The other two sites were poorly maintained and few exotic parasitoids were produced. For example, the southern Imperial Valley planting was grown on land previously planted with Sudan grass, some of which persisted, competing with okra and basil, resulting in small and chlorotic plants. From a fall sample obtained at this location, none of the 38 male parasitoid specimens identified were exotic. Overall, two of the four fields in 1998 and three of four fields in 1999 produced considerable numbers of exotic parasitoids. The southern Imperial Valley site in 1999 was moved approximately 2 miles away from the 1998 location, next to an alfalfa field. The quality of the insectary was greatly improved. The species composition 1-month post release contained 25% exotic parasitoid species. DNA testing on several exotic *Eretmocerus* specimens showed they were *Eret. emiratus* and *Eret.* sp nr. *emiratus*, which was consistent with samples taken from many other field and urban sites throughout Imperial Valley during this time.

In late summer and fall of 1998, samples were collected at the northern Imperial Valley site to assess parasitoid densities. Estimated numbers of adult parasitoids emerging were very high, representing millions over 3–4 months, a significant portion of the season (Table 15.3). The estimation of daily adult parasitoid production was based on an approximate pupal duration of 5 days during this time of year. Similar estimates were not tabulated for basil, but would be lower because *Bemisia* densities were only a fraction of those found on okra. Although okra supported high densities of whiteflies (Fig. 15.2), the

Table 15.3 Estimates of adult exotic parasitoid production from okra in a 0.11 ha summer to fall (north Imperial Valley site) insectary garden with 3,000 plants.

Date	Leaves per plant	Mean		
		Pupae per leaf	Pupae per plot (% exotic)[a]	Daily production[b] of exotic adults
		Eretmocerus spp.		
28 Aug. 1998	53	44	6,996,000 (53%)	741,576
6 Nov. 1998	12	16	576,560 (55%)	63,630
		Encarsia sophia		
28 Aug. 1998	53	3.7	588,300	117,660
6 Nov. 1998	12	36	1,296,000	259,200

[a]Percentage of all *Eretmocerus* that were exotic species versus native *Eret. eremicus*.
[b]Estimate of exotic adult parasitoid species assuming a 5-day pupal developmental period. Daily *Eretmocerus* production = [*Eret.* pupae per plot]∗[% exotic *Er.*/100]/5-day development period. Daily *Encarsia sophia* production = [*En. sophia* per plot/5 days].

insectary was not considered a source of infestation due to high levels of parasitism on this plant species. Parasitism of fourth instar whitefly on okra was typically well over 50%, commonly exceeding 80%. For example, at the northern Imperial Valley site on 28 August of 1998, we determined that fourth instar whitefly survivorship was approximately 15%, parasitism by *Eretmocerus* was 58%, parasitism by *Encarsia* was 4%, predation was 17.5% and unknown mortality was 5%, based on counts of field-collected leaves using methods described in Chapter 13.

Although this conservation approach only functioned for a portion of the year (i.e., July–November), these gardens produced large numbers of parasitoids for given whitefly densities. In comparison, the permanent insectaries used collard plants during the winter season as a bridge crop, which often produced numerous whiteflies that were not heavily parasitized from mid-spring onward. Other cold tolerant whitefly host plants were sought, but none were found to replace cole crop species during the cooler seasons.

15.4 Home Gardens

The use of urban locations for establishing exotic parasitoid species was a high priority because home sites provided perennial habitats with many evergreen tree and shrub ornamental species such as hibiscus, mulberry, orchid tree, roses, fig (*Ficus carica* L.), cape honeysuckle (*Tecomara capensis* Thunb.), lantana (*Lantana camara* L.) and snail vine (*Vigna caracalla* L.) that served as whitefly hosts. Compared to annual agricultural crops, these represented a temporally stable long-term habitat where parasitoids were provided winter habitat and food, free of insecticides. During the first years of the project, sampling of such perennials as hibiscus and mulberry revealed that perennials were typically not subject to *Bemisia* attack throughout their entire growing season (Fig.15.1). For instance, *Bemisia* densities on hibiscus were low in the spring and early summer, increasing markedly in the summer and fall. To provide additional habitat for introduced parasitoids, the home gardener project provided various other *Bemisia* host plants, most of which were common vegetable garden plants.

15.4.1 Methods

In 1996 an advertisement was placed in the local newspaper requesting participation by home owners who were interested in obtaining new parasitoid species and who could meet several requirements, including no insecticide use, minimal pruning, year-round plant propagation, and access to their property. Participating homeowners were provided with transplants in the spring, including okra, eggplant, basil, roselle, and collards or broccoli for placement in their garden. In 1997 and 1998 large pots were provided to homeowners to further ensure the propagation of transplants. Cellulose fiber pots (90 cm diameter × 60 cm height) were planted with one broccoli plant, one collard, two basil (Italian broad leaf), two okra, one rue (*Ruta graveolens* L.), and one chuparosa (*Justicia californica* Benth.). *Bemisia* and its parasitoids could be supported throughout the entire year with this combination of plant species. The use of large container pots provided consistently high quality soil and plants that could not be confused with others grown by the homeowner. The mode of parasitoid release included transplanting greenhouse insectary-grown eggplant and hibiscus plants in 10 or 15 cm diameter pots containing parasitoid pupae into gardens or planters. At the time of transplanting, individual plants had as many as 4,000 parasitoid pupae; however 1,000 to 2,000 were more common.

15.4.2 Recoveries

From mid-October to early December1996, exotic parasitoids were detected at two of 10 home sites where parasitoids had been released (2% and 18% of the *Eretmocerus* were exotic, respectively). This corresponded closely with information indicating that the exotic species were showing signs of area-wide establishment at that time. During spring of 1997, exotic *Eretmocerus* were found in four samples collected at 12 home sites where they had been released in 1996. An average of 14% of the *Eretmocerus* found at those sites during the early stages of this classical biological control project was exotic. Home garden releases in Arizona were also found to be a successful means of establishing and evaluating exotic biological control agents (Chapter 12).

The use of large container pots proved ineffective in colonization of parasitoids compared to releases in existing home gardens. Some homeowners failed to follow watering instructions, while others were ambivalent about having the planter in their yard, leading to abandonment. Overall, few parasitoids were produced with this method. The utilization of such planters in a region with a more moderate climate, where these large pots would need less watering and the level of care would be lower, is worthy of consideration.

15.5 Alfalfa as a Reservoir

Alfalfa comprises a large proportion of irrigated acreage in many southwestern desert valleys, and represents a large percentage of all host crops impacted by *B. tabaci*. In Imperial County in 1996, 85% of the acreage of whitefly-host crops and over one-third of all production acreage was alfalfa (Birdsall 1992, 2002). It may serve as an important bridge crop between melons and cotton in the summer and fall–winter vegetable crops such as broccoli and cauliflower. Although whitefly populations only reached high numbers on alfalfa in late summer, when massive migrations from melons and cotton take place (Yee et al. 1997), parasitized fourth instar whiteflies were found in alfalfa samples in low numbers by late spring (authors' data). The substantial acreage of alfalfa therefore represents a potentially large reservoir for natural enemies of *B. tabaci*.

The frequent cutting cycles typically practiced in desert alfalfa grown for forage subjects high populations of whiteflies to removal (Yee and Toscano 1996). Cutting removes all stems and foliage at heights above 15.25 cm (and often shorter), although substantial amounts of alfalfa escape cutting along the edges of raised section borders, within furrows and along field edges (authors' observations). Surveys of commercial alfalfa fields and experimental plots were conducted in 1995–1997 to evaluate the role of alfalfa as a reservoir for parasitoids. The influence of cutting on the levels of parasitism and on the distribution of whiteflies and parasitoids on alfalfa above and below the cut level was examined in plots that were cut at different schedules, at 20–28 day intervals (standard commercial practice), at double the previous interval (40–56 days), and not cut at all (for seed production). In each cutting regime, samples of stems were clipped at ground level, and in the laboratory separate counts were made of whiteflies and parasitoids on stem portions below and above 15.25 cm.

Yee et al. (1997) reported that parasitism of whitefly in alfalfa was virtually absent (<1%) at their study sites, but results from the authors' surveys showed significant levels of parasitism (more than 10% at some sites) by late fall as whiteflies increased in abundance on alfalfa. Levels of parasitism remained high into early winter, probably because cold temperatures slowed growth rates of *B. tabaci* and parasitoids. From samples of parasitized whiteflies reared to emergence, most of the naturally occurring parasitism was due to the native *Eret. eremicus*, with *Enc. luteola* Howard, another native species, as the second most abundant parasitoid. Representative distribution data from one experimental site is given in Table 15.4.

Table 15.4 Distribution of *B. tabaci* and parasitoids in an experimental plot of alfalfa in Imperial Co. sampled during 1995–1996. The total numbers of apparently healthy 4th instar whitefly and parasitized 4ths (in parentheses) per 40-stem sample on each sample date are shown. Three cutting regimes are compared: regular cut (at intervals of every 20–28 days), alternate cut (intervals of 40–56 days), and no cut (alfalfa cut only once each season). Two tallies were made on each stem: from ground level to 15.25 cm, and above15.25 cm, meant to represent the portion of stems below the cutting level and the harvested portion above the cut. Not all regimes could be sampled during the same periods.

	Regular cut		Alternate cut		No cut	
Sample date(s)	0–15.25 cm	>15.25 cm	0–15.25 cm	>15.25 cm	0–15.25 cm	>15.25 cm
Sept. 1995	38 (3)	284 (15)	–	–	4 (0)	126 (8)
Nov. 1995	205 (20)	477 (72)	–	–	124 (32)	1,293 (353)
Dec. 1995–Jan. 1996	194 (19)	1,745 (308)	318 (53)	895 (215)	24 (3)	155 (47)
Feb.–Mar. 1996	–	–	100 (5)	433 (40)	5 (3)	37 (10)
May–June 1996	10 (6)	90 (56)	47 (21)	141 (82)	8 (3)	85 (32)
Aug. 1996	0 (0)	485 (2)	12 (0)	245 (10)	2 (1)	97 (2)
Sept. 1996	66 (1)	162 (5)	13 (1)	222 (10)	2 (0)	217 (36)

Despite the occurrence of the majority of *B. tabaci* on stems above the cutting level, it is evident that substantial parasitism may occur during at least a few months of the year in alfalfa cut on a normal rotation. On alfalfa stems with longer cutting intervals, or not cut at all, a larger proportion of parasitized whiteflies may remain until adult parasitoids can emerge.

Field cage evaluations at Brawley determined that exotic *Eretmocerus* and *Encarsia* species were also capable of successful development in *B. tabaci* reared on alfalfa (Chapter 8). An agreement was reached with the US Fish & Wildlife Service to allow parasitoid releases and surveys to monitor their establishment in alfalfa fields located on the Salton Sea National Wildlife Refuge. Since the refuge was maintained to provide seasonal food for migrating waterfowl, it was not treated with insecticides, except against alfalfa weevils in the windrows after cutting and raking. The refuge was surrounded on three sides by commercial agriculture and provided a good site for establishing potentially large populations of natural enemies that could migrate outwards. During summer and fall 1996, releases of 250,000–300,000 *Eret. mundus*, *Eret. hayati* Zolnerowich and Rose (M95012 *ex* Pakistan), *Eret. emiratus*, *Enc. bimaculata* Heraty and Polaszek, *Enc. sophia*, and *Enc.*sp. nr. *pergandiella* (M95055 *ex* Brazil) were made in alfalfa at several locations on the refuge. In 1997, further releases of 400,000 *Eretmocerus* (*emiratus, hayati* and *mundus*) were made. Parasitoids were released as pupae on leaves harvested from insectary eggplants (1996 only) or as free pupae placed in small plastic cups tied to plant stems. Follow-up surveys recovered very few introduced parasitoids from these alfalfa fields, although a number of recoveries were made at nearby locations.

15.6 Other Commercial Crops

From 1994 to 1999, large-scale releases of exotic parasitoids were made into fields of spring melons to develop and demonstrate the potential for seasonal inoculative biological control as part of the overall effort on whitefly biological control in the Imperial Valley (Chapter 16). A pesticide-free pollination period of several weeks coupled with the use of imidacloprid (which had little direct impact on parasitoids) provided favorable conditions for the establishment of large populations of parasitoids in the melon crop (Chapter 16). While the main purpose of this demonstration project was to test the efficacy of whitefly control by inoculative release of parasitoids, fields also served as insectaries where large populations of exotic parasitoids developed, greatly enhancing the probability of establishing new species.

In the San Joaquin Valley of central California, citrus was used for large-scale colonization of newly imported *Bemisia* parasitoids (Chapter 14). During the late 1990s *B. tabaci* invaded central California and high numbers were reported on citrus fall foliage. In addition to homeowners' yards in central California, this crop represented the most stable system for making releases in this vast agriculturally productive region of the state. From fall to early spring little if any insecticides are used in citrus (Pickett et al. unpublished data). Released parasitoids moved between citrus and nearby weedy host plants and have been recovered near the four citrus sites 3 years following their last release (Chapter 14, Pickett unpublished data).

15.7 Conclusions

Large releases of exotic parasitoid species and the use of multiple release strategies maximized the efforts to colonize newly imported parasitoids. The limited monitoring of pupal survivorship showed that parasitoids released as pupae were commonly subject to high levels of mortality in the desert climate of Imperial Valley. Releasing small numbers of parasitoids in only a few locations would have greatly diminished the likelihood of establishing new species, and increased the time required in their spread. The time spent coordinating activities with growers, homeowners and facility managers was considerable but it ensured proper habitat for releases. The usefulness of field insectaries to propagate parasitoids to facilitate their establishment was demonstrated. Long-term insectary habitat provided the opportunity to observe parasitoids over a wide range of seasonally related changes such as weather and flora. In contrast, summer–fall insectaries provided the opportunity to reliably and more easily promote the mass production of parasitoids at multiple locations.

References

Antony, B., M.S. Palaniswami, and T.J. Henneberry. 2003. *Encarsia transvena* (Hymenoptera: Aphelinidae) development on different *Bemisia tabaci* Gennadius (Homoptera: Aleyrodidae) instars. Environmental Entomology 32: 584–591.

Ball, J.C. 1998. Host stage preference and development of two exotic parasitoids of silverleaf whitefly. In M.S. Hoddle, ed., California Conference on Biological Control: Innovation in biological control research (Proceedings), pp. 131–133. June 10–11, 1998, University of California, Berkeley, CA.

Bellows, T.S., Jr. and K. Arakawa. 1988. Dynamics of preimaginal populations of *Bemisia tabaci* (Homoptera: Aleyrodidae) and *Eretmocerus* sp. (Hymenoptera: Aphelinidae) in Southern California cotton. Environmental Entomology 17: 483–487.

Birdsall, S.L. 1992. Imperial County Agriculture Crop & Livestock Report. Agricultural Commissioner's Office, Imperial County, California.

Birdsall, S.L. 2002. Imperial County Agriculture Crop & Livestock Report. Agricultural Commissioner's Office, Imperial County, California.

Bugg, R.L. and C.H. Pickett. 1998. Introduction: enhancing biological control. In C.H. Pickett and R.L. Bugg, eds., Enhancing Biological Control – Habitat Management to Promote Natural Enemies of Agricultural Pests pp. 1–15. University of California, Berkeley and Los Angeles.

Byrne, D.N., R.J. Rathman, T.V. Orum, and J.C. Palumbo. 1996. Localized migration and dispersal by the sweet potato whitefly, *Bemisia tabaci*. Oecologia 105: 320–328.

Coll, M. 1998. Parasitoid activity and plant species composition in intercropped systems. In C.H. Pickett and R.L. Bugg, eds., Enhancing Biological Control – Habitat Management to Promote Natural Enemies of Agricultural Pests, pp. 85–119. University of California, Berkeley and Los Angeles.

DeBach, 1974. Biological Control by Natural Enemies. p.216. Cambridge University Press, London, 333 pp.

Dyer, L.E. and D.A. Landis. 1996. Effects of habitat, temperature, and sugar availability on longevity of *Eriborus terebrans* (Hymenoptera: Ichneumonidae). Environmental Entomology 25: 1192–1201.

Greenberg, S.M., B.C. Legaspi, Jr., W.A. Jones, and A. Enkegaard. 2000. Temperature-dependent life history of *Eretmocerus eremicus* (Hymenoptera: Aphelinidae) on two whitefly hosts (Homoptera: Aleyrodidae). Environmental Entomology 29: 851–860.

Headrick, D.H., T.S. Bellows, Jr., and T.M. Perring. 1996. Behaviors of female *Eretmocerus* sp. nr. *californicus* (Hymenoptera: Aphelinidae) attacking *Bemisia argentifolii* (Homoptera: Aleyrodidae) on cotton, *Gossypium hirsutum* (Malavaceae) and melon, *Cucumis melo* (Cucurbitacea). Biological Control 6: 64–75.

Simmons, A.M., S. Abd-Rabou, and G.S. McCutcheon. 2002. Incidence of parasitoids and parasitism of *Bemisia tabaci* (Homoptera: Aleyrodidae) in numerous crops. Environmental Entomology 31: 1030–1036.

Yee, W.L. and N.C. Toscano. 1996. Ovipositional preference and development of *Bemisia argentifolii* (Homoptera: Aleyrodidae) in relation to alfalfa. Journal of Economic Entomology 89: 870–876.

Yee, W.L., N.C. Toscano, J.C. Palumbo, M.J. Blua, and H.A. Yoshida. 1997. Seasonal population trends of *Bemisia argentifolii* (Homoptera: Aleyrodidae) on alfalfa in southern California and Arizona. Environmental Entomology 26: 241–249.

Chapter 16
Integrating Parasitoid Releases with Traditional Control Methodologies: Experience in the Spring Melon Production System in the Southwestern USA

G. Simmons[1], K. Hoelmer[2], and Eric Natwick[3]

Abstract During the outbreak of *Bemisia tabaci* biotype B during the 1990s in the southwestern USA efforts were made to integrate parasitoid releases with existing chemical control strategies in concert with a classical biological control program. Spring melons grown in the desert vegetable producing areas of the southwestern USA were among those crops most severely affected by whitefly infestations. High rates of parasitism of *B. tabaci* by native parasitoids were observed in melons near the end of the melon production cycle, therefore demonstration projects were conducted in the Imperial Valley of California during 1994–1998 to inoculate these crops with introduced parasitoids earlier in the season in order to increase rates of parasitism. Since spring melon fields would act as field insectaries this had the added benefit of increasing the chances of permanently establishing newly imported species of parasitoids. The project took advantage of mass production systems developed for the importation and establishment of new species of *Eretmocerus* to provide the numbers needed for large, replicated field releases. Initial tests compared the native *Eret. eremicus* with two populations of introduced *Eret. mundus* and showed that *Eret. mundus* was more effective and increased parasitism five to sixfold. Subsequent field trials utilized *Eret. hayati,* introduced from Pakistan, *Eret. emiratus* from the United Arab Emirates, and *Eret.* sp. nr. *emiratus* from Ethiopia. Various parasitoid release strategies were tested in melons grown using conventional pest control methods relying on broad spectrum insecticides, in imidacloprid treated fields, and in fields where both imidacloprid and broad spectrum pesticides were used. Releases of *Eretmocerus* were shown to be compatible with the use of imidacloprid. By comparing whitefly densities in nearby untreated fields to imidacloprid treated fields, these

[1] USDA-APHIS-PPQ Center for Plant Health Science and Technology Laboratory, 3645 East Wier Avenue, Phoenix, AZ 85040

[2] USDA-ARS Beneficial Insects Introduction Research Unit, 501 South Chapel Street, Newark, DE, 19713-3814

[3] ANR UCCE, University of California, Desert Research & Extension Center, 1050 East Holton Rd., Holtville, CA 92250-9615

J. Gould et al. (eds.), *Classical Biological Control of* Bemisia tabaci *in the United States.*

studies showed that it was possible to reduce the number of whitefly by a factor ranging from one half to 300-fold. Estimates of the number of parasitoids needed to increase parasitism by 50% or greater and to reduce whitefly densities by as much as one half over several seasons ranged from 49,400 to 197,600/ha. Releasing parasitoids in organic fields early in the season when whitefly density was low also provided significant reductions in whitefly numbers. Emergence rates using pupae released in cups placed among foliage vs. pupae mixed with vermiculite and spread mechanically by tractor with a drop-box were similar and ranged from 55 to 82%.

16.1 Introduction

During the initial stages of the outbreak of *Bemisia tabaci* (Genn.) biotype B (= *B. argentifolii* Bellows and Perring) in the 1990s in the southwestern USA, efforts were made to integrate parasitoid releases with existing chemical control strategies as part of a program to import and establish new species of parasitoids. The most severely infested areas were places where winter and early spring vegetables and melons were grown, providing year-round production of whitefly susceptible crops that allowed high whitefly populations to persist through most of the year (Chu et al. 2001; Ellsworth and Martinez-Carrillo 2001; Palumbo et al. 2001). Because high rates of parasitism of *Bemisia* by native parasitoids were observed on certain crops near the end of the production cycle (Gerling 1967; Bellows and Arakawa 1988) it was reasoned that inoculating susceptible crops with parasitoids early in the season could lead to earlier and increased parasitism providing greater natural control. A second advantage of this strategy was that increased parasitism would result in reduced migration of whitefly to successive crops, a key requirement of reducing the season-long regional whitefly population. Lastly, this strategy had the added benefit of increasing the probability of permanent establishment of the newly imported species of parasitoids by making use of spring melon fields as large field insectaries.

This project took advantage of the mass production systems developed for the importation and establishment of new species of whitefly parasitoids (Chapter 10) to provide the high numbers of parasitoids needed for large scale, replicated field releases. Both the native *Eretmocerus eremicus* Rose and Zolnerowich and several exotic species of *Eretmocerus* were used in releases (Goolsby et al. 1998). *Eretmocerus eremicus,* the dominant parasitoid in Southwestern low desert agroecosystems, did not achieve high rates of parasitism on the new whitefly biotype infesting winter cole crops and spring melons. It was reasoned that an exotic with similar biology but more adapted to the new whitefly would be a more effective parasitoid. When initial studies showed that many of the exotic species produced greater numbers of progeny and achieved higher levels of parasitism than the native *Eret. eremicus* (Hoelmer 1998, 2007, Chapter 8) continued studies were performed using only exotic species. *Eretmocerus* species were chosen for these studies because the exotic

Encarsia species exhibited male hyperparasitism due to their autoparasitic life history, which greatly complicated mass rearing procedures.

For a parasitoid release strategy to be implemented in commercial crops, it would need to be integrated with existing control practices, primarily pesticide usage. Because of the severity of the whitefly infestations, several new and more selective pesticides were introduced, often under emergency registrations. While the general use of pesticides may be viewed as an obstacle to using augmentative or inoculative releases of parasitoids, the use of selective pesticides could provide an opportunity to improve the efficacy of biological control by reducing whitefly populations with minimal harm to the parasitoid and permitting lower release rates. The increasing economic importance of organic melon production provided further motivation to develop an augmentative biological strategy for whitefly in melons, as organic growers had few effective tools for whitefly control.

This chapter reviews conventional and IPM methods of pest control in whitefly susceptible crops, with obstacles and opportunities for integrating parasitoids; describes our work in implementing parasitoid releases with existing technology, and discusses the remaining knowledge gaps needed for full implementation of an augmentative release strategy.

16.2 Conventional and IPM Strategies for Whitefly Control in Spring Melons

Spring melons grown in the desert vegetable producing areas of the southwestern USA were among the crops most severely affected by whitefly infestations; reductions in fruit quality due to sooty mold, reduced fruit size and sugar content, as well as complete destruction of entire fields, would occur during heavy infestations (Perring et al. 1993; Brown et al. 1995; Ellsworth and Martinez-Carrillo 2001). Whitefly control measures consisted of applications of broad spectrum pesticides such as pyrethroids (e.g., Capture) and endosulfan (Thiodan) up to 7 days before harvest. During especially severe infestations as many as five to six applications were needed for adequate control (E. Natwick, personal observation). After imidacloprid (Admire) became available in 1995, its use as a systemic (either injected into the soil or applied by drip-irrigation) became the standard whitefly control practice for melon production. Because of its systemic action, imidacloprid was expected to have less impact on beneficial species because adults would not come in direct contact with pesticide residues (Palumbo et al. 2001). This was viewed as another favorable factor in implementing a parasitoid release program in spring melons. Imidacloprid applied to young plants gave control for 45–60 days. Depending on the severity of the whitefly infestation and other factors, such as planting date and proximity to nearby infested crops, an additional two to three applications of pyrethroids or endosulfan were often needed to control whitefly until harvest (E. Natwick, personal observation).

Early in the season before flowering, pesticides are rarely needed. Spring melons require a 4–6- week pollination period; limited or no pesticide applications are made while bees are placed in the field. This created a pesticide-free window for inoculative releases of parasitoids early in the season, allowing parasitoids to attack the first generation of whitefly present in the fields and increase the population of parasitoids.

16.3 Parasitoid Releases in Melons

Parasitoid release strategies in which parasitoids were released either by hand in small cups or with tractor drop-box systems were tested in melons grown using conventional pest control methods relying on broad spectrum insecticides, in imidacloprid treated fields, and in fields where both imidacloprid and broad spectrum pesticides were used. Parasitoid releases were also tested in organic melon production, where only approved and generally less toxic and shorter-lived pesticides were used.

The general sampling methods were the same for all experiments in each year. Release plots were sampled at 2-week intervals, with at least 30 leaves collected per plot. Each sample consisted of a single leaf the same distance from the crown of the plant as those which contained the whitefly stages of the appropriate age for observing parasitism (determined by prior observation) and selected from a randomly chosen plant. All samples were examined in the laboratory using a dissecting microscope, and whitefly pupae and parasitoids were counted on either the entire leaf or a 2.7 cm^2 disk taken from the base of each leaf. The proportion of whiteflies parasitized on each leaf was calculated as the total number of parasitoid pupae and parasitoid pupae with exit holes divided by the total number of parasitoid pupae, parasitoid pupae with exit holes, *Bemisia* pupae and *Bemisia* pupal exuvia.

16.4 Testing Parasitoid Release in Small Plots, 1994–1995

The first release trials were conducted in small plots embedded within larger commercial fields of cantaloupe (*Cucumis melo* var. *reticulatus* L.) and honeydew (*Cucumis melo* var. *inodorus* L.). The fields ranged from 2.4 to 28.4 ha in size, with release plots measuring 0.2 ha. Parasitoids were released as pupae on leaves harvested from insectary plants, or as loose pupae distributed in 0.47 l paper containers and allowed to eclose in the field. The containers were placed under the leaves of melon plants to provide shade and were spaced in alternating rows every 1.5 m. This resulted in a single release point every 1.5 m^2. The first parasitoid releases were made as soon as whitefly nymphs were observed on melon leaves, which for spring planted melons was usually within 3–4 weeks after planting. Releases were made approximately weekly until the end of blooming period, which resulted in a total of eight to ten releases.

16.4.1 Small Plot Studies, 1994

Two species of parasitoids were tested in 1994. The native *Eret. eremicus* and one of the first available exotic parasitoids, *Eret. mundus* Mercet (USDA-APHIS Mission Biological Control Laboratory culture M92019 ex. Padappai, India), were released into paired 0.2 ha plots within 11 larger fields of cantaloupe (varieties Topmark and Packstart) in a randomized complete block experiment at the rates of 1.1 million/ha for *Eret. eremicus* and 294,000/ha for *Eret. mundus*. Parasitoid release resulted in significant increases in the percentage parasitism relative to control plots (repeated measures ANOVA, $F = 7.36$, d.f. = 2, 14, $P = 0.0066$) (Fig.16.1). *Eretmocerus mundus* produced the highest parasitism, averaging 47% at the peak vs. 28% by *Eret. eremicus*. The higher rate of parasitism achieved by releasing *Eret. mundus*, despite a nearly four times higher release rate of *Eret. eremicus*, was the first field evidence suggesting an exotic parasitoid could be more effective than the native species. Further evaluations of both the native and exotic parasitoids supported these early results, which showed that several of the imported exotic species were more effective parasitoids (Chapter 8).

During the course of the experiment, whitefly density increased in all treatments (Fig. 16.1). There were 27% fewer whiteflies in release plots compared with no-release plots although these differences were not significant (repeated measures ANOVA, $F = 0.02$, d.f. = 2, 14, $P = 0.983$).

16.4.2 Small Plot Studies, 1995

Eretmocerus mundus populations from Murcia, Spain (MBCL culture M92014) and Padappai, India (M92019) were tested in 1995. Releases were made into 0.2 ha plots of cantaloupe within ten larger fields ranging from 2 to 30 ha in size. Three of ten fields were treated with imidacloprid and seven were untreated. Within each field there were three treatments: release of each of the two *Eret. mundus* populations and a no-release control plot, distributed in randomized complete blocks with a distance of at least 244 m between plots. While *Eretmocerus* species are clearly capable of moving across the distance of these buffers, mark–release–recapture studies have shown that the majority of released parasitoids only move a few meters (Simmons 2000; Bellamy et al. 2001; Hagler et al. 2002), and the distance between plots was considered adequate to reduce the possibility of confounding effects caused by parasitoid movement to a minimum. Ten weekly releases were made from 21 March to 19 May with a total release of 254,519/ha of the Indian population and 210,039/ha of the Spanish population.

The peak mean rates of parasitism achieved at harvest were 41% for the Indian and 34% for the Spanish populations of *Eret. mundus* in plots not treated with imidacloprid, both significantly greater than the 7% parasitism by the native *Eret. eremicus* seen in control plots (repeated measures ANOVA, $F = 18.63$, d.f. = 2, 7, $P = 0.0016$)

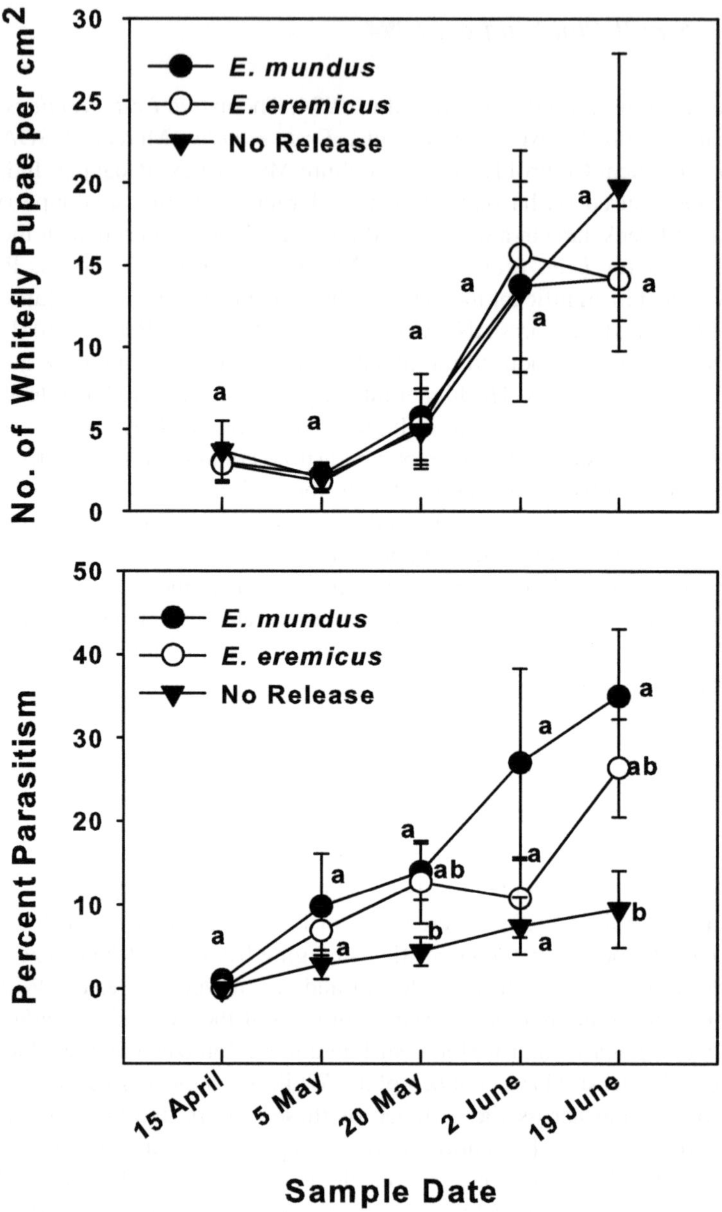

Fig. 16.1 Percentage parasitism and density of whitefly following release of *Eret. mundus* and *Eret. eremicus* in spring melons in 1994. Means followed by the same letters are not significantly different by repeated measures ANOVA at $P = 0.05$, n = 9 plots.

(Fig. 16.2). Rates of parasitism in single plots were as high as 60% in the Indian *Eret. mundus* release plots, indicating that substantial rates of parasitism were possible. Reductions in the density of whitefly pupae for the Indian *Eret. mundus* were as high as 59% relative to the no-release control plots but this was not significant (repeated measures ANOVA, $F = 0.80$, d.f. = 2, 7, $P = 0.486$) (Fig. 16.2).

In fields treated with imidacloprid, whitefly nymphs were present throughout the season at much lower densities than those observed in untreated melons (compare Figs. 16.2A and B). Some whiteflies survived imidacloprid treatment possibly because of uneven distribution of imidacloprid in plant tissues or poor placement of material in the field, which can reduce the effectiveness of uptake and distribution within the plant (Palumbo et al. 2001). These resident whitefly populations as well as the progeny of whitefly migrants later in the season increased exponentially during the last weeks before harvest (Fig. 16.2).

Releases of the two *Eret. mundus* populations into imidacloprid treated plots resulted in rates of parasitism as high as 43% compared to an average of 7% in the no-release control plots and decreased the density of the whiteflies by as much as 57%

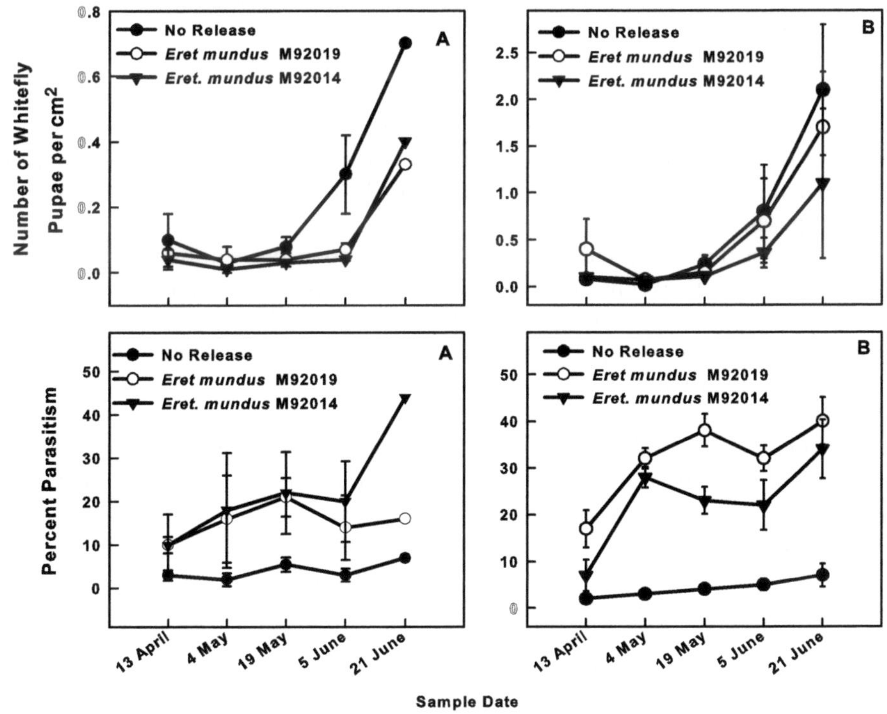

Fig. 16.2 Percentage parasitism and density of whitefly pupae in imidacloprid treated and untreated melon fields following release of two populations of *Eret. mundus* in 1995. (**a**) Imidacloprid treated fields (n = 3, error bars are one std. error of the mean). (**b**) Untreated fields (n = 7, error bars are one standard error of the mean).

(Fig. 16.2). Although there were not enough replicates of imidacloprid treated plots to show statistical significance, these results suggested that parasitoid release was compatible with imidacloprid treatment and that large reductions in whitefly density were possible. This was one of the first studies to suggest that parasitoid release was compatible with imidacloprid treatment and set the stage for further investigations on the potential for an IPM strategy combining imidacloprid with parasitoid release.

16.4.3 Parasitoid Release with Imidacloprid + Bifenthrin, 1996

Within one growing season, imidacloprid became the standard strategy for whitefly control in desert melon production systems (E. Natwick, personal observation). Because of the positive results from our limited study in 1995 combining parasitoid release and imidacloprid, a replicated experiment comparing parasitoid release to applications of broad spectrum pesticides in imidacloprid treated melons was conducted. Cantaloupe variety 'Topmark' was planted on 26 March in a 1.4 ha field at the University of California research station in Holtville, CA. This field was divided into 36 plots four rows wide on 1.5 m centers by 18.3 m long. There was a 9.1 m long and a 6.1 m wide buffer of bare ground between plots to reduce pesticide drift into unsprayed plots and to reduce the number of parasitoids moving from release plots into no-release plots. Based on results of several dispersal studies with *Eretmocerus* species (Simmons 2000; Bellamy and Byrne 2001; Hagler et al. 2002) this was a sufficient distance to ensure that few parasitoids would move from one plot to another, possibly confounding results.

Imidacloprid was applied at planting 5.1 cm under the seed line at the rate of 1.2 l/ha to all plots except in the untreated control plots. There were four treatments: (1) untreated control, (2) imidacloprid treated, (3) imidacloprid plus bifenthrin (as Capture) application, and (4) imidacloprid plus parasitoid release. There were nine replicates of each treatment arrayed in a randomized complete blocks design.

Biweekly sampling for whitefly and parasitoids began on 18 April and continued until 14 June for a total of five separate sampling dates. On each sample date random samples were selected from 20 plants from the center two rows of each plot. One sample leaf was collected from each plant.

Eretmocerus hayati Zolnerowich and Rose (MBCL M95012 *ex.* Multan, Pakistan) was released in parasitoid release plots beginning on 11 April and continuing until 24 May. Six weekly releases of loose pupae were made at the rate of 247,000 parasitoids/ha using methods described previously. On 8 May, spraying of pesticides by ground began in the imidacloprid plus bifenthrin plots and continued at 4–7- day intervals until 28 May, for a total of five applications. Bifenthrin was used in the first four applications, in the second application it was combined with endosulfan, and on the last application only endosulfan was used.

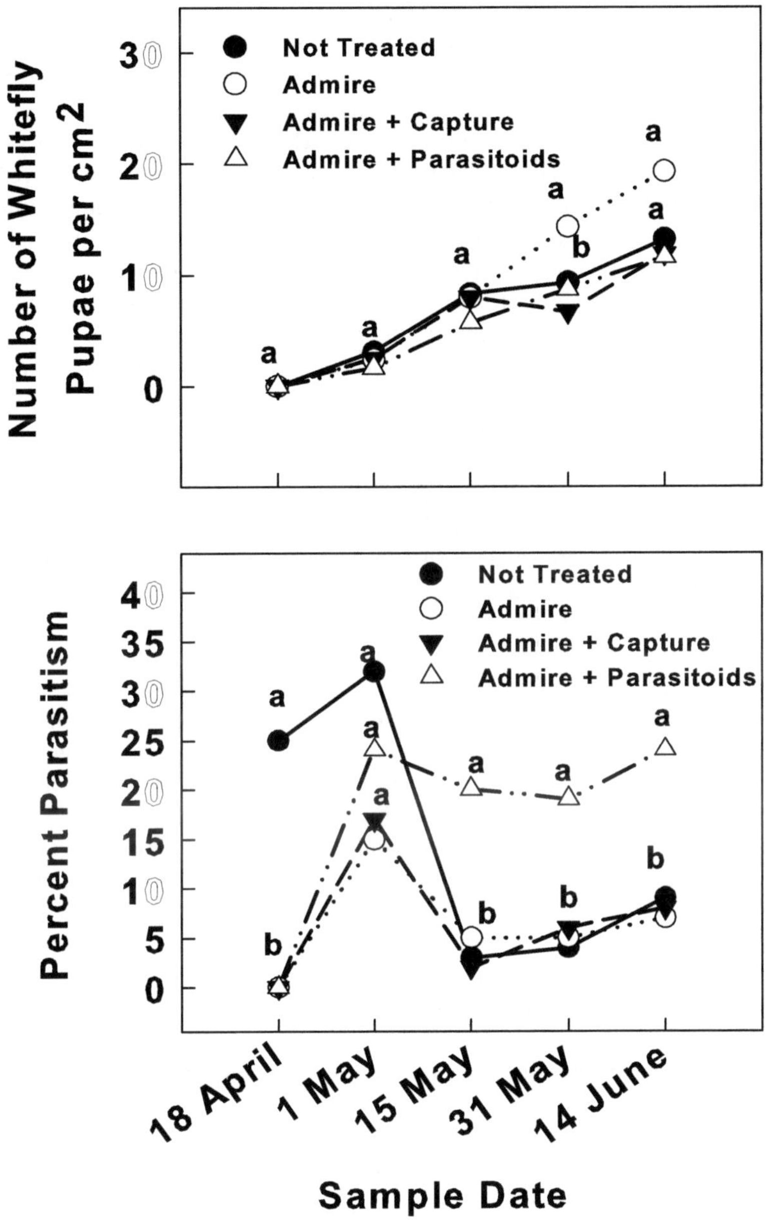

Fig. 16.3 Percentage parasitism and density of whitefly pupae in melons receiving either (1) releases of *Eretmocerus hayati* plus an application of imidacloprid (Admire®) or (2) bifenthrin plus imidacloprid in 1996. Results are the means of nine replicate plots for each treatment. Values marked with different letters are significantly different by Tukey's mean separation test.

The highest percentage of parasitism was found in the untreated plots on the first two sample dates and peaked at 33% on 1 May, but was not significantly different from the other treatments on these dates (ANOVA, $F = 1.88$, d.f. = 3, 21, $P = 0.5636$) (Fig. 16.3). Because the first releases of parasitoids were made just a few days before the first sample was taken, all of the parasitized whitefly found on the first sample date and much of the parasitism found on the second sample date would have been caused by naturally occurring parasitoids. Early colonization of whitefly infested melons by naturally occurring parasitoids might be retarded by the imidacloprid treatment, as there were no parasitoids present in any of the other imidacloprid plots on the first sample date (Fig. 16.3). The absence of parasitoids could be either the result of direct lethal effects or indirect effects such as the unavailability of suitable stages of whitefly nymphs as hosts. As the experiment progressed parasitism increased in the release plots, peaking at 25% on the last sample date; this was significantly higher than the parasitism of 4–7% observed in the no-release plots (ANOVA, $F = 12.46$, d.f. = 3, 5, $P = 0.0093$, Tukey's HSD $P < 0.05$) (Fig. 16.3).

Whitefly density steadily increased in all treatments. The highest density of whitefly pupae were found in the imidacloprid treated plot on the last two sample dates, 14 and 19 pupae/cm^2 of leaf surface respectively (Fig. 16.3). In the other treatments, the number of whitefly pupae ranged from 5 to 10 pupae/cm^2, but these differences were significant only on 31 May (ANOVA, $F = 6.60$, d.f. = 3, 24, $P = 0.0021$, Tukey's HSD $P < 0.05$) (Fig. 16.3). One surprising result was that whitefly densities in the untreated control plots were about the same as in the imidacloprid plus parasitoids and the imidacloprid plus bifenthrin treated plots, and were significantly lower than the imidacloprid only treatment on 31 May (Fig. 16.3). The untreated vines were in decline and could not support higher densities of whitefly as well as the healthier, imidacloprid treated vines. In this case, the imidacloprid-only treatment should be regarded as the control for the treatments either of imidacloprid plus parasitoid release or imidacloprid plus bifenthrin application.

Due to an early extreme whitefly infestation, spraying of the imidacloprid plus bifenthrin plots began earlier than anticipated. When the applications of bifenthrin began, it was clear that many whiteflies had survived the imidacloprid treatment, indicating that control in all imidacloprid treated plots had been poor. Figure 16.3 shows that the number of whitefly pupae continued to increase even after applications of bifenthrin began in the first week of May. These results demonstrated that parasitoid release can increase parasitism in imidacloprid treated melons, but the overall levels of parasitism achieved were lower than necessary for good control. It is unclear why higher levels of parasitism were not achieved; large numbers of migrating whitefly may have overwhelmed the ability of the released parasitoids to provide adequate control. Although whiteflies were not well controlled in any of the treatments, it appears that imidacloprid plus parasitoid release was about equal to imidacloprid plus bifenthrin, as the densities of whitefly were similar in these two treatments (Fig. 16.3). This was the desired result, suggesting that replacing bifenthrin treatment with parasitoid release may be possible.

16.5 Commercial Scale Releases: Demonstration Projects, Release Rate Studies, and Test of Parasitoid Release Systems

16.5.1 Demonstration Releases, 1997

During the spring of 1997 a large-scale parasitoid release demonstration project was conducted on four farms totaling 20.2 ha of spring cantaloupe. Because positive results from the studies in smaller plots showed there was potential for providing effective control of whitefly with parasitoid releases, the question arose of whether parasitoid release would be more effective if entire fields were treated. Previous studies in cotton showed that immigration into small release plots from the surrounding fields could decrease the effectiveness of parasitoids (Simmons 2000), and it was reasoned that treating entire fields would be more effective. Releases of *Eret. emiratus* Zolnerowich and Rose (MBCL M95104 *ex.* United Arab Emirates) were made in two cantaloupe fields under organic production and two fields in which imidacloprid was applied for whitefly control. *Eretmocerus emiratus* was chosen for release because recent evaluations showed that it was one of the most effective new species of parasitoids (Chapter 8). The organic cantaloupe fields included a 4.5 ha block (Early Organic) of an early season variety ('PackStart') and a 6.9 ha block (Late Organic) of mid-season cantaloupe ('Top Mark'). The imidacloprid-treated fields were two mid-season planted blocks of cantaloupes ('Top Mark'), consisting of a 3.6 ha block (Imidacloprid 1) and a 5.7 ha block (Imidacloprid 2). Nearby fields of similar size and growing practice were selected as non-release control fields. These fields used were not considered replicates due to differences in planting times, size and location; true replication was not feasible due to the size of the fields and the large numbers of parasitoids required. The results of these experiments should be viewed as a demonstration of the potential for whitefly control with parasitoid release over an entire field.

Releases were made over a 4-week period beginning as soon as an average of one whitefly nymph per plant was observed in the field. Because release rates were determined by both the availability of parasitoids and by the density of whitefly, two release rates were used for the organic fields: 106,000 parasitoids/ha in the Early Organic field; in the Late Organic field, 326,000 parasitoids/ha were released because of a higher whitefly density. The imidacloprid treated fields had parasitoid release rates of 319,000/ha (Imidacloprid 1) and 193,000/ha (Imidacloprid 2). The point of these experiments was to show that there was a positive effect of parasitoid release regardless of the number released, not to refine release rates.

High levels of parasitism and reductions of whitefly were achieved in the organic fields (Fig. 16.4). The Late Organic field had peak levels of parasitism ranging from 60 to 100%, which resulted in a reduction of whitefly density by as much as 94% relative to the control field. Peak levels of parasitism in the Early Organic field ranged from 60% to 70%, which resulted in a reduction of whitefly levels by 65% compared to the control fields.

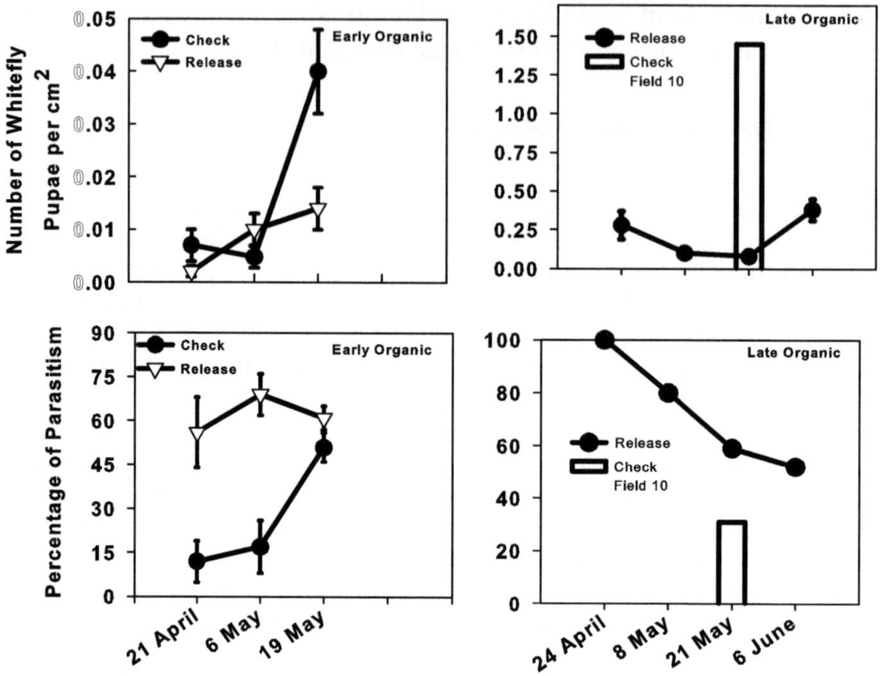

Fig. 16.4 Percentage parasitism and density of whitefly pupae following releases of *Eret. emiratus* in organic melon fields in 1997. Early Organic: early season planting, 4.5 ha. Late Organic 2: mid-season planting, 6.9 ha.

In the imidacloprid treated fields, peak levels of parasitism ranged from 80% to100% (Fig. 16.5) resulting in reductions of whitefly by as much as 26–74% in comparison to nearby control fields (Fig. 16.5). Whitefly density was much lower in the imidacloprid treated fields than control fields, and the increased parasitism eliminated the need for late season applications of pyrethroids that are often needed for whitefly control. This demonstration showed that parasitoids released over large areas can have field-wide impacts in both organic and imidacloprid treated fields.

16.5.2 Release Rate Studies, 1998

Several release-rate experiments in large fields were conducted to determine the minimum number of parasitoids needed for efficacious control after the positive results from the small plot trials in 1994–1996 and the larger demonstration fields in 1997. Studies were conducted in the spring of 1998 in four cantaloupe ('Topmark')

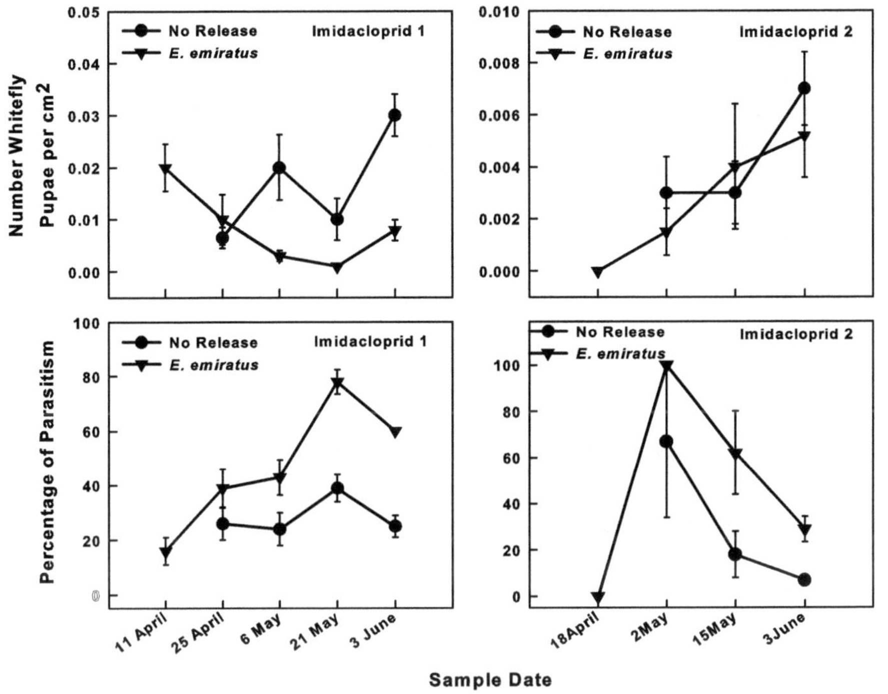

Fig. 16.5 Percentage parasitism and density of whitefly pupae following releases of *Eret. emiratus* in imidacloprid treated fields in 1997. Imidacloprid 1: mid-season planting, 3.6 ha. Imidacloprid 2: mid-season planting, 5.7 ha.

fields totaling 71 ha. Two fields each were grown following either organic production practices or standard imidacloprid application practices. Organic 1, a 22 ha field, was planted on 26 January and Organic 2, a 3.6 ha field, was planted on 5 March. Imidacloprid 1 was planted on 14 January and was 31 ha in size; Imidacloprid 2 was planted on 25 January and was 24 ha in size. Both imidacloprid fields and Organic 1 contained two replicate plots ranging from 0.7 to 2.8 ha for each parasitoid release rate and two control plots. Organic 2 contained only one release plot for each release rate and one control plot.

Pupae of *Eret. emiratus* were distributed at five rates of 0, 24,700, 49,400, 98,800, and 197,600/ha. In the organic fields, parasitoids were released by placing pupae in small paper cups along every other row. Two releases of parasitoids were made in Organic 1, spaced 10 days apart; a single release was made in Organic 2. Parasitoid pupae were mixed with vermiculite and released over every other row with a tractor mounted drop box system (Bug-Dropper, AgAttack Inc, Visalia, CA) in the imidacloprid treated fields. The drop boxes dropped the pupae directly over the plant canopy so that pupae would either rest on the leaves or drop to the soil below in the shade of the plants. The parasitoids needed for each field were released in two groups, 10to 15 days apart.

Samples of 100–500 pupae from each release field were collected for all release dates to determine the percentage of adult emergence, which was used to estimate the actual number of live parasitoids released on each date. Samples of 30 leaves per release plot were randomly collected at 2-week intervals until harvest. The selected leaves contained the appropriate whitefly stages that could be visually examined to estimate parasitism. The rates of parasitism and whitefly density from each release field were analyzed as single experiments with non-linear regression analysis by regressing the response variables against the number released of parasitoids released per acre.

16.5.3 Results of Imidacloprid Treatment

16.5.3.1 Imidacloprid 1

Actual release rates ranged up to 148,200/ha based on an average emergence rate of 75% over both release dates (Table 16.1). The percentage of parasitism in the Imidacloprid 1 field varied over the course of the experiment but was generally higher with increased release rates. The highest rates of parasitism were achieved on 6 May and then declined (Fig. 16.6). On 6 May, the 148,200/ha release plot had the highest parasitism with a mean parasitism rate of 43%. None of the non-linear regressions of the proportion of parasitism versus the number of parasitoids released were significant (Fig. 16.7), although the curve fit to the data for some sample dates showed a trend of saturation of parasitism at release rates greater than 86,000–98,800 parasitoids released/ha (Fig. 16.7, 18 June). The density of whitefly pupae increased during the course of the experiment (Fig. 16.6) and generally decreased with increased parasitoid release (Fig. 16.6). Overall whitefly densities were an order of magnitude lower than seen in previous years, which may have contributed to the variation in rates of parasitism. On some sample dates, whitefly densities were close to zero (Fig. 16.6). There were no significant regressions of the density of whitefly with release number, although on some sample dates the fitted curves suggested that whitefly declined with increased parasitoid release rate (Fig. 16.8).

16.5.3.2 Imidacloprid 2

There were two releases of parasitoids approximately 2 and 3 weeks apart in release blocks 1 and 2 respectively. The average emergence rate for both releases for block 1 was 64%, resulting in release numbers up to 130,000 parasitoids/ha; for block 2, the average emergence rate was 58%, resulting in releases up to 123,000/ha) (Table 16.1). This gives an average of 126,500 parasitoids released per hectare for the two replicate

Table 16.1 Release dates, parasitoid emergence rates, and the number of released parasitoids for each release plot in the 1998 release rate experiment.

			Release date		% Emergence		No. released (1,000s)/ha				
Release field	Release method	Block	Release 1	Release 2	Release 1	Release 2					
Imidacloprid 1	Drop-box	1	4/15	5/6	68%	82%	0	20	37	74	148
		2	4/15	5/6	68%	82%	0	20	37	74	148
Imidacloprid 2	Drop-box	1	3/25	4/8	70%	62%	0	17	32	64	130
		2	4/1	4/23	61%	55%	0	15	30	57	114
Organic 1	Cups	1	3/19	3/30	67%	77%	0	17	35	72	143
		2	3/19	3/31	67%	77%	0	17	35	72	143
Organic 2	Cups	1	4/17	–	79%	–	0	20	37	76	153

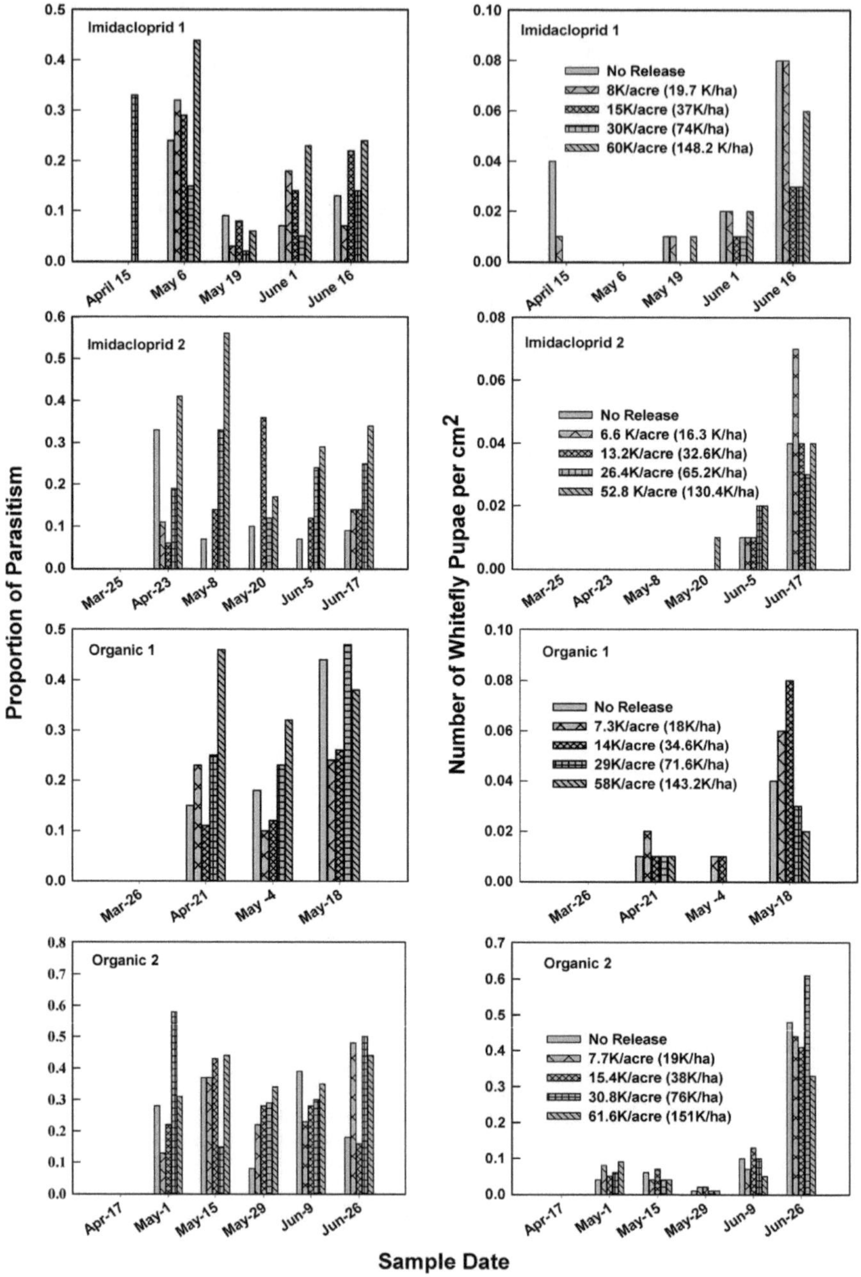

Fig. 16.6 Proportion of parasitism and density of whitefly pupae following releases of five different rates of *Eret. emiratus* in spring melons in 1998.

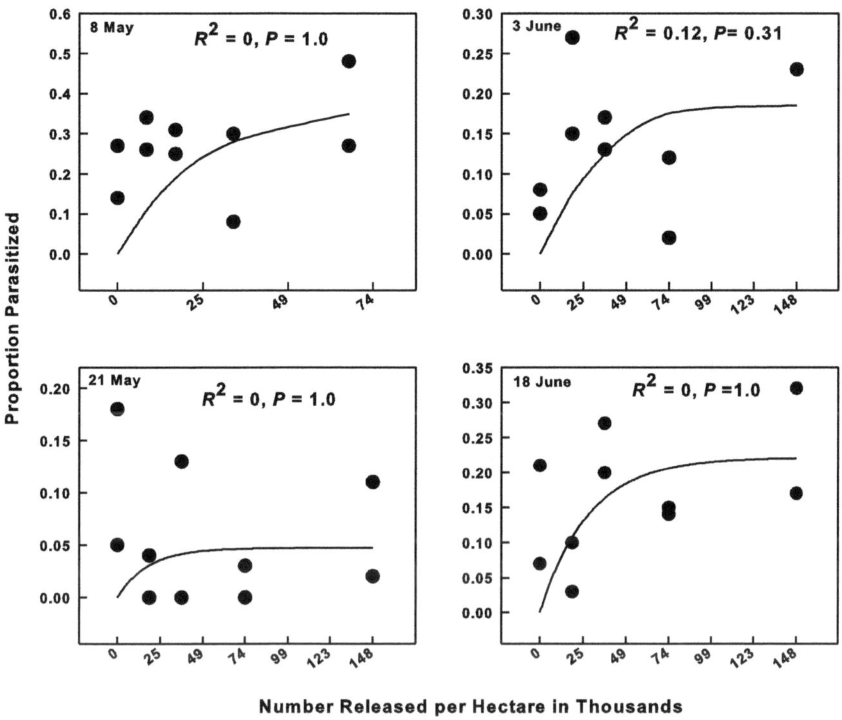

Fig. 16.7 Proportion of parasitized *B. tabaci* versus the release rate of *Eret. emiratus* released for each sample date in 1998, Imidacloprid 1 field. Solid lines are fitted regression lines.

blocks. On most sample dates the percentage of parasitism was highest in the high release plots and peaked on the 8 May sample date, averaging 53% (Fig. 16.6). The regressions of the proportion parasitized were significant with *P* values ranging from 0.03 to 0.007 (Fig. 16.9) on three sample dates, 8 May, 5 June, and 17 June, and with R^2 ranging from 0.48 to 0.68 (Fig. 16.9). Inspection of the fitted regression curves suggests that early in the experiment increasing the parasitoid release rate resulted in a linear increase in parasitism (Fig. 16.9, 8 May). As the experiment progressed, overall parasitism declined and the rate of parasitism appeared to saturate above releases of 49,000 parasitoids/ha (Fig. 16.9, 5 June, 17 June), suggesting that releases above this did not increase the final rate of parasitism. Because the increase in the rate of parasitism appeared to be linear on early sample dates (Fig. 16.9, 8 May) it would be useful to test higher release rates to determine if saturation of parasitism rates would still occur. The density of whitefly pupae increased in all plots over the course of the experiment (Fig. 16.6). The regression of whitefly density on release rate showed a trend of decline in whitefly density with increased parasitoid release, although only the 23 April sample demonstrated a significant regression (Fig. 16.10).

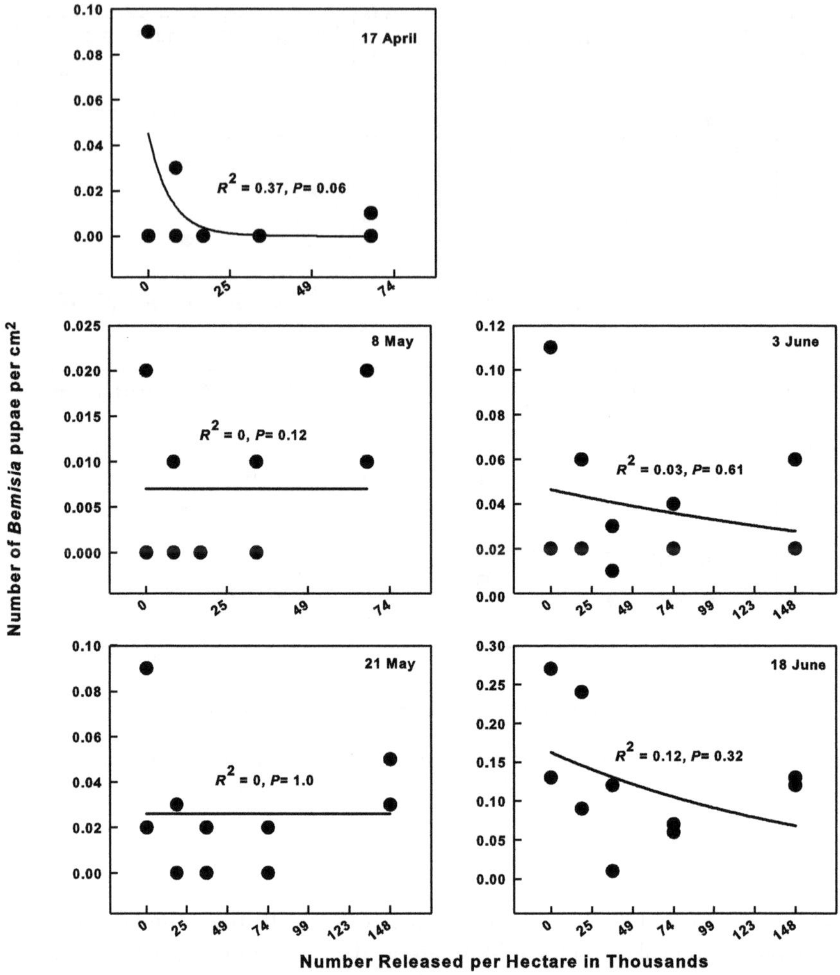

Fig. 16.8 Density of whitefly pupae versus the release rate of *Eret. emiratus* for each sample date in 1998 release rate experiment, Imidacloprid 1 field. Solid lines are fitted regression lines.

16.6 Organic Treatment

16.6.1 Organic 1

There were two releases of parasitoids into the two replicates, 10 and 11 days apart. The average emergence rate for both releases was 72% resulting in a release of up to 143,200 parasitoids/ha (Table 16.1). The highest rates of parasitism were in the

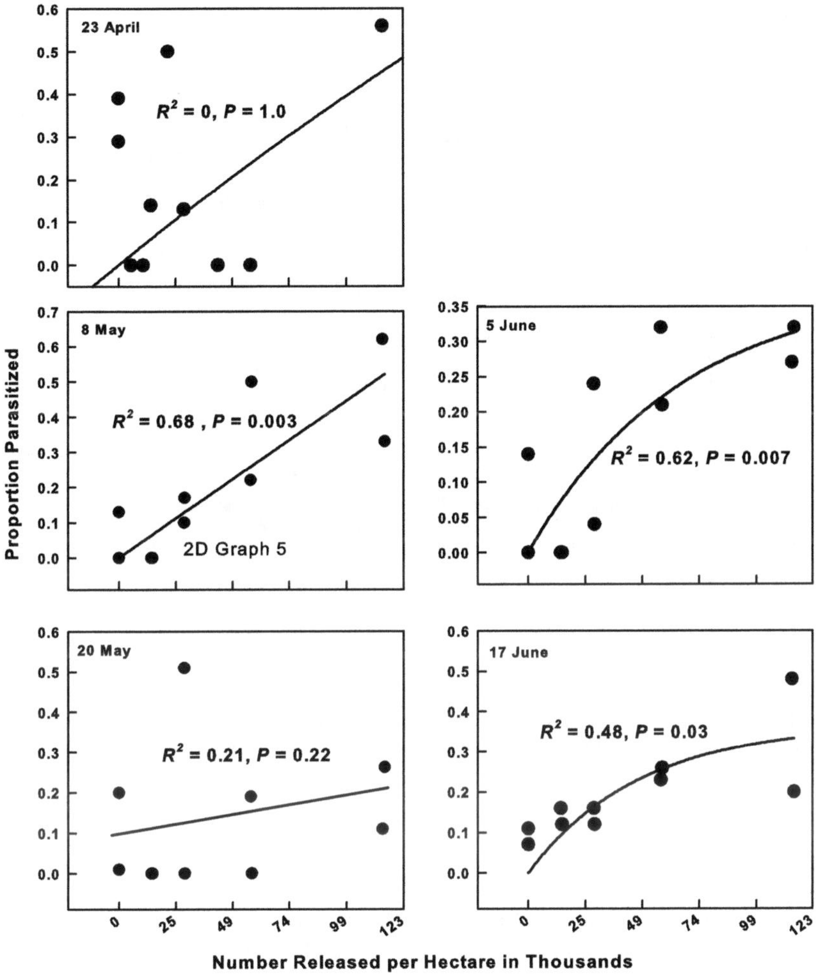

Fig. 16.9 Proportion of parasitized whitefly pupae versus the release rate of *Eret. emiratus* for each sample date in 1998, Imidacloprid 2 field. Solid lines are fitted regression lines.

two highest release rate plots and ranged from 40% to 48% (Fig. 16.6). None of the regressions of rates of parasitism against the number of parasitoids released were significant. The regression curves fitted to data suggest that saturation in parasitism occurs above release rates of about 49,000 parasitoids/ha (Fig. 16.11).

Whitefly density declined with greater parasitoid releases, with average reductions between the control and high release plots as great as 50% (Fig. 16.6). On the 23 April sample, there was a statistically significant regression showing a negative

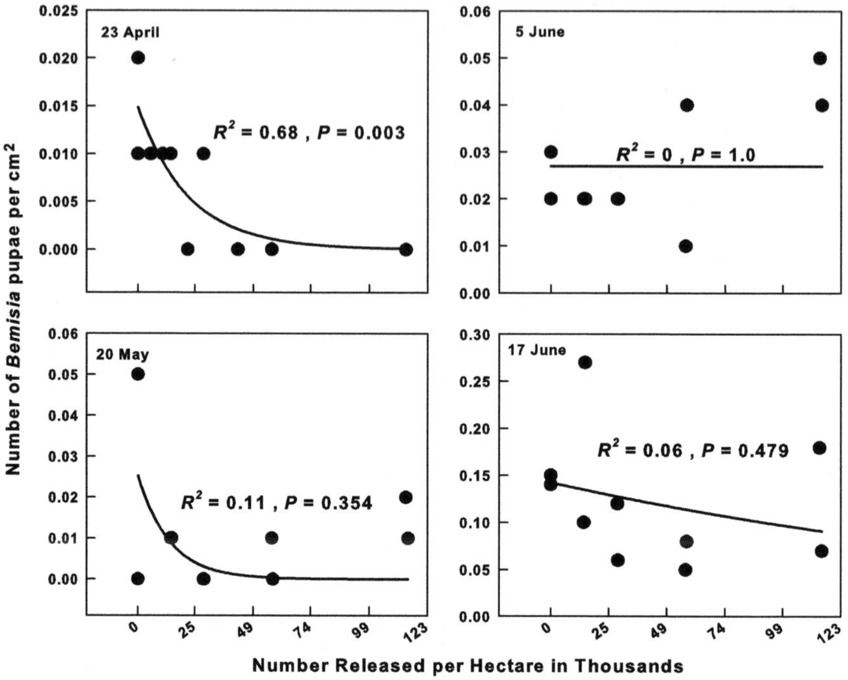

Fig. 16.10 Density of whitefly pupae versus the release rate of *Eret. emiratus* for each sample date in 1998, Imidacloprid 2 field. Solid lines are fitted regression lines.

relationship between whitefly density and the parasitoid release rate. The regressions on the other dates sampled were not statistically significant, though the trend of the fitted regression lines suggest a negative correlation between parasitoid release rate and whitefly density (Fig. 16.12).

16.6.2 Organic 2

One release of parasitoids was made. The number of parasitoids released ranged up to 151,000/ha based on the average emergence rate of 79% (Table 16.1). The highest rate of parasitism was 59% on the 1 May sample (Fig. 16.6) in the second highest release plot, 71,600/ha. Average rates of parasitism increased in all release plots during the course of the experiment, and on most sample dates the high release plots had higher rates of parasitism (Fig. 16.6). For the 31 May sample there was a statistically significant fit to the data for the regression of the rate whitefly parasitism versus the number of released parasitoids with an R^2 of 0.81 (Fig. 16.13). There

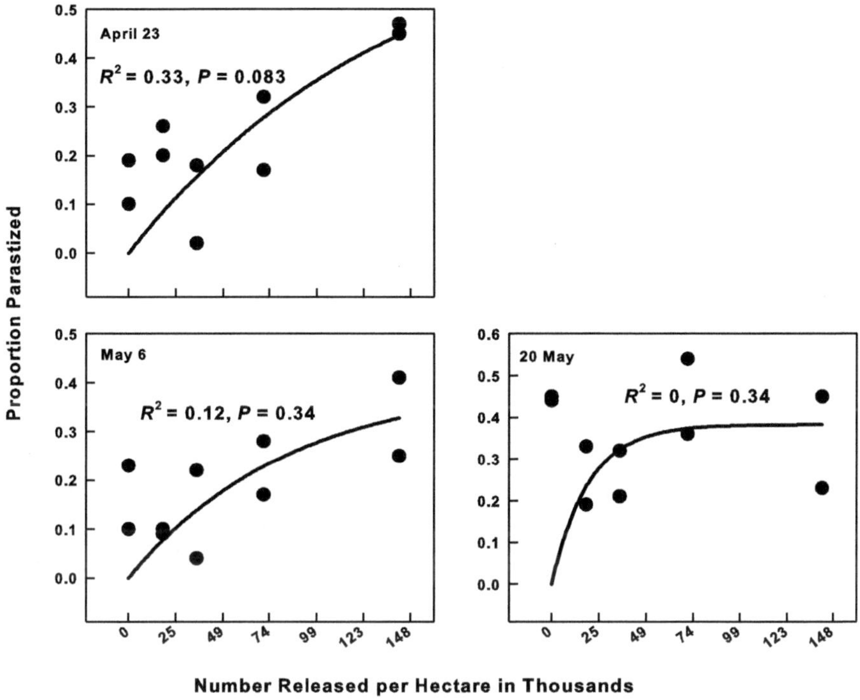

Fig. 16.11 Proportion of parasitized whitefly pupae versus the release rate of *Eret. emiratus* for each sample date in 1998, Organic 1 field. Solid lines are fitted regression lines.

were no significant regressions for other sample dates, possibly because there were no replications of release plots in this field (d.f. = 1,3). The significant regression on 31 May suggests that saturation of parasitism rates may occur at release rates higher than about 49,400 parasitoids/ha.

Whitefly density increased in all release plots over all sample dates. The lowest whitefly density was usually observed in the high release plots with reductions in whitefly density as great as 40–50% observed on the last two sample dates (Fig. 16.6). None of the regressions for whitefly density versus the number of released parasitoids were statistically significant, though the trend of fitted curves suggests a negative relationship between parasitoid release rate and the density of whitefly (Fig. 16.14).

16.7 Discussion of Release Rate Trials, 1998

Regression analysis of rates of parasitism and whitefly density versus the number of released parasitoids showed that for some of the fields there was a significant relationship to the number of released parasitoids. In some of the other cases, where

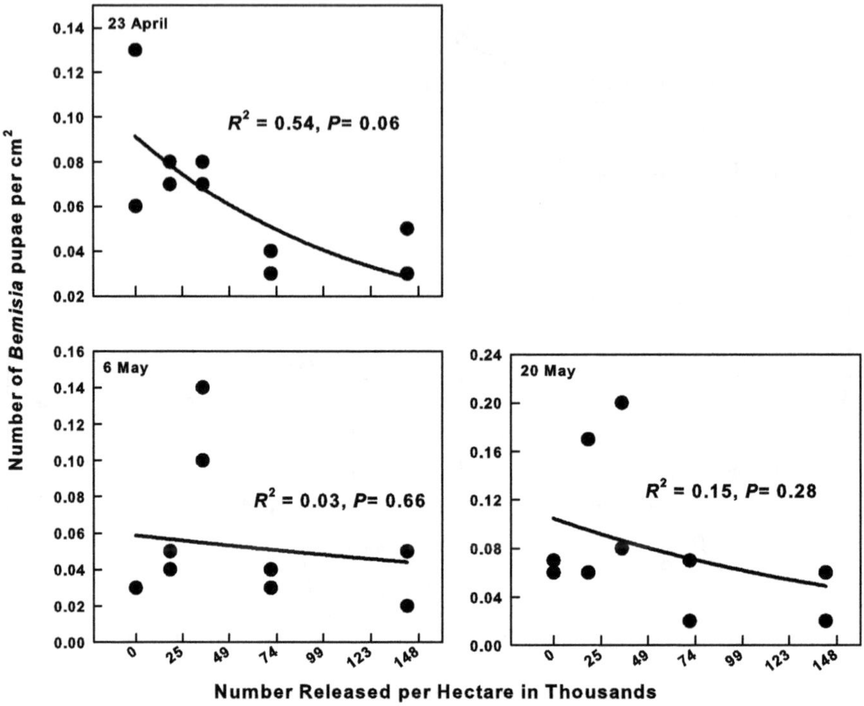

Fig. 16.12 Density of whitefly pupae versus the release rate of *Eret. emiratus* for each sample date in 1998, Organic 1 field. Solid lines are fitted regression lines.

no significant regressions were found, the shape of the fitted curves suggests the same relationship. More replications may have improved the statistical power of the regressions. Though release plots were relatively large, parasitoids from adjacent higher release rate plots may have migrated to other plots, weakening the relationship between release rate and parasitism. However, examination of the fitted regression curves show that releases above about 49,000 parasitoids/ha resulted in saturation of parasitism rates and a flattening of the curve showing a decrease in whitefly density, suggesting that higher release rates would not achieve greater control and that this rate would be the most effective release rate. A previous study in spring melons suggested higher release rates, 98,800–197,500/ha, were needed, but the whitefly densities were up to 25 times higher than in the present studies which may explain the higher numbers of parasitoids needed (Simmons et al. 2002). Reductions in whitefly density between the control and higher release rates were as high as 50%, which was as not as high as seen in some years but was within the range of previous years' observations. It is unclear why higher levels of parasitism or greater reductions were not achieved compared to other years. The observation that saturation in rates of parasitism occurred above 49,000/ha suggests that the number of

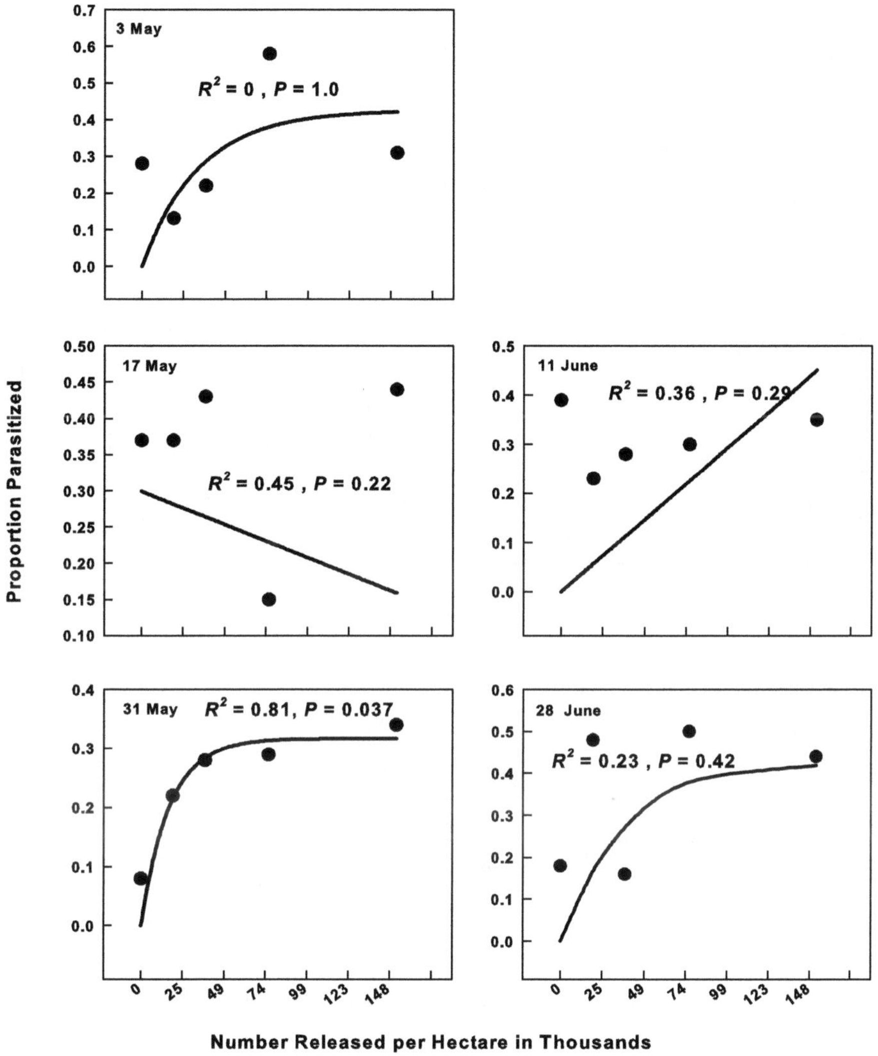

Fig. 16.13 Proportion of parasitized whitefly pupae versus the release rate of *Eret. emiratus* for each sample date in 1998, Organic 2 field. Solid lines are fitted regression lines.

parasitoids released was not a limiting factor. Because whitefly densities were quite low compared to previous years, it is unlikely that whitefly immigration from surrounding fields was responsible for lower rates of parasitism or lower relative reductions in whitefly density between no-release plots and the highest release plots. It is possible that whitefly densities were low enough that the parasitoids dispersed from the release area because of low numbers of hosts. Bellamy et al. (2004)

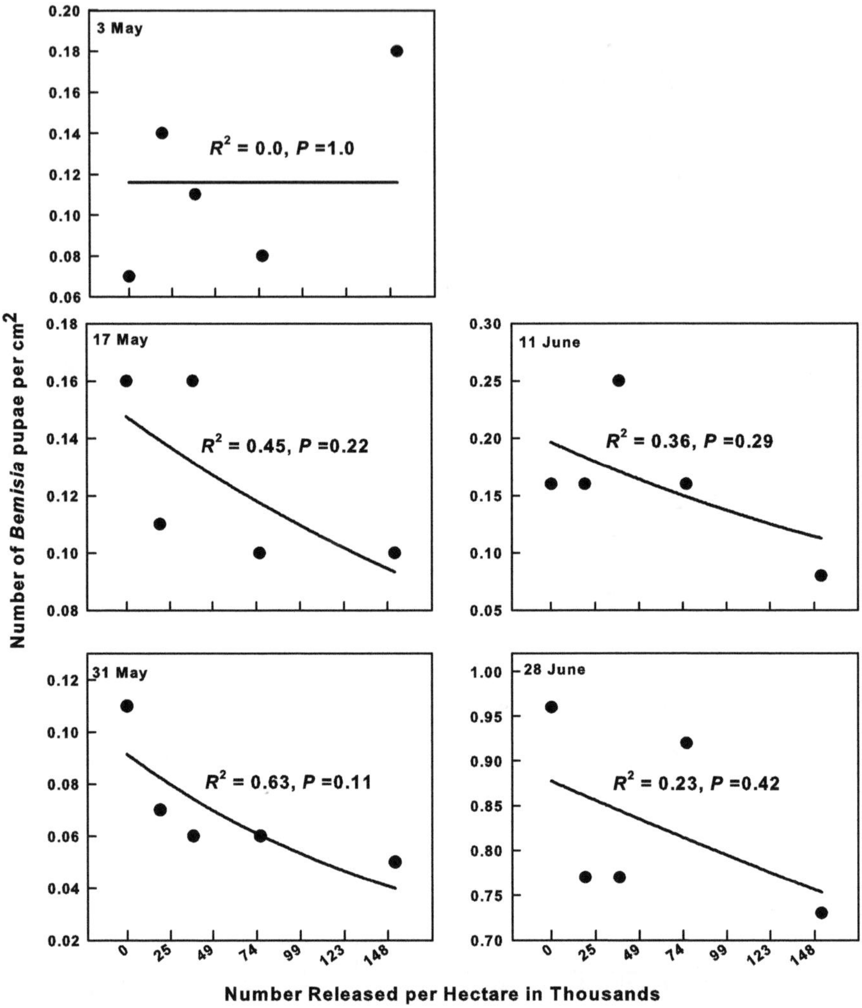

Fig. 16.14 Density of whitefly pupae versus the release rate of *Eret. emiratus* for each sample date in 1998, Organic 2 field. Solid lines are fitted regression lines.

suggest that dispersal of the native *Eret. eremicus* may be inverse density dependent resulting in greater movement when host densities are low; it is unknown if this is a factor in dispersal patterns of *Eret. emiratus*.

Emergence rates using the release cup method and the parasitoids mixed with vermiculite and spread with the drop-box equipment were similar and ranged from 55% to 82%. These emergence rates are similar to those seen in other studies where pupae were released in cups (Pickett et al. 2004), which suggests that pupal emergence rates are not affected by the drop-box machine. This agrees with studies of other beneficial insects spread with this equipment and shows that this release technique causes little harm (Gardner and Giles 1996).

16.8 Prospects for the Future

Estimates of the number of parasitoids needed to increase parasitism by 50% or greater and to reduce whitefly densities by as much as one half over several seasons ranged from 49,400 to 197,600/ha. Current costs for bulk purchases of *Eretmocerus* species parasitoids are estimated at 6.00–8.00 USD per 1,000 (D. Cahn, Syngenta Bioline, Oxnard, CA, personal communication), which at the above rates would cost growers $296–1185/ha. During the years of the highest whitefly infestations in the Imperial Valley melon growers spent as much as $740 or more per hectare for whitefly control (J. Benson, Benson Farms, Brawley, California, personal communication). If it were assumed that parasitoid release could provide satisfactory whitefly control, at current costs the release of parasitoids could be considered an economically feasible form of control.

In the 1998 release rate studies and in the previous year's experiments parasitoid release was shown to be compatible with the use of imidacloprid. By comparing whitefly densities in nearby untreated fields to imidacloprid treated fields, these studies showed that it was possible to reduce the number of whitefly by a factor ranging from one half to 300-fold (see Figs. 16.2, 16.5 and 16.6). With the price of imidacloprid (as Admire) at $122/l, an application rate of 1.2–1.75 l/ha the cost per hectare is $144–217 and using the low estimate of a release of 49,000 parasitoids/hectare, the combined cost for whitefly control would range from $442 to $514/ha which would be within the range of acceptable cost for whitefly control. Organic growers are limited in the types of pesticides they can use to control whitefly, most of which are minimally effective. The strategy for organic melon production in the Imperial Valley has been to shift production to early planting in December and January so the harvest is completed by mid May, largely escaping the period of highest whitefly density that begins later in the spring (A. Bornt, Bornt Farms, Calexico, California, personal communication). Releasing parasitoids in organic fields earlier in the season when whitefly density is lower can cause significant reductions in whitefly, especially in years with early high numbers of whiteflies, extending the organic melon production season.

This economic argument is speculative and several issues need to be addressed before parasitoid release can be proposed as an economically viable form of whitefly control. Some of these are determination of the economic threshold for control of whitefly in melons using augmentative releases, refinement of release techniques, and development of predictive population models to determine the optimal timing and frequency of parasitoid release relative to the whitefly density. Refinements of release techniques that reduce the number of parasitoids needed for release would lower costs. Average emergence rates for parasitoids released in cups and by drop box were in the 70–80% range, but improved emergence rates close to 100% can be achieved when parasitoids are delivered to the field in the form of pupae on transplants (banker plants) (Goolsby 1999; Pickett et al. 2004). Other advantages of the banker plant technique include placing parasitoids in the field early enough so that they are timed to attack the first generation of whitefly that become established in the field and providing a low level of whiteflies on the transplants ensuring that hosts will be available for the first generation of parasitoids. Using transplants to deliver parasitoids to the field is more costly than other release methods, but it permits use of lower release rates. Since some growers use transplants as standard growing practice, a system for introducing parasitoids onto plants in an existing transplant production nursery could be developed at low cost (Goolsby and Ciomperlik 1999; Pickett et al. 2004).

Reduction of the whitefly population in desert valleys caused by growers adopting more efficient region-wide control measures, such as the widespread use of imidacloprid, timely destruction of crop residues after harvest, and alteration of cropping schedules of whitefly susceptible plants may lead to more economical biological control (Ellsworth and Martinez-Carrillo 2001; Palumbo et al. 2001). These practices improve the prospects for using parasitoid release as a control measure by suppressing the regional whitefly population, thereby delaying the whitefly infestation of the fields and reducing the number of whitefly emigrating from other crops, and resulting in lower whitefly populations that must be controlled. Using augmentative biological control may also help reduce grower's costs at harvest time by eliminating the need to observe cumbersome worker safety regulations and lengthy worker reentry periods that are required by the use of restricted use pesticides.

Lastly, improvements in rearing technology and commercial-scale production of parasitoids should translate into lower costs. In order for this to happen, more research on mass-rearing is needed, as well as help in developing markets for exotic whitefly parasitoids. European growers of greenhouse and protected crops have adopted the use of the Palearctic species *Eret. mundus* (Lara et al. 2002; Urbaneja et al. 2003; D. Cahn, Syngenta-BioLine, Oxnard, CA, personal communication) which is highly effective against *B. tabaci* (see Chapter 8), while North American growers have preferred to use the Nearctic species *Eret. eremicus*. While *Eret. eremicus* is not as effective against *B. tabaci* as the introduced *Eretmocerus* species (see Chapter 8), it will parasitize both *B. tabaci* and *Trialeurodes vaporariorum* (Westwood) (D. Cahn, personal communication). Further development of whitefly biological control for melons will depend on tapping the existing markets and production systems for whitefly biological control to provide melon growers access to the more effective Palearctic *Eretmocerus* species at reasonable prices.

References

Bellamy, D.E. and D.N. Byrne. 2001. Effects of gender and mating status on self-directed dispersal by the whitefly parasitoid *Eretmocerus eremicus*. Ecological Entomology 26: 571–577.

Bellamy, D.E., M.K. Asplen, and D.N. Byrne. 2004. Impact of *Eretmocerus eremicus* (Hymenoptera: Aphelinidae) on open-field *Bemisia tabaci* (Homoptera: Aleyrodidae) populations. Biological Control 29: 227–234.

Bellows, T.S., Jr. and K. Arakawa. 1988. Dynamics of preimaginal populations of *Bemisia tabaci* (Homoptera: Aleyrodidae) and *Eretmocerus* sp. (Hymenoptera: Aphelinidae) in southern California cotton. Environmental Entomology 17: 483–487.

Brown, J.K., D.R. Frohlich, and R.C. Rosell. 1995. The sweetpotato or silverleaf whiteflies: biotypes of *Bemisia tabaci* or a species complex? Annual Review of Entomology 40: 511–534.

Chu, C.C., T.J. Henneberry, E.T. Natwick, D. Ritter, and S.L. Birdsall. 2001. Efficacy of CC traps and seasonal activity of adult *Bemisia argentifolii* (Homoptera: Aleyrodidae) in Imperial and Palo Verde Valleys, California. Journal of Economic Entomology 94: 47–54.

Ellsworth, P.C. and J.L. Martinez-Carrillo. 2001. IPM for *Bemisia tabaci*: a case study from North America. Crop Protection 20: 853–869.

Gardner, J. and K. Giles. 1996. Handling and environmental effects on viability of mechanically dispensed green lacewing eggs. Biological Control 7: 245–250.

Gerling, D. 1967. Bionomics of the whitefly-parasite complex associated with cotton in southern California (Homoptera: Aleurodidae; Hymenoptera: Aphelinidae). Annals of the Entomological Society of America 60: 1306–1321.

Goolsby, J.A. and M.A. Ciomperlik. 1999. Development of parasitoid inoculated seedling transplants for augmentative biological control of silverleaf whitefly (Homoptera: Aleyrodidae). Florida Entomologist 82: 532–545.

Goolsby, J.A., M.A. Ciomperlik, B.C. Legaspi, J.C. Legaspi, and L.E. Wendel. 1998. Laboratory and field evaluation of exotic parasitoids of *Bemisia tabaci* (Gennadius) (Biotype "B") (Homoptera: Aleyrodidae) in the lower Rio Grande Valley of Texas. Biological Control 12: 127–135.

Hagler, J.R., C.G. Jackson, T.J. Henneberry, and J.R. Gould. 2002. Parasitoid mark-release-recapture techniques – II. Development and application of a protein marking technique for *Eretmocerus* spp., parasitoids of *Bemisia argentifolii*. Biocontrol Science and Technology 12: 661–675.

Hoelmer, K.A. 1998. Comparative field cage evaluations of top-performing introduced parasitoids in desert cantaloupes. In T.J. Henneberry, N.C. Toscano, R.M. Faust, and D. Kopp, eds., Silverleaf Whitefly: National Research, Action and Technology Transfer Plan, 1997–2001; First Annual Review of the Second Five-Year Plan, pp. 68. US Department of Agriculture, Agricultural Research Service, ARS 1998–01.

Hoelmer, K.A. 2007. Field cage evaluation of introduced *Eretmocerus* species (Hymenoptera: Aphelinidae) against *Bemisia tabaci* strain B (Homoptera: Aleyrodidae) on cantaloupe. Biological Control 43: 156–162.

Lara, L., J. Blom, P. Stansly, D. Beltrán, J. Calvo, and A. Urbaneja. 2002. Bemipar *(Eretmocerus mundus)* y Bemimix *(E. mundus + E. eremicus):* nuevos productos para el control biológico de la mosca blanca *Bemisia tabaci*. Almería en Verde 0: 19–20.

Palumbo, J.C., A.R. Horowitz, and N. Prabhaker. 2001. Insecticidal control and resistance management for *Bemisia tabaci*. Crop Protection 20: 739–765.

Perring, T.M., A.D. Cooper, R.J. Rodriguez, C.A. Farrar, and T. S. Bellows. 1993. Identification of a whitefly species by genomic and behavioral studies. Science 259: 74–77.

Pickett, C.H., G.S. Simmons, E. Lozano, and J.A. Goolsby. 2004. Augmentative biological control of whitefly using transplants. BioControl 49: 665–688.

Simmons, G.S. 2000. Studies on dispersal of a native parasitoid *Eretmocerus eremicus* and augmentative biological control of *Bemisia tabaci* infesting cotton. [University of Arizona]. Dissertation Abstracts International 61: 1189.

Simmons, G.S., K. Hoelmer, C. Pickett, E. Natwick, and E. Andress. 2002. Augmentative biological control of silverleaf whitefly infesting melons with releases of exotic *Eretmocerus* species: a review of the progress. In T.J. Henneberry, N.C. Toscano, T.M. Perring, and R.M. Faust, eds., Silverleaf Whitefly: National Research, Action & Technology Transfer Plan 1997–2001: Fifth Annual Review of the Second Five-Year Plan, pp. 168. US Department of Agriculture, Agricultural Research Service, ARS 2002–01.

Urbaneja, A., J. Calvo, and P.A. Stansly. 2003. Calibration of release rates of *Eretmocerus mundus* Mercet (Hymenoptera: Aphelinidae) for control of *Bemisia tabaci* in sweet pepper and tomato. Bulletin OILB/SROP (IOBC/WPRS) 26: 69.

Chapter 17
Multivariate Analysis of *Bemisia tabaci* Biotype B and Associated Parasitoid Populations within the Imperial Valley Agricultural System

Earl Andress[1], Mark Quinn[2], and Juli Gould[3]

Abstract Multivariate analysis techniques were used to separate the effects of various agroecosystem variables on populations of *Bemisia tabaci* and its native and exotic parasitoids in the Imperial Valley of California. The effects detected were significant and complex. Parasitoid numbers increased with higher *B. tabaci* density, but positive density-dependence was not observed. Parasitism consistently accounted for a significant portion of *B. tabaci* mortality, and exotics accounted for a large portion of the mortality clearly attributable to parasitoids especially in alfalfa and broccoli. Percentage parasitism was greatest at sites associated with high regional coverage of alfalfa, and alfalfa seems to be an excellent reservoir for parasitoids. Regional coverage of cantaloupe also seems to enhance biocontrol of *B. tabaci* at sample sites, including biocontrol from exotics. Populations of parasitoids at sample sites were enhanced by regional coverage of cotton. Populations of parasitoids were lower at sites exposed to relatively high regional insecticide loads, but not necessarily local pesticide loads. This study was initiated only 2 years after the first releases of *Eret.* sp. nr. *emiratus* (*ex.* Ethiopia), the species that eventually became the most numerous exotic parasitoid. It is likely that exotic parasitoid populations will continue to adapt for some time to come, and that their impact has not yet reached its full potential. Once sufficient time has elapsed for populations of introduced species of parasitoids to reach their full potential, multivariate studies of this type can be used determine the impact of natural enemy populations on *B. tabaci* populations after accounting for the effects of surrounding crops and insecticide use.

[1]USDA-APHIS-PPQ Pink Bollworm Rearing Facility, 3645 E. Chipman Road, Phoenix, AZ 85040

[2]Department of Crop and Soil Sciences, Washington State University, P.O. Box 646420, Pullman, WA 99164-6420

[3]USDA-APHIS-PPQ, Center for Plant Health Science and Technology Laboratory, Building 1398, Otis ANGB, MA 02542-5008

17.1 Introduction

The impact of released natural enemies on some whitefly pests has been well documented (Quezada 1974; Dowell et al. 1979; Gould et al. 1992). Designing methodologies to evaluate the impact of parasitoids released against *Bemisia tabaci* biotype B (Genn.) (=*B. argentifolii* Bellows and Perring) (Hemiptera: Aleyrodidae), however, is not straightforward. Evaluating the success of classical biological control programs for *B. tabaci* was complicated by the whitefly's broad host range, coupled with the influence of farming practices, cropping patterns, and climate. To further complicate matters, these factors may operate at different spatial and temporal scales and are likely interrelated. For these reasons, data on parasitoid releases in the Imperial Valley of California were collected and analyzed using multivariate statistics to determine the impact of introduced whitefly parasitoids.

Host crops are available to *B. tabaci* in the Imperial Valley of California throughout the year. In favorable weather conditions gravid females can cause a population explosion if suppressive measures are not in place. Alfalfa (*Medicago sativa* L.) is the major crop in the Imperial Valley averaging between 71,000 ha and 91,000 ha over a 20-year period (Finnell 1980–1986; Birdsall 1987–2001). Although effectively found year round, *B. tabaci* populations are generally considered low in alfalfa compared to cantaloupe (*Cucumis melo* L. var. *reticulatus* (Naudin.)) or cotton (*Gossypium hirsustum* L.). Alfalfa can therefore provide a more stable habitat for *B. tabaci* parasitoids. Landscaping plants at residences and businesses scattered throughout the valley also provide perennially stable habitats and refuge to both whiteflies and parasitoids.

Seasonal crops are exceptionally good hosts for *B. tabaci* but for much shorter periods of time. During the fall and winter, cole crops (*Brassica* spp.) are present, and if overwintering *B. tabaci* populations on these crops are not controlled, they may quickly colonize spring crops. Shortly after the cole crop season ends and the fields are plowed under, spring cantaloupes and cotton emerge, providing some of the most favored *B. tabaci* host material in the Imperial Valley. Cantaloupe fields planted for spring harvest are available into June and cotton fields contain host plants through September or October. A monitoring program using traps conducted jointly by the Imperial County Agricultural Commissioner's office and the United States Department of Agriculture, Animal and Plant Health Inspection Service (USDA-APHIS) shows that peak *B. tabaci* populations historically occur in August or September (Fig. 17.1). Under current crop production patterns, cotton is the dominant host crop during these months, supporting peak populations of *B. tabaci*. During the height of the of *B. tabaci* outbreaks in the late 1980s and early 1990s, cantaloupe fields planted for fall harvest were available from July through November; fall cantaloupe acreage is now greatly reduced (Fig. 17.2) because of pressure from *B. tabaci* populations and associated control costs.

When parasitoids were being imported to the USA from all over the world to control *B. tabaci* in the 1990s, other control strategies also were being developed and implemented. New chemical tools became available that proved especially useful in

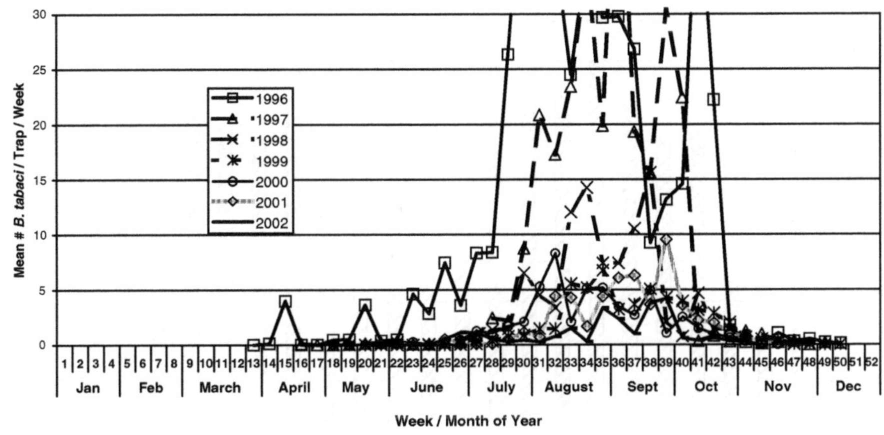

Fig. 17.1 Mean numbers of *B. tabaci* captured in 30 traps throughout the Imperial Valley from 1996 to 2002, illustrating annual seasonal peak in late summer–early fall (Courtesy of Imperial County Agriculture Commissioner).

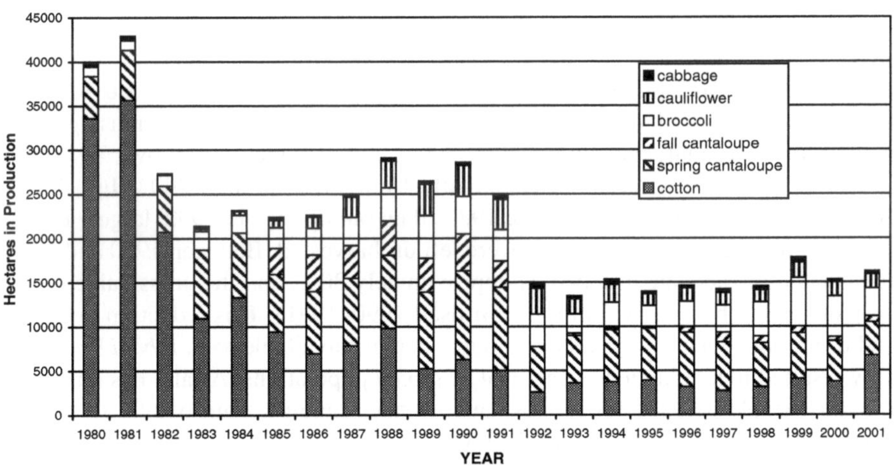

Fig. 17.2 Principle *B. tabaci* host plant crops (exclusive of alfalfa) available in the Imperial Valley, listed by gross acreage and year. Spring and fall cantaloupes were not reported separately prior to 1985 (From Finnell 1980–1986 and Birdsall 1987–2001).

controlling *B. tabaci* in cantaloupe and cole crops, reducing extreme population increases observed in these crops during the peak outbreak years. Imidacloprid, applied to the soil, was and continues to be widely used in cole crops as well as in cantaloupe (State of California 1990–2003); tank mixes of acephate and fenpropathrin were used on cotton (State of California 1990–2003). Other factors that

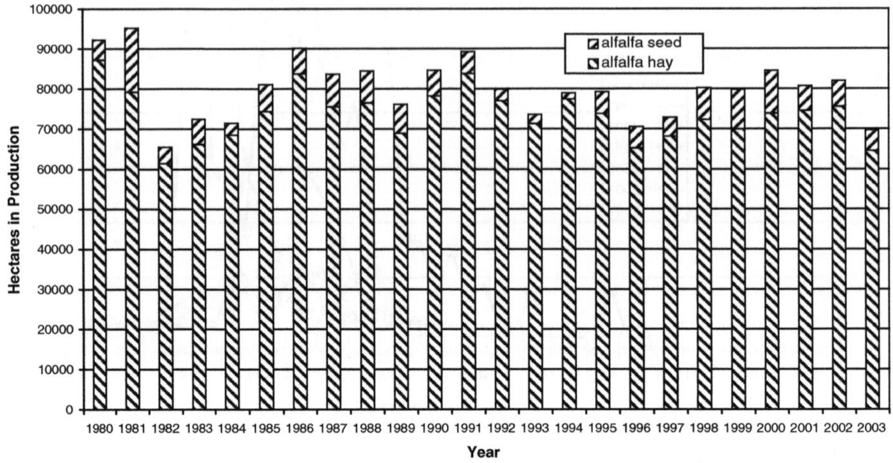

Fig. 17.3 Gross acreage of alfalfa from 1980 through 2001 (From Finnell 1980–1986 and Birdsall 1987–2001).

influenced *B. tabaci* populations were undergoing change. Control costs for pink bollworm, boll weevil, and more recently *B. tabaci* biotype B and increasing competition in world markets, combined to reduce cotton acreage from over 40,000 ha in the 1970s to less than 8,000 ha annually from 1989 through 2002 (Fig. 17.2). Acreage of cantaloupe peaked in the 1980s then plummeted in 1992, largely due to *B. tabaci* damage, and many fields were left un-harvested. Between 1990 and 1991 fall cantaloupe acreage dropped by approximately 30%, and the gross value of the crop decreased by over $15 million (Birdsall 1990–1991). This reduction in cotton and cantaloupe (especially fall cantaloupe) acreage available to *B. tabaci* has probably resulted in a reduction of regional *B. tabaci* populations. Alfalfa has remained the most stable acreage for the last two decades ranging from 65,000 to 89,000 harvestable hectares annually (Fig. 17.3).

17.2 Evaluating the Impact of Biological Control of *B. tabaci*

Classical biological control has been a very successful strategy for controlling several species of non-native pest whiteflies (DeBach and Rose 1976; Onillon 1990). Indigenous natural enemies, such as *Eretmocerus eremicus* (Hymenoptera: Aphelinidae), failed to adequately regulate *B. tabaci* biotype B populations in the Imperial Valley, as evidenced by the epidemic outbreak populations of the 1980s and early 1990s. Releases of imported *B. tabaci* natural enemies in a classical biocontrol program began in the mid-1990s in the Imperial Valley (Hoelmer 1996) and continued into 2000. In the mid-1990s, *B. tabaci* densities began dropping noticeably, and in 2002 Imperial Valley experienced its fifth year in a row of low *B. tabaci* populations (Fig. 17.4). Exotic parasitoids have become established, and their prevalence is increasing in the Imperial Valley of California (Roltsch et al. 1999; Goolsby et al. 2005; Roltsch personal communication).

The goal of all classical biological control programs is a reduction in pest populations caused by the activity of the introduced natural enemy. Establishment of natural enemies is essential for effective classical biological control and is easy enough to prove by recovering the introduced species for several years after the last inoculative release. However establishment alone does not necessarily indicate a change in the target pest population; other factors active in the system may mislead one to credit biocontrol with all of the reduction of pest populations. The problem was to determine whether the observed reduction in *B. tabaci* populations in the Imperial Valley of California was, in fact, a result of introduced natural enemies or caused by other factors.

A typical analysis of successful biocontrol by parasitoids shows high densities of the pest before the introduction of exotic natural enemies. After the introduction, mortality attributed to natural enemies increases, and the density of the pest falls and remains at low levels. Differences in pest density before and after natural enemy release can be correlated to natural enemy activity and interpreted as the effect of the natural enemy. This type of analysis does not work well when the density of the pest from year to year or from site to site is highly variable and influenced by factors such as migration, changing cropping patterns, year-specific weather patterns, and/or changing pesticide use. All of these factors affect the

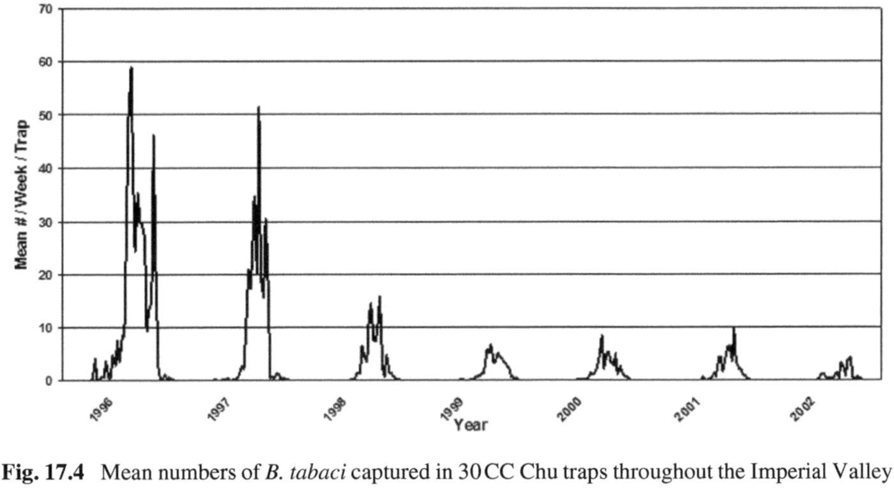

Fig. 17.4 Mean numbers of *B. tabaci* captured in 30 CC Chu traps throughout the Imperial Valley from 1996 to 2002, plotted to illustrate consecutive annual decreasing trend (Courtesy of Imperial County Agricultural Commissioner).

dynamics of *B. tabaci*. One factor that contributes heavily to high sample variability is that the sampling units are not stable through time. Other successfully evaluated whitefly species are pests of perennial plants that can be sampled repeatedly. *Bemisia tabaci* attacks annual crops such as melons, cotton, and cole crops that are not always grown in the same field each year, and important factors that influence *B. tabaci* density (nearness to sources of whiteflies, pesticide use for other pests, etc.) change accordingly.

One approach to evaluating and predicting the success of biological control agents is to use multivariate statistical methods to define habitat and crop production factors associated with the establishment, efficacy, and population/community structure of the natural enemies. Sample sites or habitat patches can be described, differentiated, and grouped based on landscape analysis (Menalled et al. 1999). For example, Quinn et al. (1991, 1993, 1995) used multivariate analyses to define several plant community and soil variables associated with generalist predators of grasshoppers on rangeland. Neuenschwander (1996) and D'Almeida et al. (1998) identified several regional landscape and life-history variables associated with populations of the spiraling whitefly, *Aleurodicus dispersus* Russell. Similarly, Brazzle et al. (1997) used a multivariate approach to identify several production, life-history, and landscape variables associated with the abundance of *B. tabaci* in the Imperial Valley of southern California. Multivariate methods should be particularly suitable for evaluating the relationships among parasitoid population parameters, such as percentage parasitism and abundance, host characteristics, landscape characteristics at various spatial scales, and crop production factors.

Studies were conducted from 1999 through 2001 to determine relationships among crop and insecticide use patterns, *B. tabaci* and parasitoid populations, and biocontrol of *B. tabaci* in the Imperial Valley agroecosystem using multivariate and landscape ecology methods. These methods are particularly suitable for evaluating the efficacy of *B. tabaci* parasitoids in the Imperial Valley because of the complex spatial and temporal dynamics of host crops in the region, the availability of pesticide use information collected by the Imperial County Agricultural Commissioner's office on all agricultural fields in the valley, and the availability of a digitized map of all fields in the valley.

17.3 Methods

17.3.1 Insect Sampling

Insects were sampled on four different crops, alfalfa, broccoli (*Brassica oleracea* L. *italica* group), cantaloupe, and cotton in two different years (Table 17.1). Although parasitoid releases continued in Imperial County as late as the summer of 2000, care was taken to ensure that none of the data used in this study were from fields in which releases occurred. Samples were collected in 1999 through 2001 as

Table 17.1 *B. tabaci* host crops studied, dates of sample collection, and number of fields and sites sampled in the Imperial Valley, California, 1999–2001.

Crop	Dates sampled	Number of fields sampled	Total sites
Broccoli	16 November–30 December 1999	30	50
	2 January–18 January 2001	21	42
Cantaloupe	5 April–3 May 2000	22	44
	21 May–13 June 2001	21	42
Alfalfa	31 July–14 September 2000	20	42
	27 September–23 October 2001	22	43
Cotton	5 June–12 June 2000	24	45
	6 July–14 August 2001	20	40

soon as sufficient numbers of *B. tabaci* were widely available on target host crops as determined by periodic examination of randomly selected fields. The number of fields sampled ranged from 20 to 30 per crop type, and one to four sites were sampled within a contiguous stand of a single crop. Locations of fields of each crop type were chosen to ensure representation of all regions in the Imperial Valley and on the availability of cooperators. Large fields were divided into multiple quadrants of approximately 10 ha, and insects were sampled in one to four randomly chosen quadrants per field. *Bemisia tabaci* populations peaked earlier in 2000 than in 2001 (Fig. 17.1) so sampling times were correspondingly earlier in 2000 (Table 17.1).

At each sample site, one leaf from each of 40 randomly selected plants was sampled to estimate population levels of *B. tabaci* and their parasitoids and mortality from parasitoids, predators and other sources. Leaves were collected from the portion of the canopy empirically determined to have the highest populations of 3rd and 4th-instar *B. tabaci* and brought to the lab for analysis. Depending on initial estimates of *B. tabaci* densities, insect counts were made from either whole leaves (leaf area was measured using a Li-Cor LI-300A leaf area meter (Li-Cor Biosciences, Lincoln, Nebraska) or from single, 3.5-cm^2 leaf disks cut from the leaves. Host population levels were estimated by counting the number of *B. tabaci* eggs, 1st–3rd instar nymphs, 4th-instar pupae, and exuviae and dividing by the area to obtain numbers per square centimeter. Numbers of older *B. tabaci* were calculated as the sum of 4th-instar pupae and *B. tabaci* exuviae; this constitutes the cohort of *B. tabaci* in which parasitism is detectable using morphological methods. Population levels of *Eretmocerus* and *Encarsia* (Hymenoptera: Aphelinidae) species were estimated by counting the number of parasitized pupae and exuviae with parasitoid emergence holes. The total number of parasitoids was calculated as the sum of parasitoid pupae, parasitoid exuviae, and immature whitefly nymphs with displaced mycetomes (indicating a parasitoid larva inside the whitefly nymph). Mortality due to non-parasitoid factors was determined by counting dead nymphs that showed no evidence of parasitism, and included *B. tabaci* with signs of predation.

Percentage parasitism and parasitoid origin (exotic or indigenous) were determined by DNA hybridization (squash blot). When sample numbers allowed, 100 second through fourth instar *B. tabaci* nymphs were randomly collected from each sample and were subjected to DNA hybridization. At lower densities, the entire sample of second through fourth instars was analyzed. A squash blot hybridization technique using probes for highly abundant but species-specific DNA was used to identify species of *Eretmocerus* even at early stages of development (Heilmann 1997, 1998 and Heilmann et al. 2002). Insect squashing and stabilization of DNA was done at the USDA-APHIS laboratory in Brawley, California and the hybridization with a series of species specific probes labeled with ^{32}P was conducted at the Department of Biochemistry, North Dakota State University.

Immature *B. tabaci* were squashed onto nylon membranes that had been treated first with 0.4M NaOH/1.5M NaCl and then with 1M Tris-HCl, pH 7.0/1.5M NaCl. The DNA was bound to the membrane by UV light treatment or by baking at 80°C (Sambrook et al. 2001). Hybridization was accomplished using the method of Church and Gilbert (1984). Prehybridization was performed at 65°C for 12–18h in a solution of 0.5M $NaPO_4$, pH 7.0, 7% sodium dodecylsulfate (SDS) and 1% bovine serum albumin (BSA). Purified probe DNA was labeled with ^{32}P by the random primer method (Feinberg and Vogelstein 1983). About 2×10^6 cpm/ml was added to the pre-hybridization mix and the incubation continued for another 18–24h. The hybridization solution was removed and the membranes washed twice at 65°C with 50mM $NaPO_4$, 5% SDS and 0.5% BSA and then twice at 65°C with 50mM $NaPO_4$, 1% SDS. The filters were finally rinsed briefly in room temperature 2X SSC (0.3M NaCl, 0.03M Na citrate), wrapped wet in plastic wrap and exposed to x-ray film at 80°C. If the species specific labeled probe had bound with the insect DNA on the membrane then the x-ray revealed a dark blot on the membrane. The same membrane had to be subjected to a series of labeled probes and subsequently x-rayed to determine which parasitoid species, if any, had attacked the *B. tabaci* nymph.

17.3.2 Insecticide Load Assessment

Insecticide loads were assessed using five variables: (1) total number of insecticide applications applied to the sampled field (since alfalfa is a perennial it can receive pesticide treatments over the course of several years; this study used data only from the calendar year in which the sample was taken); (2) number of insecticide applications applied to a sampled field in the 10 weeks prior to collecting insects; (3) total number of insecticide applications in an area approximately 1 km around each sampled field; (4) number of insecticide applications in a 2km diameter region in the 10 weeks prior to collecting insects; and (5) the number of days between the last insecticide application to a sampled field and the insect sampling date. Regional insecticide loads were examined because of their potential effects on dispersing populations of whitefly and natural enemies. Data based on the use of spinosad and *Bacillus thuringensis* insecticides were excluded from the analyses.

17.3.3 Agroecosystem Landscape Analysis

A digitized map indicating the placement of crop fields, roads, urban areas, riparian habitats, and other landscape features was obtained from the Imperial Irrigation District and imported into an ARCView GIS database (ESRI 1992–1997). All crops within 2 km of each sample site were identified by visiting each field and these data were incorporated into the database; due to logistical problems, the fields between 1 and 2 km of the sample sites for the second broccoli crop were not inspected. The GIS program was used to calculate the coverage of all crop types, riparian habitat, desert, and urban areas within 1 and 2 km of each sample site. The number of different crop types within these distances was also recorded as a measure of crop diversity.

17.3.4 Statistical Analysis

At any given time in the Imperial Valley, the agroecosystem landscape consists of a mosaic of one or more host crops (e.g., cotton, broccoli, melons, alfalfa), non-host crops, non-crop vegetation, urban areas, and desert habitat. Because of the complexity of the agroecosystem, it was expected that many of these variables would be correlated. Factor analysis of landscape and insecticide load variables was used to determine relationships among the agroecosystem components and to create a reduced set of landscape and crop production variables, or factors, to describe the agroecosystem. A preliminary factor analysis was conducted with the variables shown in Tables 17.2 and 17.3, and included coverage of broccoli, cotton, alfalfa, cantaloupe, combined annual host crops (i.e., cotton, broccoli, and cantaloupe), non-host crops, urban areas, desert habitat, and number of crop types. These variables were measured at two spatial scales: between 0 and 1 km from sample sites, and between 1 and 2 km from sample sites (except for the second broccoli crop). Results from the preliminary factor analysis indicated that landscape variables measured at the two spatial scales had similar factor loadings, indicating that coverage within 1 km of sample sites correlated highly with coverage between 1 and 2 km of sites. A second factor analysis was conducted on the same landscape variables, omitting the measurements at the largest spatial scale, which allowed us to include data associated with the crop of broccoli sampled for insects in 2001. For both analyses the SAS procedure PROC FACTOR was used to derive varimax rotated factor patterns (SAS Institute1996). All factors with eigenvalues (amount of variation explained by each factor) greater than 0.75 were included in subsequent analyses. Spearman correlation analysis (Sokal and Rohlf 1981) was used to determine relationships between insect variables and the factor loadings from the four principal factors derived from factor analysis of the landscape and insecticide use data.

The Wilcoxon sign rank test and the Kruskal-Wallis test (Sokal and Rohlf 1981; PROC NPAR1WAY, SAS Institute, 1996) were used to determine the effect of the

Table 17.2 Imperial Valley agroecosystem landscape variables associated with *B. tabaci* host crops sampled for insects, 1999–2001.

	Host crop sampled (number of hectares)			
Landscape component	Broccoli	Alfalfa	Cotton	Cantaloupe
Broccoli coverage – 0–1 km	55.1 ± 3.50	1.1 ± 0.65	1.0 ± 0.56	1.6 ± 0.83
Broccoli coverage – 1–2 km	49.3 ± 9.26	6.5 ± 1.76	4.8 ± 1.57	4.8 ± 1.51
Cotton coverage – 0–1 km	0.8 ± 0.48	9.0 ± 2.04	54.1 ± 3.21	1.7 ± 0.88
Cotton coverage – 1–2 km	4.3 ± 2.01	18.0 ± 3.55	32.5 ± 4.73	7.0 ± 1.81
Cantaloupe coverage – 0–1 km	0.0 ± 0.00	0.7 ± 0.52	3.8 ± 1.32	50.4 ± 3.22
Cantaloupe coverage – 1–2 km	0.0 ± 0.00	3.7 ± 1.33	13.2 ± 3.50	29.4 ± 4.47
Other melon coverage – 0–1 km	0.9 ± 0.50	0.0 ± 0.00	0.0 ± 0.00	1.7 ± 0.71
Other melon coverage – 1–2 km	9.3 ± 2.48	0.0 ± 0.00	0.6 ± 0.43	4.3 ± 1.73
Primary host crops – 0–1 km	63.0 ± 3.68	11.2 ± 2.25	59.3 ± 3.46	58.5 ± 3.75
Primary host crops – 1–2 km	73.6 ± 9.75	29.9 ± 4.57	52.4 ± 5.45	54.5 ± 6.34
Alfalfa coverage – 0–1 km	89.1 ± 6.19	129.3 ± 6.59	71.0 ± 6.26	99.2 ± 6.37
Alfalfa coverage – 1–2 km	299.3 ± 19.47	244.7 ± 15.53	201.4 ± 16.80	265.1 ± 15.52
Nonhost crop coverage – 0–1 km	61.8 ± 4.57	66.8 ± 5.70	101.9 ± 7.23	78.9 ± 5.22
Nonhost crop coverage – 1–2 km	170.0 ± 11.41	263.9 ± 19.87	288.4 ± 17.04	208.2 ± 13.79
Urban area coverage – 0–1 km	1.5 ± 1.01	0.1 ± 0.01	2.8 ± 1.77	3.2 ± 1.73
Urban area coverage – 1–2 km	4.8 ± 2.56	5.5 ± 2.43	10.1 ± 4.77	12.1 ± 6.61
Desert area coverage – 0–1 km	2.2 ± 1.26	9.2 ± 3.70	4.4 ± 2.26	5.8 ± 3.03
Desert area coverage – 1–2 km	24.6 ± 10.06	37.0 ± 12.33	34.1 ± 9.74	35.0 ± 11.51
Number of crop types – 0–1 km	4.6 ± 0.14	3.6 ± 0.20	4.9 ± 0.25	5.0 ± 0.19
Number of crop types – 1–2 km	7.0 ± 0.29	5.7 ± 0.37	6.7 ± 0.39	7.2 ± 0.33

Values are mean coverage in hectares (±1 SEM) for landscape components within 1 km of insect sample sites in broccoli, alfalfa, cotton, or cantaloupe, and coverage between 1 km and 2 km of insect sample sites. Sample sizes (number of sites) were alfalfa (85), cotton (85), cantaloupe (86), and broccoli. Samples were taken at 0–1 km (92), and 1–2 km (50).

Table 17.3 Insecticide load variables associated with major whitefly host crops sampled for insects in the Imperial Valley, California, 1999–2001.

	Crop			
Insecticide load variable	Broccoli	Alfalfa	Cotton	Cantaloupe
Days since last application[a]	44.7 ± 2.87a	165.1 ± 9.82b	51.3 ± 4.02a	66.3 ± 4.61a
Total field applications[b]	10.6 ± 0.72a	2.2 ± 0.20b	2.8 ± 0.34b	0.8 ± 0.12c
Recent field applications[c]	4.3 ± 0.56a	0.3 ± 0.08b	2.0 ± 0.31c	0.6 ± 0.10b
Total regional applications[d]	68.3 ± 4.76a	24.0 ± 2.37b	22.4 ± 2.38b	24.4 ± 2.16b
Recent regional applications[e]	29.3 ± 2.61a	4.1 ± 0.59b	7.4 ± 0.84c	10.9 ± 1.41c

Values are means ±1 SEM. Sample size equals the number of sites for each crop: broccoli (92), alfalfa (86), cotton (86), cantaloupe (86), (2 years combined). Means within rows followed by the same letter are not significantly different ($P < 0.001$; Fisher's protected LSD test).

[a]Number of days between the last application of an insecticide and the insect sampling date.

[b]Total number of insecticide applications made to the insect sampling site.

[c]Number of insecticide applications made to the insect sampling sites within 10 weeks of sampling insects.

[d]Total number of insecticide applications made to all crop fields within 0.8 km of the insect sampling sites.

[e]Number of insecticide application made to all crop fields within 0.8 km of insect sampling sites, within 10 weeks of sampling insects.

four host crop types on numbers of insects and percentage mortality from parasitoids and non-parasitoid related factors. These tests were used because not all of the variables could be normalized for parametric analyses. Analysis of variance (PROC GLM, SAS Institute 1996) was used to determine the effect of host crop and landscape/production factors on the abundance of whitefly. Analysis of covariance was used to determine the effects of host crop, landscape/production factors and initial whitefly population levels on abundance of total parasitoids, percentage parasitism by native and total parasitoids, and percentage mortality from predators and abiotic factors. The four factors derived from the factor analysis were used in the analyses as measures of landscape and insecticide use patterns. The covariable used in the analysis of covariance of percent parasitism (assessed with the molecular technique) was total number of whiteflies (estimated using the morphological technique described above). Thus, percent parasitism and initial abundance of whitefly were independently estimated. All response variables were either square-root, $\log_{10}$, or arcsin transformed to ensure normality and homoscedasticity of residuals.

17.4 Results

17.4.1 The Imperial Valley Agroecosystem

The Imperial Valley agroecosystem is a very large and complex mosaic of host crops, non-host crops, and non-crop vegetation. The landscape is temporally dynamic because of the high coverage of annual crops in the region. Table 17.2 shows the landscape variables associated with insect sample sites. Sample sites in all host crops were associated with high local and regional coverage of alfalfa, the dominant crop grown in the Imperial Valley, but generally, there was only one dominant annual host crop associated with each sample site. *Bemisia tabaci* and parasitoids always have two major host crops available to them within a range of 1 km, alfalfa and an annual host crop. For example, during the winter when broccoli was sampled for insects, regional coverage of cotton, cantaloupe, other melons, and other host crops was very small. Coverage of non-host crops was always less than coverage of host crops, and urban areas represented a small proportion of area around sample sites. The number of crop types around sample sites did not change significantly with host crop ($P > 0.05$; analysis of variance).

Data from the California Department of Pesticide Regulation suggest that insects in alfalfa are typically not exposed to direct insecticide applications due to the length of time between applications (Table 17.3). The mean time between the last insecticide application in alfalfa and insect sampling dates was 165 days, which was significantly greater than the application interval in the other host crops (Table 17.3; $P < 0.001$). Mean total number of insecticide applications was only 2.2 for alfalfa fields sampled in this study, and insecticide usage in the region around the alfalfa fields was relatively low, particularly within 10 weeks of the insect sampling date. In contrast, broccoli fields received the most recent applications, the most total applications, and were in regions that received high insecticide loads. Insecticide loads in cotton and cantaloupe were more similar to insecticide loads in alfalfa than in broccoli (Table 17.3). Based only on insecticide load data, broccoli may not be a good crop for biocontrol programs given the number of field and regional insecticide applications. Alfalfa may be the best crop habitat for minimizing exposure to insecticides, but cotton and cantaloupe also appear to be good crop habitats for biocontrol agents (in terms of insecticide loads). The particular insecticide and timing of applications are also important to consider when analyzing the impact of insecticides on *B. tabaci* and parasitoid populations. For example, cantaloupe seems to have very few insecticides applied to the fields being sampled. However, the one early imidacloprid application applied to almost all cantaloupe fields in the Imperial Valley is targeted at *B. tabaci* and is effective in preventing buildup of *B. tabaci* populations, as evidenced by the average number of days between the sampling date and the previous insecticide application date.

Results from factor analysis of the landscape and production variables indicated that four factors explained 95% of the variation in the variables associated with the insect sample sites. Table 17.4 shows information on which landscape and cropping parameters are most closely associated with each other and the relative amounts of variability that the groups of variables (factors) contribute. Factor 1 is associated primarily with insecticide load variables, with high loadings or values associated with the individual insecticide application variables as well as the coverage

Table 17.4 Factor analysis of landscape and insecticide load variables associated with insect sample sites in broccoli, cotton, cantaloupe, and alfalfa fields sampled in two different years, Imperial Valley, California, 1999–2001.

Variable[a]	Factor 1	Factor 2	Factor 3	Factor 4
Application days	−0.444	−0.209	−0.258	0.064
Total local applications	0.823	−0.049	−0.121	0.130
Recent local applications	0.768	0.044	−0.008	−0.060
Total regional applications	0.576	0.045	0.089	0.501
Recent regional applications	0.620	0.145	0.159	0.397
Alfalfa coverage	−0.161	−0.570	−0.099	0.183
Broccoli coverage	0.717	−0.095	0.175	0.197
Cotton coverage	−0.019	0.180	0.115	−0.590
Cantaloupe coverage	−0.313	0.140	0.556	0.169
Other melon coverage	0.004	0.023	0.132	0.101
Annual host crop coverage	0.344	0.195	0.691	−0.149
Nonhost crop coverage	−0.115	0.692	−0.133	−0.022
Urban area coverage	−0.070	−0.050	−0.033	−0.126
Desert coverage	−0.046	0.057	−0.165	0.025
Number of crop types	0.050	0.587	0.134	0.209
Eigenvalues	0.489	0.226	0.120	0.117

[a]See Table 17.3 for an explanation of insecticide variables. Crop variables refer to coverage within 1 km of insect sample sites. Variables contributing to each factor are explained in the text.

of broccoli around sample sites. As indicated in Table 17.3, broccoli had much higher insecticide loads than the other crops, which shows that local and regional insecticide loads are related. Factor 2 has a high negative loading associated with coverage of alfalfa and high positive loadings associated with coverage of non-host crops and number of crop types. Regions with high alfalfa coverage tended to have low coverage of non-host crops and low crop diversity. Factor 3 was associated with coverage of cantaloupe and the annual host crop currently being sampled (primary host crop). Factor 4 had high loadings associated with number of total regional insecticide applications (positive loading) and coverage of cotton (negative loading).

17.4.2 Agroecosystem Structure and Biocontrol

Bemisia tabaci populations were very low in the Imperial Valley from 1998 through 2002 (Fig. 17.4). Populations did not reach a level adequate for sampling in individual broccoli fields until just prior to harvest time. The cool weather typical during the growing season for broccoli greatly extended the generation time for *B. tabaci* (Wagner 1993; Nava-Camberos et al. 2001) and probably also for associated parasitoids. This slow development time late in the cropping cycle helped ensure that the high insect densities in samples never translated into high valley-wide populations and many of these insects probably did not survive the harvest and subsequent plow down. Even at low population levels, the effect of host crop on *B. tabaci* populations was evident (Tables 17.5 and 17.6). Density of *B. tabaci* nymphs was noticeably greater in broccoli that in the other host crops. Although mean number of total *B. tabaci* was much greater in broccoli than the other crops, the differences were not significant because the higher mean abundance in broccoli reflects a few sites with a disproportionate number of insects. Density of total *B. tabaci* was significantly less in alfalfa than in cotton or cantaloupe. Results from analysis of variance indicated that Factor 2 explained a significant amount of variation in total whitefly ($F = 3.3$, $P < 0.1$; Table 17.6). Factor 2 had a high negative value (i.e., loading) associated with alfalfa coverage and high positive values associated with non-host crop coverage and number of crop types (Table 17.4), suggesting that higher whitefly populations occurred at sites associated with low regional coverage of alfalfa, high regional coverage of non-host crops, and high regional crop diversity.

Percentage mortality from all factors other than parasitoids was quite high in alfalfa (43%) and cotton (41%), and was significantly lower in broccoli and cantaloupe (Table 17.5). This may reflect mortality of natural enemies due to high pesticide loads in broccoli and the systemic insecticides used in cantaloupe and broccoli that were targeted specifically for *B. tabaci*. In alfalfa and cotton, which had long periods between sampling and the most recent insecticide application (Table 17.3), mortality not associated with production of parasitoid progeny was due mostly to the activity of predators, parasitoid host feeding, and other natural causes. Analysis

Table 17.5 Imperial Valley agroecosystem insect variables associated with major *B. tabaci* host crops sampled for insects from 1999 to 2001.

	Crop			
Insect variables	Broccoli	Alfalfa	Cotton	Cantaloupe
No. *B. tabaci* eggs	0.52 ± 0.064ac	1.47 ± 0.262b	2.76 ± 0.611c	0.24 ± 0.104d
No. *B. tabaci* 2nd–3rd instar nymphs	2.98 ± 0.447a	1.13 ± 0.209b	0.76 ± 0.210c	0.39 ± 0.087c
No. *B. tabaci* 4th instar nymphs	2.23 ± 0.363a	0.39 ± 0.090b	0.71 ± 0.204b	0.21 ± 0.048b
No. *B. tabaci* pupae	0.14 ± 0.063a	0.08 ± 0.021a	0.22 ± 0.069b	0.22 ± 0.043c
No. *B. tabaci* exuvia	0.90 ± 0.215a	0.18 ± 0.040a	0.53 ± 0.103b	0.53 ± 0.084b
No. older *B. tabaci*[a]	3.19 ± 0.551ab	0.65 ± 0.124a	1.46 ± 0.355b	0.96 ± 0.166b
No. *Eretmocerus*[c]	0.04 ± 0.014a	0.05 ± 0.008b	0.09 ± 0.022bc	0.04 ± 0.013c
No. *Encarsia*[c]	0.04 ± 0.010a	0.01 ± 0.001a	0.00 ± 0.001b	0.00 ± 0.001a
No. parasitoids[c]	0.11 ± 0.031a	0.16 ± 0.020b	0.19 ± 0.038c	0.07 ± 0.014c
% Non-parasitoid mortality[b]	10.07 ± 0.762a	42.71 ± 2.529b	40.63 ± 1.413b	17.39 ± 0.989c
% Parasitism by indigenous species[d,e]	2.86 ± 0.774a	30.94 ± 2.186b	28.23 ± 2.392b	13.59 ± 1.436c
% Parasitism by exotic species[d, f]	5.43 ± 1.370a	13.09 ± 1.627b	3.00 ± 0.584ac	2.67 ± 0.563c
Total % parasitism[d]	8.29 ± 1.847a	44.03 ± 2.800b	31.23 ± 2.652c	16.26 ± 1.736d
Exotic species: indigenous species	2.31 ± 0.588a	0.91 ± 0.253b	0.12 ± 0.023c	0.27 ± 0.068c

Values for numbers of insects are means ±1 SEM insects per cm^2 leaf area. Means within rows followed by the same letter are not significantly different ($P > 0.05$; Wilcoxon and Kruskal Wallis test).

[a]Numbers of 4th instar nymphs, pupae, and exuvia combined.

[b](Number of dead whitefly from non-parasitoid sources/total number of whitefly) × 100.

[c]Numbers of larvae, pupae, exuvia, dead parasitoids, and immatures with displaced mycetomes, combined, as determined by morphological methods.

[d]Parasitoids identified using molecular techniques (see Section 17.3).

[e]*Eretmocerus eremicus.*

[f]*Eretmocerus emiratus* Zolnerowich and Rose (APHIS culture M95104), *Eret.* sp. nr. *emiratus* ex. Ethiopia (M96076), *Eret. hayati* Zolnerowich and Rose (M95012), *Eret. mundus* Mercet (various sources), and *Encarsia sofia* (Girault and Dodd) (M95103).

of covariance indicated a significant negative relationship between mortality from non-parasitoid sources and number of whitefly (Table 17.6); percent mortality from sources other than parasitoids tended to increase as whitefly populations decreased. A significant amount of variation in non-parasitoid mortality was explained by Factor 4 ($F = 24.0$, $P < 0.001$; Table 17.6), indicating that mortality from sources other than parasitoids was associated with high regional insecticide loads and low regional coverage of cotton (Table 17.4). Since a correlation to local insecticide applications is noticeably absent, this relationship possibly results from effects on insects outside the field being sampled. Perhaps immigration of *B. tabaci* from

Table 17.6 Analysis of variance and analysis of covariance of insect variables, Imperial Valley, California, 1999–2001.

Response variable	Source of variation	df	F-value	Regression parameter estimate ± SEM
Total no. whitefly	Crop	7	13.9****	
	Factor 1	1	0.1	–0.0164 ± 0.0665
	Factor 2	1	3.3*	0.1009 ± 0.0555
	Factor 3	1	0.2	–0.0247 ± 0.0637
	Factor 4	1	2.2	0.1003 ± 0.0673
	Residual	336		
% Nonparasitoid mortality[a]	Crop	7	16.5****	
	No. *B. tabaci*	1	172.1****	–0.0989 ± 0.0075
	Factor 1	1	2.1	–0.0554 ± 0.0378
	Factor 2	1	1.0	0.0315 ± 0.0316
	Factor 3	1	0.8	–0.0327 ± 0.0362
	Factor 4	1	24.0****	0.1885 ± 0.0384
	Residual	335		
Total no. parasitoids[a]	Crop	7	11.2****	
	No. *B. tabaci*	1	117.4****	0.0342 ± 0.0032
	Factor 1	1	0.0	0.0001 ± 0.0145
	Factor 2	1	0.9	–0.0114 ± 0.0121
	Factor 3	1	3.0*	0.0239 ± 0.0138
	Factor 4	1	19.6****	–0.0649 ± 0.0147
	Residual	346		
Total % parasitism[b]	Crop	6	15.9****	
	No. *B. tabaci*	1	0.1	–0.0017 ± 0.0076
	Factor 1	1	3.4*	–0.0700 ± 0.0378
	Factor 2	1	4.8**	–0.0502 ± 0.0230
	Factor 3	1	7.0***	0.0675 ± 0.0256
	Factor 4	1	1.2	–0.0328 ± 0.0299
	Residual	248		
% Parasitism – indigenous[b]	Crop	6	17.0****	
	No. *B. tabaci*	1	0.0	0.0017 ± 0.0117
	Factor 1	1	4.2**	–0.1187 ± 0.0581
	Factor 2	1	5.4**	–0.0820 ± 0.0354
	Factor 3	1	7.2***	0.1056 ± 0.0394
	Factor 4	1	0.5	–0.0324 ± 0.0460
	Residual	248		
% Parasitism – exotic[b]	Crop	6	6.9****	
	No. *B. tabaci*	1	0.0	–0.0003 ± 0.0146
	Factor 1	1	1.4	0.0845 ± 0.0707
	Factor 2	1	0.0	–0.0001 ± 0.0390
	Factor 3	1	3.1*	0.0902 ± 0.0511
	Factor 4	1	1.1	–0.0593 ± 0.0578
	Residual	147		

* P < 0.1; ** P < 0.05; *** P < 0.01; **** P < 0.001.

[a] Determined with morphological methods.

[b] Determined with molecular methods.

surrounding fields was reduced by regional applications of insecticides while natural enemies within the sampled field were minimally affected.

Densities of *Eretmocerus*, *Encarsia*, and total parasitoids, determined by examining the leaves for insects, were quite low in all crops, reflecting the low densities of their host (Table 17.5). Total numbers of parasitoids and *Eretmocerus* were significantly lower in broccoli than the other crops. Analysis of covariance indicated that the total number of parasitoids was positively correlated with number of whitefly ($F = 117.4$, $P < 0.001$; Table 17.6). Both Factor 3 (cantaloupe coverage and primary host crop) and Factor 4 (cotton coverage and regional insecticide application) explained a significant amount of variation in number of parasitoids ($F = 3.0$, $P <$ 0.1, and $F = 19.6$, $P < 0.001$, respectively). The positive relationship between Factor 3 and number of parasitoids (Table 17.6) and the high Factor 3 values associated with coverage of cantaloupe and annual host crops (Table 17.4) suggests that parasitoid populations at sample sites were affected positively by high regional coverage of cantaloupe and annual host crops. Similarly, the significant Factor 4 variable suggests that parasitoid populations were enhanced at sample sites associated with a high regional coverage of cotton and low regional insecticide loads (Table 17.4).

Total percentage parasitism, determined by molecular techniques, was affected significantly by host crop ($F = 15.9$, $P < 0.001$; Tables 17.5 and 17.6). Mean percentage parasitism was greatest in alfalfa (44%), followed by cotton (31%), cantaloupe (16%), and broccoli (8%). Percentage parasitism was not significantly related to numbers of whiteflies (Table 17.6). The abundance of whiteflies used in the analysis of covariance was estimated independently from percentage mortality from parasitoids (see Section 17.3). Factors 1, 2, and 3 explained a significant amount of variation in total percent parasitism (Table 17.6), indicating a complex relationship between the agroecosystem landscape and biocontrol from parasitoids. Factor 1, an insecticide load and broccoli coverage variable (Table 17.4), was negatively associated with total percent parasitism, suggesting that biocontrol from parasitoids was negatively affected by insecticide applications and regional coverage of broccoli. Factor 2 was also negatively associated with percent parasitism. The large negative loadings associated with regional coverage of alfalfa and positive loadings associated with non-host crop coverage and crop diversity (Table 17.4) suggests that these landscape variables influence parasitoid biocontrol, which is more effective in areas with high coverage of alfalfa and low coverage of non-host crops and low crop diversity. Factor 3, defined by high coverage of cantaloupe and annual host crops (Table 17.4), was positively associated with total percent parasitism.

Percentage parasitism by indigenous species (*Eret. eremicus* Rose and Zolnerowich) was highest in alfalfa (31%) and cotton (28%), and lowest in broccoli (3%) (Table 17.5) and was similar to levels for all parasitoid species combined (Tables 17.5 and 17.6). Percentage parasitism by exotic species was greatest in alfalfa (13%), and was generally lower than parasitism by indigenous species. Broccoli supported a significantly greater proportion of exotic parasitoids, whereas cotton and cantaloupe were strongly favored by indigenous species. Factor 3 was positively related to percentage parasitism by exotic species, suggesting greater mortality by exotic parasitoids at sites associated with high coverage of cantaloupe and other annual host crops. We speculate that the effect of cantaloupe on parasitism by exotic species in surrounding fields were greater than the effects of broccoli because of different cropping seasons and higher numbers of parasitoids produced.

The mean levels of parasitism observed fall within the ranges observed in earlier field studies. Augmentation of field populations with mass reared parasitoids has resulted in relatively high parasitism levels. In field release studies in the Imperial Valley Simmons et al. (1997) reported mean levels of parasitism up to 40% and levels in individual fields up to 60%. Ciomperlik and Goolsby (1996) found exotic species parasitizing 10–18% of *B. tabaci* in cotton, and Minkenberg et al. (1995) found 13.7% parasitism in field studies on melons and 21.5% in caged studies. Early reports on parasitism by indigenous parasitoids range from a low of 5% on cole crops (Roltsch and Pickett 1995) and 7% on cantaloupe (Simmons et al. 1997) to a moderate 30% on ornamentals (Ball and Weddle 1993) and cantaloupe (Roltsch and Pickett 1995).

17.5 Summary

Using this multivariate system of analysis led to successful separation of the effects of various agroecosystem variables on whitefly and parasitoid field populations and examination of the relationship between parasitoids and *B. tabaci* during the sampling period. The effects on *B. tabaci* and parasitoid populations due to the variables in the agroecosystem were detectable, significant and complex. Parasitoid numbers increased as *B. tabaci* numbers increased, but there was not an observable positive density-dependent response. Higher percentage parasitism in response to increasing host numbers that occurs with a positive density-dependent response has been shown to contribute to population regulation (Debach and Rosen 1991). The release of exotic natural enemies and their subsequent establishment and spread was probably not the leading cause of the reduction in whitefly populations in the Imperial Valley during the years we sampled. A negative correlation was evident with non-parasitoid mortality factors and may relate to other natural enemies. Parasitism consistently accounted for a significant portion of *B. tabaci* mortality, and exotics accounted for a large portion of the mortality clearly attributable to parasitoids especially in alfalfa and broccoli. Percentage parasitism was greatest at sites associated with high regional coverage of alfalfa, as alfalfa seems to be an excellent reservoir of parasitoids. Regional coverage of cantaloupe also seems to enhance biocontrol of *B. tabaci* at sample sites, including biocontrol from exotics. Populations of parasitoids at sample sites were enhanced by regional coverage of cotton. Populations of parasitoids were lower at sites exposed to relatively high regional insecticide loads, but not necessarily local pesticide loads.

At the time that this study was conducted, introduced parasitoid species were still increasing and only beginning to be consistently recovered throughout the agricultural system (Roltsch 2000). It is likely that exotic parasitoid populations will continue to adapt for some time to come, and that their impact has not yet reached its full potential. Once sufficient time has elapsed for populations of introduced species of parasitoids to reach their full potential, multivariate studies of this type can be used to determine the impact of natural enemy populations on *B. tabaci* populations after accounting for the effects of surrounding crops and insecticide use.

References

Ball, J.C., and D. Weddle. 1993. Untitled report. In Henneberry, T.J., N.C. Toscano, R.M. Faust, and J. R. Coppedge, eds., Sweetpotato Whitefly: 1993 Supplement to the Five Year National Research and Action Plan – First Annual Review, Tempe, Arizona, January 18–21, 1993. US Department of Agriculture, Agricultural Research Service, ARS-112, p. 93.

Birdsall, S. 1987–2001. Imperial County agricultural crop and livestock report (Annual Report Series). Imperial County Agricultural Commissioner's Office, Imperial County, California.

Brazzle, J.R., K.M. Heinz, and M.P. Parrella. 1997. Multivariate approach to identifying patterns of *Bemisia argentifolii* (Homoptera: Aleyrodidae) infesting cotton. Environmental Entomology 26: 995–1003.

Church, G.M. and W. Gilbert 1984. Genomic sequencing. Proceedings of the National Academy of Science 81: 1991–1998.

Ciomperlik, M. and J.A. Goolsby. 1996. Field impact evaluations of exotic parasitoids in the Lower Rio Grande Valley on selected crops. In Henneberry, T.J., N.C. Toscano, R.M. Faust, and J. R. Coppedge, eds., Silverleaf Whitefly: 1996 Supplement to the Five Year National Research and Action Plan – 4th Annual Review, February 4–6, 1996, San Antonio, Texas. US Department of Agriculture, Agricultural Research Service, ARS-1996–01, p. 107.

D'Almeida, Y.A., J.A. Lys, P. Neuenschwander, and O. Ajuonu. 1998. Impact of two accidentally introduced *Encarsia* species (Hymenoptera: Aphelinidae) and other biotic and abiotic factors on the spiraling whitefly *Aleurodicus dispersus* (Russell) (Homoptera: Aleyrodidae) in Benin, West Africa. Biocontrol Science and Technology 8: 163–173.

DeBach, P. and M. Rose. 1976. Biological control status of woolly whitefly. Citrograph 61: 162–184.

DeBach, P., and D. Rosen. 1991. Biological Control by Natural Enemies, 2nd Edition. Cambridge University Press, Cambridge.

Dowell, R.V., G.E. Fitzpatrick, and J.A. Reinert. 1979. Biological control of citrus blackfly in southern Florida. Environmental Entomology 8: 595–597.

ESRI. 1992–1997. ArcView GIS 3.2. Environmental Systems Research Institute, Redlands, California.

Feinberg, A.P. and B. Vogelstein. 1983. A technique for radiolabeling DNA restriction endonuclease fragments to high specific activity. Analytical Biochemistry 132: 6–13.

Finnell, C.M. 1980–1986. Imperial County agricultural crop & livestock report (Annual Report Series). Imperial County Agricultural Commissioner's Office, Imperial County, California.

Goolsby, J.A., P. De Barro, A.A. Kirk, R. Sutherst, M. Ciomperlik, P. Ellsworth, J. Gould, K. Hoelmer, S. Naranjo, M. Rose, W. Roltsch, R. Ruiz, C. Pickett, and D. Vacek. 2005. Post-release evaluation of the biological control of *Bemisia tabaci* biotype "B" in the USA and the development of predictive tools to guide introductions for other countries. Biological Control 32: 70–77.

Gould, J.R., T.S. Bellows, Jr., and T.D. Paine. 1992. Population dynamics of *Siphoninus phillyreae* in California in the presence and absence of a parasitoid, *Encarsia partenopea*. Ecological Entomology 17: 127–134.

Heilmann, L.J. 1997. Use of satellite DNA sequences as species-specific probes for *Eretmocerus* and *Encarsia* parasitoids. In Henneberry, T.J., T.M. Perring, and R.M. Faust, eds., Supplement to the Five-Year National Research and Action Plan: Progress Review, Technology Transfer, and New Research and Action Plan (1997–2001), Fifth Annual Review, San Diego, California. January 28–30, 1997. US Department of Agriculture, 1997–02, 272 pp.

Heilmann, L.J. 1998. Satellite DNAs as identification probes for *Encarsia* and *Eretmocerus* wasps. In Henneberry, T.J., N.C. Toscano, T.M. Perring, and R.M. Faust, eds., Silverleaf Whitefly: National Research, Action, and Technology Transfer Plan, 1997–2001 (Formerly Sweetpotato Whitefly, Strain B): First Annual Review of the Second 5-Year Plan, Charleston, South Carolina, February 3–5, 1998. US Department of Agriculture, 1998–01, 187 pp.

Heilmann, L.J. Z. Jianzhong Z., A.E. Oleson, E.R. Andress, J.R. Gould, and D.R. Nelson. 2002. A DNA test for parasitization of whiteflies by native and exotic *Eretmocerus* species. Proceedings of the 4th International Symposium on Molecular Insect Science in Tucson, Arizona, May 28–June 6, 2002.

Hoelmer, K.A. 1996. Whitefly parasitoids: can they control field populations of *Bemisia*? In D. Gerling, ed., *Bemisia*: 1995 Taxonomy, Biology, Damage, Control and Management, pp. 451–476. Intercept, Andover, UK.

Menalled, F.D., P.C. Marino, S.H. Gage, and D.A. Landis. 1999. Does landscape structure affect parasitism and parasitoid diversity? Ecological Applications 9: 634–641.

Minkenberg, O.P.J.M., H.J. Henter, J. Kaltenbach, C. Leonard, R. Malloy, J. Tovar-Soto, K. Zeigweid, R. Greatex, and K.T. Alcock. 1995. Evaluation of *Eretmocerus* nr. *californicus* ex. Arizona for augmentative biological control of silverleaf whitefly on field and greenhouse crops. In Henneberry, T.J., N.C. Toscano, R.M. Faust, and J.R. Coppedge, eds., Silverleaf Whitefly: 1995 Supplement to the Five Year National Research and Action Plan – 3rd Annual Review, January 28–31, 1995, San Diego, California, p. 131. US Department of Agriculture, Agricultural Research Service, ARS-1995–2.

Nava-Camberos, U., D.G. Riley, and M.K. Harris. 2001. Temperature and host plant effects on development, survival, and fecundity of *Bemisia argentifolii* (Homeptera: Aleyrodidae). Environmental Entomology 30: 55–63.

Neuenschwander, P. 1996. Evaluating the efficacy of biological control of three exotic homopteran pests in tropical Africa. Entomophaga 41: 405–424.

Onillon, J.C. 1990. The uses of natural enemies for the biological control of whiteflies. In D. Gerling, ed., Whiteflies, their Bionomics, Pest Status and Management, pp. 226–313. Intercept, Andover, UK.

Quezada, J.R. 1974. Biological control of *Aleurocnthus woglumi* [Homoptera, Aleyrodidae] in El Salvador. Entomophaga 19: 243–254.

Quinn, M.A., R.L. Kepner, D.D. Walgenbach, R.N. Foster, R.A. Bohls, P.D. Pooler, K.C. Reuter, and J.L. Swain. 1991. Effect of habitat characteristics and perturbation from insecticides on the community dynamics of ground beetles (Coleoptera: Carabidae) on mixed-grass rangeland. Environmental Entomology 20: 1285–1294.

Quinn, M.A., R.L. Kepner, D.D. Walgenbach, R.N. Foster, R.A. Bohls, P.D. Pooler, K.C. Reuter, and J.L. Swain. 1993. Grasshopper stages of development as indicators of nontarget arthropod activity: implications for grasshopper management programs on mixed-grass rangeland. Environmental Entomology 22: 532–540.

Quinn, M.A., D.D. Walgenbach, R.N. Foster, K.C. Reuter, and J.P. McCaffrey. 1995. Tritrophic associations between spider wasps (Hymenoptera: Pompilidae), wolf spiders (Araneae: Lycosidae), and grasshoppers (Orthoptera: Acrididae) on mixed-grass prairie. Journal of the Kansas Entomological Society 68: 7–19.

Roltsch, W. 2002. Establishment of introduced parasitoids of the silverleaf whitefly in Imperial Valley, CA. In T.J. Henneberry, N.C. Toscano, T.M. Perring, and R.M. Faust, eds., Silverleaf Whitefly: National Research, Action, and Technology Transfer Plan, 1997–2001 – Fourth

Annual Review of the Second 5-year Plan and Final Report for 1992–2002, 11–13 February, 2001, San Diego, California. US Department of Agriculture, Agricultural Research Service, ARS-2002–01, p. 167.

Roltsch, W. and C. Pickett. 1995. Silverleaf whitefly natural enemy refuges in the Imperial Valley. In T.J. Henneberry, N.C. Toscano, R.M. Faust, and J.R. Coppedge, eds., Silverleaf Whitefly: 1995 Supplement to the Five Year National Research and Action Plan - 3rd Annual Review, January 28–31, 1995, San Diego, California. US Department of Agriculture, Agricultural Research Service, ARS-1995–2, p. 135.

Roltsch, W., G. Simmons, and K. Hoelmer. 1999. Establishment of introduced parasitoids of the silverleaf whitefly in Imperial Valley, CA. In T.J. Henneberry and R.M. Faust, eds., Silverleaf Whitefly: National Research, Action, and Technology Transfer Plan, 1997–2001 – Second Annual Review of the Second 5-year Plan, January 31-February 2, 1999. Albuquerque, New Mexico. US Department of Agriculture, Agricultural Research Service, ARS-1999–01, p. 86.

Sambrook, J. and D.W. Russell. 2001. Molecular Cloning: A Laboratory Manual, 3rd Edition. Cold Spring Harbor Laboratory Press, Woodbury, NY.

SAS Institute. 1996. SAS/STAT user's guide for personal computers, version 6.12. SAS Institute, Cary, NC.

Simmons, G.S., K.H. Hoelmer, R. Staten, T. Boratynski, and E. Natwick. 1997. Seasonal innoculative biological control with parasitoids of *Bemisia* infesting cantaloupe in the Imperial Valley of California: A report on three years of investigations. In T.J. Henneberry, N.C. Toscano, T.M. Perring, and R.M. Faust, eds., Silverleaf Whitefly: 1997 Supplement to the Five Year National Research and Action Plan-Progress Review, Technology Transfer, and New Research and Action Plan (1997–2001), Fifth Annual Review, January 28–30, 1997, San Diego, California. US Department of Agriculture, Agricultural Research Service, ARS-1997–2, p. 170.

Sokal, R.R. and F.J. Rohlf. 1981. Biometry. W.H. Freeman, New York.

State of California. 1990–2003. Pesticide use report database (annual report series). State of California EPA, Division Pesticide Regulation, Environmental Monitoring and Pest Management Branch.

Wagner, T.L. 1993. Temperature-dependent development of sweetpotato whitefly, *Bemisia tabaci* (Homoptera: Aleyrodidae), biotype "B". Beltwide Cotton Conferences, pp. 714–716. National Cotton Council.

Chapter 18
Indigenous Parasitoids of *Bemisia* in the USA and Potential for Non-Target Impacts of Exotic Parasitoid Introductions

Kim A. Hoelmer[1], David J. Schuster[2], and Matthew A. Ciomperlik[3]

Abstract Surveys to document the presence and species composition of native natural enemies were conducted prior to the introduction of non-indigenous agents against sweetpotato whitefly, *Bemisia tabaci* biotype B, in the USA. Agricultural officials surveyed for the presence and natural enemies of *B. tabaci* in eight southeastern states, and researchers in Florida, Puerto Rico, South Carolina, Mississippi, Texas, and California conducted separate surveys of regional crops and other whitefly infested plants. General survey procedures in each area were similar and involved the periodic collection of *Bemisia*-infested foliage from a wide range of crops, weeds, ornamentals, and other native plants. In California deserts, surveys also included other whitefly species on host plants sharing habitats with host plants of *B. tabaci*. The greatest diversity of native parasitoid species attacking *B. tabaci* was reported in Florida, due perhaps to the diversity of invasive whitefly species established in Florida. Only two or three parasitoid species were responsible for the majority of parasitism of *B. tabaci* within any given region of the USA. The predominant species attacking *B. tabaci* prior to the introduction of new Palearctic parasitoid species were *Eretmocerus tejanus* (in Texas), *Eretmocerus eremicus* (Arizona and California), *Eretmocerus* sp. (undescribed, southeast USA), *Encarsia pergandiella*/*Enc. tabacivora* (southeastern USA and Texas), and *Encarsia luteola* (southwestern USA). *Trialeurodes abutiloneus* was the only other whitefly species regularly found on the same herbaceous, annual host plants utilized by *B. tabaci,* and it is a natural reservoir for many of the indigenous parasitoids that attack *B. tabaci* in much of the southern USA. Parasitoids reared from other whitefly species found on native desert vegetation were not the same species as those occurring on

[1]USDA-ARS, Beneficial Insects Introduction Research Unit,
501 South Chapel Street, Newark, DE 19713-3814

[2]University of Florida, Institute of Food & Agricultural Sciences,
Gulf Coast Research & Education Center, 14625 CR 672, Wimauma, FL 33598

[3]USDA-APHIS-PPQ, Center for Plant Health Science and Technology Laboratory, 22675 N. Moorefield Rd, Edinburg, TX 78451-9398

J. Gould et al. (eds.), *Classical Biological Control of* Bemisia tabaci *in the United States.*

B. tabaci and *T. abutiloneus*. At the time of latest surveys in 2001 (California) and 2003 (Texas), the exotic species that were introduced have remained limited to their intended target, *B. tabaci*.

18.1 Introduction

Prior to the introduction of any non-indigenous biological control agents into an area it is essential to know which natural enemies are already present so it is clear that establishment of an agent resulted from the deliberate introduction. Surveys to document the presence and species composition of native natural enemies that have adapted to a newly introduced pest species are vital in any classical biological control program before the introduction of new natural enemies. Such surveys were conducted prior to the introduction of non-indigenous agents against sweetpotato whitefly, *Bemisia tabaci* (Gennadius) biotype B (= silverleaf whitefly, *B. argentifolii* Bellows and Perring), in the USA. The potential for unanticipated non-target impacts of introduced biological control agents on native species is also a concern. Knowledge of the indigenous species is also critical to detect and evaluate such impacts. Environmental assessments prepared in support of releases of *Eretmocerus* and *Encarsia* against *B. tabaci* in the USA (USDA 1995a, b) recognized that species of *Eretmocerus* are known to attack only whiteflies, and that while a broader range of biologies are encompassed within the genus *Encarsia,* some individual species are relatively specific in the choice of hosts they attack. Only nonindigenous species obtained from foreign collections of *B. tabaci* were considered for release in the USA (Goolsby et al. 1998). The host specificity, and thus the potential for non-target impact, of newly introduced natural enemies was addressed in the southwestern USA by incorporating surveys of selected indigenous whiteflies found in habitats where *B. tabaci* occurred before, during and following the release programs to document any evidence of reproduction by exotic agents in these species.

The first reports of *B. tabaci* biotype B in the USA came from Florida where heavy infestations occurred in 1986 on ornamental crops in glasshouses and plastic-covered saranhouses (Schuster et al. 1989). The outbreak was determined to be a new biotype of *B. tabaci* (Perring et al. 1993), which was designated biotype B to distinguish it from *B. tabaci* biotype A. Biotype A had been found occasionally in Florida since the late 1800s but was not considered a pest (Hamon and Salguero 1987). Because there was no prior information about native natural enemies attacking *Bemisia* in Florida, surveys were conducted between 1988 and 1992 to document the native parasitoid species attacking the whitefly and to gain an understanding of their possible contribution to an integrated approach for managing the whitefly. Follow-up sampling was conducted during the spring of 1994. Sampling in Florida was expanded to include Puerto Rico beginning in 1990 because of *B. tabaci* outbreaks there.

Infestations of *B. tabaci* biotype B rapidly spread west and north from Florida, impacting vegetable production in the southeastern USA. Surveys of native parasitoids in South Carolina were initiated in 1992 (Simmons 1993), and in 1991 a multi-state survey for the occurrence of *B. tabaci* and its parasitoids was initiated by APHIS utilizing PPQ officers in various southeastern and southern states (Hennessey 1993). Similar surveys were also conducted in Mississippi, primarily in cotton-growing regions, in anticipation of future releases of natural enemies against the expected invasion of cotton by *B. tabaci* biotype B (Smith et al. 1997).

The subtropical Lower Rio Grande Valley of Texas, with its mixed cropping of vegetables and cotton, was one of the areas most heavily impacted by *B. tabaci* in the mid-southern USA (Riley and Sparks 1993). In 1992 the USDA-APHIS-PPQ Mission Biological Control Laboratory (MBCL; now called the Center for Plant Health Science and Technology Laboratory, Edinburg, TX), APHIS-PPQ Central Region, and Texas A&M University began a collaborative project to determine whitefly population dynamics and identify native natural enemy species and their seasonal abundance in Texas.

In the southwestern USA, the mild desert winters, hot summers, and year-round production of susceptible crops in the desert agricultural valleys of California and Arizona led to massive outbreaks of *B. tabaci* biotype B by the early 1990s (Perring et al. 1991). As was the case in Florida, *B. tabaci* biotype A was already present in the region and sporadically reached pest levels on cotton (*Gossypium hirsutum* L.) (Johnson et al. 1982). In desert valleys as in other regions of the USA, *B. tabaci* populations were sustained not only on a variety of crops, but also on many weeds, native plants, and widely planted ornamental species. A survey of desert host plants of *B. tabaci* was initiated in 1992 in the Imperial Valley of southeastern California by the county agricultural commissioner's office in cooperation with USDA-APHIS MBCL, and provided the first survey data on parasitoids of whitefly on these hosts. This survey was expanded by the USDA-APHIS Phoenix Plant Protection Center and the California Department of Agriculture to include *B. tabaci* on crops, ornamentals, and native desert host plants, and several native whitefly species on desert vegetation.

Studies in southern California and Arizona cotton fields conducted during outbreaks of *B. tabaci* biotype A in the 1960s reported several aphelinid parasitoids of *B. tabaci* on cotton, including *Eretmocerus eremicus* Rose and Zolnerowich (as *Eret. californicus* Howard or *Eret. haldemani* Howard), *Encarsia meritoria* Gahan, and *Enc. luteola* Howard (as *Enc. formosa* Gahan 'desert form') (Gerling 1966, 1967). Gerling (1967) also found *B. tabaci* on cheeseweed, *Malva parviflora* L., but did not mention rearing parasitoids from whiteflies on this host plant. There were no records of *B. tabaci* parasitoids from other regions of the USA prior to the introduction of biotype B in the late 1980s. However, several introductions of non-indigenous *Eretmocerus* and *Encarsia* were made between 1985 and 1987 in the Imperial Valley in an attempt to reduce damage caused by *B. tabaci* biotype A. *Eretmocerus mundus* Mercet and *Encarsia sophia* (Girault and Dodd) (as *Enc. transvena* (Timberlake)) were obtained from Rawalpindi, Pakistan and released in large numbers at widely separated locations throughout the Imperial Valley (Nuessly 1990). *Encarsia sophia* reproduced in field cages placed in cotton prior to the removal of cages. However, subsequent surveys in 1988 gave no evidence of their establishment nor were they found during surveys for indigenous parasitoids that were begun several years later in California following the introduction of *B. tabaci* biotype B.

18.2 Survey Methods

Each group developed its own procedures for surveying local crops and other whitefly infested plants. However, general survey procedures in each area were similar and involved the periodic collection of *Bemisia*-infested foliage from a wide range of crops, weeds, ornamentals, and other native plants. In California deserts, surveys included other whitefly species on host plants sharing habitats with host plants of *B. tabaci*. Because all parasitoids of *B. tabaci* and other whitefly species emerge from late 4th instars ('pupae'), sampling targeted foliage containing this stage. Leaf samples were held in containers in the laboratory until emergence of all parasitoids was complete. Specimens were identified according to existing keys in literature and unpublished information provided by specialists at local institutions, or by consultation with appropriate taxonomists. When *B. tabaci* biotype B was first noted in Florida, the taxonomic knowledge of parasitoids of *Bemisia* was inadequate to provide species identification of many of the native and exotic species encountered in surveys and foreign exploration. Fortunately, this situation improved rapidly during the course of the interagency project, as taxonomic resources were devoted to revising principal groups of *Encarsia* and *Eretmocerus*, describing new species, and preparing and publishing new identification keys (e.g., Heraty and Polaszek 2000; Rose and Zolnerowich 1997; see also Chapters 4 and 5).

18.3 Regional Surveys

18.3.1 Southeastern States (APHIS Survey)

The multi-state survey conducted by APHIS-PPQ officers in eight southern states (Alabama, Florida, Georgia, Louisiana, Mississippi, North Carolina, South Carolina, and Texas) found *B. tabaci* in all of these states. Five species of *Encarsia* (*luteola, opulenta* (Silvestri), *pergandiella* Howard, sp. nr. *strenua,* and an unidentified species) and two undetermined species of *Eretmocerus* (probably including *Eret. tejanus* Rose and Zolnerowich) were reported. All were generally distributed in these states except for *Enc. opulenta*, which was reported only from southern Florida (Hennessey 1993). Further surveys in some of these states were conducted by other parties and are reported below.

18.3.2 Florida and Puerto Rico

Foliage was collected from 111 different crop and weed species during the survey. Collections in Florida were made weekly, biweekly, or monthly depending upon the time of year and location (Schuster et al. 1998). Sample sites were located in nearly the every region of the state, including Quincy in the panhandle; the University of Florida at Gainesville in the north central area; Leesburg and Sanford in the central area; Ruskin, Parrish and Bradenton in the west central region; Immokalee in the southwest area; 10 sites along the Florida Turnpike running from the central area to the extreme southeast coastal area; and the Florida Keys. Collections in Puerto Rico were made by individuals visiting the commonwealth for other reasons, and were, therefore, completed on an irregular basis. The number of leaves collected varied by plant species and location, depending upon the density of the whitefly population and the size of the leaves of the host plant. At some sites sampling was standardized by limiting the number of leaves collected to ten from the crops and weeds present, or to ten from either okra (*Abelmoschus esculentus* (L.) Moench) or collards (*Brassica oleracea* L.), planted specifically for monitoring parasitoid presence. Foliage with late instar nymphs was collected, examined for the presence of other insects, and placed in cylindrical paper cans (0.24, 0.47 or 0.95 l). Adult whiteflies and parasitoids were sorted to species and counted after they had emerged and died, which permitted the processing of large numbers of samples at a time. Because erroneous host records may have resulted if more than one host species was present in the sample, microscope slide preparations were made of the hosts from the Gainesville and Florida Turnpike collection sites to identify or confirm species identifications when necessary. Representative adults of each parasitoid species from each host from each location were mounted on microscope slides to confirm parasitoid identifications and to preserve a collection of voucher specimens. Vouchers are housed at the Florida State Collection of Arthropods, Florida Department of Agriculture and Consumer Services in Gainesville, Florida.

At least 13 species of parasitoids were recovered from *B. tabaci* in Florida, ten species were found in Puerto Rico, and one hyperparasitoid (*Signiphora aleyrodis* Ashmead) was recovered (Table 18.1). *Encarsia pergandiella* Howard and *Enc. tabacivora* Viggiani are sympatric and were placed in synonymy by Polaszek et al. (1992) and Schauff et al. (1996); G. Evans and J. Woolley and R. Johnson (personal communications) think this synonymy is incorrect. Evans (personal communication) believes that *Enc. tabacivora* predominates in Florida and, possibly, Puerto Rico; because of the different opinions we give both names in Table 18.1. *Encarsia meritoria* is a complex that includes *Enc. haitiensis* Dozier, *Enc. brasiliensis* (Hempel) and *Enc. hispida* DeSantis. The latter species was placed into synonymy with *Enc. meritoria* by Viggiani (1989), removed from synonymy by Polaszek et al. (1992), and returned to synonymy by Schauff et al. (1996). Polaszek (personal communication) still believes the two are separate species and one of the authors (D.S.) believes that the species reared from *B. tabaci* in this region is *Enc. hispida* and not *Enc.*

Table 18.1 Percentage species composition of parasitoids reared from survey samples of *Bemisia* spp. in Florida (n = 49,258) and Puerto Rico (n = 1531) from 1988 to 1994 (From Schuster et al. 1998).

Parasitoid sp.	Florida	Puerto Rico
Hym.: Platygasteroidea: Platygastridae		
Amitus bennetti Viggiani and Evans	<0.1	2.2
Hym.: Chalcidoidea: Aphelinidae		
Encarsia formosa Gahan	<0.1	6.4
Encarsia lanceolata Evans and Polaszek	<0.1	1.0
Encarsia luteola Howard	5.4	2.1
Encarsia meritoria Gahan/*hispida* De Santis	<0.1	8.8
Encarsia nigricephala Dozier	17.4	1.2
Encarsia pergandiella Howard/*tabacivora* Viggiani	61.7	71.5
Encarsia pseudocitrella Evans and Polaszek	<0.1	0.0
Encarsia quaintancei Howard	1.0	0.3
Encarsia strenua group	0.6	0.0
Encarsia sophia (Girault and Dodd) [=*transvena* (Timberlake)]	1.6	0.0
Encarsia sp.	0.2	0.6
Eretmocerus spp.	11.9	4.6
Hym.: Chalcidoidea: Signiphoridae		
Signiphora aleyrodis Ashmead	0.1	1.4

meritoria; because of the lack of agreement, both names are included in Table 18.1. The *Enc. strenua* group comprises at least six species known to attack *B. tabaci* and was summarized by Heraty and Polaszek (2000).

Encarsia pergandiella/tabacivora was the parasitoid most frequently recovered from *B. tabaci* in both Florida and Puerto Rico, comprising 62% and 72%, respectively (Table 18.1). *Encarsia nigricephala* Dozier and an undescribed *Eretmocerus* species were the next most abundant parasitoids in Florida and accounted for an additional 29% of the parasitoids recovered from *B. tabaci*. In Puerto Rico *Enc. meritoria/hispida*, *Enc. formosa* Gahan and *Eretmocerus* spp. were the next most abundant parasitoids recovered, accounting for about 19% of the specimens.

Although there is a rich diversity of parasitoids attacking *B. tabaci* in Florida and Puerto Rico, most species are too rare to have an impact on whitefly populations. The most abundant parasitoid found in Florida, *Enc. pergandiella/tabacivora*, is an autoparasitoid, where females are primary endoparasitoids of whiteflies and males are secondary endoparasitoids of conspecific female larvae or female larvae of other whitefly parasitoids. It is not clear how this complicated life history influences the level of impact parasitoids are able to exert on whitefly populations. However, *Enc. pergandiella/tabacivora*, in combination with other parasitoid species, can cause high levels of whitefly mortality in Florida. In a mixed vegetable organic farm in southwest Florida, nearly 80% of the whitefly nymphs were parasitized on tomato and >80% of the nymphs were parasitized on eggplant (*Solanum melongena* L.) (Brewster et al. 1997; Stansly et al. 1997). Whitefly mortality due to parasitism and predation on weeds on the perimeter of tomato (*Lycopersicon esculentum* Mill.) fields in west central Florida ranged from 40% to 90% in 1991 and 10% to 70% in 1992 (Schuster et al. 1992). Up to 100% of whitefly nymphs were parasitized on unsprayed peanuts (*Arachis hypogaea* L.) in north central Florida (McAuslane et al. 1993, 1994).

18.3.3 South Carolina

Surveys were conducted in Charleston County on snap beans (*Phaseolus vulgaris* L.) and cucumber (*Cucumis sativus* L.) in 1992 (Simmons 1993), and on sweet potato (*Ipomaea batata* (L.) Lam.) at five coastal locations in 1993 and 1994 (Simmons and Elsey 1994; Simmons 1995). Samples reared from 1992 to 1994 contained an *Eretmocerus* sp., *Enc. nigricephala, Enc. strenua* (Silvestri), *Enc. quaintancei* Howard, *Enc. pergandiella,* and an unidentified *Signiphora* species. Parasitoid abundance varied during the year but was highest in September and October. The predominant species when highest parasitism occurred were *Enc. pergandiella* and *Enc. nigricephala.*

18.3.4 Mississippi

Surveys were made in 1995 and 1996 in four distinct agroecosystems (Mississippi delta, northeastern Mississippi hills, coastal plains, and Black Prairie (east-central Mississippi)) and included 35 host plant species in 1995 and 53 in 1996, comprising field and vegetable crops and ornamental plants (Smith et al. 1997). Four whitefly species or biotypes were found in these surveys, including *Trialeurodes abutiloneus* (Haldeman), *T. vaporariorum* (Westwood), and both *B. tabaci* biotypes A and B. The two *Bemisia* strains were distinguished by placing adults from samples onto squash leaves for oviposition and watching for development of squash silverleaf, a growth disorder linked specifically to feeding by *B. tabaci* biotype B nymphs (Yokomi et al. 1990).

Parasitoids reared from the whiteflies reported by Smith et al. (1997) included *Amitus* sp., *Enc. americana* (DeBach and Rose), *Enc. cubensis* Gahan, *Enc. meritoria, Enc. nigricephala, Enc. pergandiella, Enc. quaintancei*, and unidentified *Eretmocerus* sp., *Signiphora* sp., and *Metaphycus* sp. All species were found in each year of the survey. *Eretmocerus* sp. was the most abundant in both years (42% and 48% of specimens), while all species of *Encarsia* accounted for 33–34% of the samples. Unidentified specimens comprised about 13% each year. However, parasitoid recoveries were not reported separately by whitefly host species, and it is very likely that some of the species reported were reared only from the *Trialeurodes* species and not from *B. tabaci.*

18.3.5 Texas

Surveys were initiated in 1992 by USDA-ARS (Jones 1993) and were included in larger multi-agency surveys organized in 1993. Twenty eight field sites were sampled along four transects across the lower Rio Grande Valley (LRGV), an area of intensive agriculture in south Texas, in an east–west direction from 1993 to 1995. The 4 transects extended from Rio Grande City to Harlingen (Highway 83), Alton to Rio Hondo (Highway 107), McCook to Willamar (Highway 490), and Granjeno to Carricitos (Highway 281). Samples were collected from field sites along the transects to determine the occurrence and abundance of *B. tabaci* and the native parasitoid complex and to evaluate the relative importance of host plants for whitefly reproduction, including crops and weedy plant species. Every 2 weeks, ten leaf samples per host plant per site were collected to determine numbers of immature whiteflies (eggs and nymphs) and parasitoids. All leaf samples were standardized by leaf age (3rd and 6th leaf node for nymphs and eggs, respectively, in cotton, cucurbits, and solanaceous crops, or 3rd or 4th leaf from the soil surface for adults and nymphs in *Brassica* crops). The ten leaf samples containing parasitized and unparasitized whitefly nymphs on leaves were brought to the laboratory to count numbers of immature *B. tabaci*, which were ranked into numerical categories (0 = none, 1 = 1–10, 2 = 11–100, 3 = 101–10,000, 4 = >10,000 whiteflies per sample). Percentage of parasitism was determined and samples were held for parasitoid emergence to determine species composition. Once emerged, adult parasitoids were sorted according to species, counted and sexed, and were forwarded to the Mission Biological Control Laboratory to verify species identification and for vouchering. Parasitoid voucher specimens were sent to Michael Schauff at the USDA-ARS Systematics Entomology Laboratory, Beltsville, Maryland for further taxonomic verification.

The 1992 survey recorded at least four species of parasitoids on crop and non-crop hosts. *Eretmocerus tejanus* (as nr. *californicus*) dominated early in the season and in the fall on crucifers, but *Enc. pergandiella/tabacivora* became dominant on most host plants by summer. An unidentified *Encarsia* species was most abundant on fall melons (*Cucumis melo* L.) and *Enc.* nr. *strenua* was the fourth species reared (Jones 1993).

A total of 35 crops and eight weed species were sampled for whitefly and parasitoids from 1993 to 1995 (Riley and Ciomperlik 1997). Although *B. tabaci* was found on all of these crops and weeds, some plant species supported greater numbers than others. The crops with notably higher numbers of eggs included cantaloupe melons and cotton, and high numbers of nymphs occurred on melons, cotton, cucumbers, squash (*Cucurbita* spp.), tomatoes, and beans. Six weed species were found to harbor low to moderate whitefly densities. Two widely distributed weeds, annual sowthistle (*Sonchus oleraceus* L.) and redroot pigweed (*Amaranthus retroflexus* L.), supported the largest number of whitefly nymphs.

The five parasitoid species routinely collected and identified during this expanded survey were, in order of decreasing abundance, *Eret. tejanus*, *Enc. pergandiella/ tabacivora*, *Enc. meritoria*, *Enc.* nr. *strenua*, and *En. formosa/luteola*. The first two species were found in almost every sample of parasitized whitefly and at all collection locations, suggesting that both species are widely distributed across the LRGV of Texas. *Eretmocerus tejanus* was generally more abundant early in the season, and various *Encarsia* species became very abundant later in the season (Moomaw et al. 1994). The total number of parasitoids collected during sampling from October 1993 to December 1995 is given in Table 18.2.

Crops with the highest number of parasitoids per sample (without regard to the total leaf area sampled), in descending order, were okra, cucumber, and eggplant. The largest numbers of parasitoids on weeds were collected from sowthistle, wild sunflower (*Helianthus annuus* L.) and redroot pigweed. Although more than 500 plant species are recorded as hosts of *B. tabaci* (Mound and Halsey 1978) not all hosts are equally suitable for whitefly reproduction (van Lenteren and Noldus 1990). The density of *B. tabaci* on various host plants may likewise affect parasitoid populations. A relatively large number of parasitoids and whitefly were collected from annual sowthistle in relation to the number of times it was sampled, suggesting it may serve as a good overwintering host for native natural enemies. Conversely, redroot pigweed produced relatively few parasitoids in relation to the number of times it was sampled, suggesting that it may serve as a better reservoir for whitefly than native parasitoid species.

This survey indicated that several crops and weed plant species, including cotton, cole crops (*Brassica* spp.), cucurbits (*Cucurbita* spp.), sowthistle and redroot pigweed may support large numbers of *B. tabaci*. Differences in host plant suitability for whiteflies and parasitoids, along with large acreages planted with highly suitable host crop plants in the LRGV may help explain why *B. tabaci* reached significant pest status there. Parasitoid populations were low in winter months on cole crops, increased on spring melons and other cucurbits, reached peak levels on summer cotton, and declined slightly on fall melons and cucurbits. Although the proportion of *B. tabaci* attacked by parasitoids reached high levels on some crops late in the season (Moomaw et al. 1994) it was evident that native natural enemies were incapable of maintaining *B. tabaci* populations below injury levels (Ciomperlik et al. 1995).

During follow-up collections in 2002–2003 in south Texas, bandedwinged whitefly, *T. abutiloneus,* was found on some of the seasonal plant hosts sampled in the absence of primary *B. tabaci* host crops. The only parasitoids reared from *T. abutiloneus* on these hosts were *Eret. tejanus* and *Enc. tabacivora/pergandiella*

Table 18.2 Percent species composition of parasitoids attacking *Bemisia tabaci* (n = 39,877) in the Lower Rio Grande Valley, Texas during October 1993–December 1995.

Parasitoid species	%
Encarsia formosa Gahan/*luteola* Howard	0.7
Encarsia meritoria Gahan/*hispida* De Santis	0.3
Encarsia pergandiella Howard/*tabacivora* Viggiani	42.4
Encarsia strenua group	2.1
Eretmocerus tejanus (Rose and Zolnerowich)	54.5

(P. DeBarro and J. Goolsby, personal communication). This suggests that the native parasitoid complex in south Texas is still present in the LRGV agroecosystem, and remains associated with *T. abutiloneus,* whereas the introduced *Eretmocerus* species seem to be restricted to *Bemisia*. Despite the dominance of the introduced *Eretmocerus* species, *Eret. tejanus* is still found occasionally in samples of *B. tabaci.*

18.3.6 California/Arizona Desert Valleys

18.3.6.1 Native Desert Plants and Weeds

Surveys of vegetation in the desert areas of Imperial Valley were begun in 1992 to identify hosts of *B. tabaci,* to determine which native parasitoids were attacking *B. tabaci* biotype B on desert plants, to investigate the role of desert plants as biological control refuges for the release of natural enemies, to collect background information to support evaluations of subsequent release/refuge programs, and to document host ranges of native and introduced parasitoids that will help evaluate potential non-target impacts of the exotic species (Bellows et al. 1994; Wendel et al. 1995; Hoelmer and Culver 1997). Survey locations for desert vegetation included more than 50 sites along the eastern and western desert margins of the Imperial Valley as well as occasional collections from adjacent desert regions of Arizona. Sample locations included the counties of Imperial, Riverside, and San Diego in California, and Yuma, Maricopa, Pima, and LaPaz in Arizona. Plant voucher specimens were collected, identified, and stored in the Imperial County Agricultural Department Herbarium.

Desert hosts of *B. tabaci* were mostly annual plants rather than perennial shrubs and trees. Of the 120 native or naturalized species of desert plants that were inspected at survey sites, 30% were recorded as hosts of *B. tabaci* and the most common are listed in Table 18.3. Typically these hosts had low whitefly densities and only for a short period of time during the year, corresponding to the availability of water. The species most often found with whitefly nymphs included *Datura discolor* Bernh. (desert thorn-apple), *Dicoria canescens* Gray (desert twin bugs, bugseed), and *Argythamnia neomexicana* Muell.-Arg. (New Mexico silverbush, Mexican ditaxis). Host plants of *B. tabaci* from which the greatest numbers of parasitoids were reared included *A. neomexicana, D. discolor*, *D. canescens*, *Heterotheca subaxillaris* (Lam.) Britt. and Rusby (camphor weed), *Pluchea purpurascens* (Sw.) DC. (annual fleabane), and *Xanthium strumarium* L. var. *canadense* (P. Mill.) Torr. and Gray (cocklebur). Imperial County survey records showed that *B. tabaci* typically appeared on hosts in desert areas only after whitefly populations had already increased to large numbers on crops in adjacent irrigated areas. Parasitoids that emerged from *B. tabaci* pupae on desert hosts included *Eret. eremicus*, *Enc. luteola, Enc. meritoria,* and *Enc. coquilletti* Howard.

Table 18.3 Plant hosts of selected whiteflies surveyed in urban areas and adjacent desert regions of Imperial Co., California. All plant names are according to the USDA Natural Resources Conservation Service Database (2005).

Desert plant hosts surveyed	Urban ornamental plant hosts surveyed
Albizia sinaloensis Britton and Rose (Sonoran silktree)	*Fraxinus* spp. (ash)
Ambrosia ilicifolia (Gray) Payne (hollyleaf bur ragweed)	*Ceratonia siliqua* L. (St. John's bread, carob)
Argythamnia neomexicana Muell.-Arg (New Mexico silverbush)	*Lantana camara* L. (lantana)
Chamaesyce polycarpa (Benth.) Millsp. ex Parish (smallseed sandmat)	*Rosa* spp. (rose)
Cucurbita palmata S. Wats. (coyote gourd)	*Cupaniopsis anacardioides* (A. Rich.) Radlkf. (carrotwood)
Datura discolor Bernh. (desert thorn-apple)	*Morus alba* L. and *nigra* L. (mulberry)
Dicorea canescens Gray (bugseed)	
Fouquieria splendens Englem. (ocotillo)	**Whiteflies sampled on urban ornamental plants**
Heterotheca subaxillaris (Lam.) Britt. and Rusby (camphor weed)	*Bemisia tabaci* (Genn.) biotype B
Hyptis emoryi Torr. (desert lavender)	*Siphoninus phillyreae* (Haliday)
Hymenoclea salsola Torr. and Gray ex Gray (cheesebush, burrobrush)	*Aleuropleurocelus* spp.
Jatropha cinerea (Ortega) Muell.-Arg. (Arizona nettlespurge)	*Aleuropleurocelus coachellensis* Drews and Sampson
Mentzelia multiflora (Nutt.) Gray var. *longiloba* (Adonis blazing star)	*Tetraleurodes caulicola* Nakahara
Nicotiana obtusifolia Mertens and Galiotti var. *obtusifolia* (desert tobacco)	*Tetraleurodes acaciae* (Quaintance)
Palafoxia arida B.L. Turner and Morris var. *gigantea* (giant Spanish needle)	*Tetraleurodes mori* (Quaintance)
Pectis papposa Harvey and Gray (cinchweed)	
Pluchea sericea (Nutt.) Coville (arrowweed)	
Pluchea purpurascens (Sw.) DC. (annual fleabane)	
Proboscidea parviflora (devils claw)	
Prosopis glandulosa Torr. (mesquite)	
Psorothamnus schottii (Torr.) Barneby (indigo bush)	
Psorothamnus spinosus (Gray) Barneby (smoketree)	
Sphaeralcea ambigua Gray (desert mallow)	
Tiquilia plicata (Torr.) A. Richards (fanleaf crinklemat)	
Xanthium strumarium L. var. *canadense* (P. Mill.) Torr. and Gray (cocklebur)	
Whiteflies sampled on desert host plants	
Bemisia tabaci (Genn.) biotype B	
Trialeurodes abutiloneus (Haldeman)	
Aleuropleurocelus coachellensis Drews and Sampson	
Aleuropleurocelus spp.	
Tetraleurodes caulicola Nakahara	

Predators of *Bemisia* occasionally observed on desert plants during survey site visits included *Geocoris* and *Orius* sp. (Hemiptera: Geocoridae, Anthocoridae), *Semidalis* sp. nr. *flinti* (Neuroptera: Conoipterygidae), *Chrysopa* sp. (Neuroptera: Chrysopidae), *Hippodamia convergens* (Guerin-Meneville) (Coleoptera: Coccinellidae), and unidentified ants.

Other species of whiteflies found on Lower Colorado desert plants included *T. abutiloneus, Aleuropleurocelus coachellensis* Drews and Sampson, several *Aleuropleurocelus* species that were new and undescribed or described as 'near' known species, and *Tetraleurodes caulicola* Nakahara. The taxonomic relationship between indigenous desert *Aleuropleurocelus* and *Tetraleurodes* species and the morphological characters defining the two genera are unclear (R. Gill, personal communication), so the identity of some of the whiteflies found on native desert vegetation remains tentative.

Trialeurodes abutiloneus was the only other whitefly species regularly found on the same herbaceous, annual host plants utilized by *B. tabaci*. Parasitoids reared from this whitefly included the same four species associated with *B. tabaci* and two additional *Encarsia* species that were found less commonly (Table 18.4). Other whitefly species were found only on native desert vegetation or in low numbers on a few ornamental shrubs and trees (Table 18.3), and some of the desert plants were host to both *Aleuropleurocelus* species and *T. caulicola*. Parasitoids reared from the native desert vegetation did not include any of those occurring on *B. tabaci* and *T. abutiloneus* (Table 18.4). Many of the parasitoids reared from *Aleuropleurocelus* and *Tetraleurodes* are undescribed species (M. Rose and J. Heraty, personal communications) and occurred in low numbers due to the sporadic occurrence of their hosts. Parasitoids included several different species of *Eretmocerus* from cheesebush or burrobrush (*Hymenoclea salsola* Torr. and Gray ex Gray), Sonoran silktree (*Albizia sinaloensis* Britton and Rose), and Arizona nettlespurge or white lomboy (*Jatropha cinerea* (Ortega) Muell.-Arg.), an undetermined number of *Encarsia* species from several hosts, and one or more species of *Neopomphale* from many of these hosts (Hymenoptera: Eulophidae). One or more unidentified species of Signiphoridae (Hymenoptera: Aphelinidae) were reared from several *Aleuropleurocelus* species; these were presumed to be hyperparasitoids of primary whitefly parasitoids, as they are typically hosts of signiphorids.

By fall of 2001, when desert surveys were discontinued, none of the introduced parasitoids released against *B. tabaci* had been recovered from any indigenous species of whiteflies sampled on desert plants, even though exotic *Eretmocerus* had been recovered from *B. tabaci* at many locations, including a number of isolated desert survey sites. Conceivably this situation could change if populations of one or more introduced parasitoids continue to increase; therefore periodic monitoring surveys are necessary in order to detect host shifts to non-target species.

Table 18.4 Parasitoids reared from *Bemisia tabaci* and selected indigenous whiteflies in desert areas near agricultural production in southwestern Arizona (Yuma, Maricopa, Pima, La Paz counties) and southeastern California (Imperial, Riverside, San Diego counties)[a].

From *Bemisia tabaci* on numerous desert host plants:
Eretmocerus eremicus Rose and Zolnerowich
Encarsia luteola Howard
Encarsia meritoria Gahan
Encarsia coquilletti Howard (rarely)
From *Trialeurodes abutiloneus* on various hosts:
Eretmocerus eremicus
Encarsia luteola
Encarsia meritoria
Encarsia coquilletti
Encarsia sp. nov.
Encarsia nr. *porteri* (probably species 'Z' of Gerling, 1967)
From *Aleuropleurocelus coachellensis* on arrowweed (*Pluchea sericea*):
Encarsia sp. nov.
Neopomphale sp. nov.
From *Aleuropleurocelus* sp. on cheesebush (*Hymenoclea salsola*):
Neopomphale sp. nov.
Eretmocerus sp. nov.
Encarsia sp. nov.
Amitus sp.
From *Tetraleurodes caulicola* on smoketree (*Dalea* (*Psorothamnus*) *spinosa*):
Encarsia sp. nov.
Neopomphale sp. nov.
From *Aleuropleurocelus* sp. nr. *coachellensis* and *Tetraleurodes caulicola* on indigo bush (*Dalea (Psorothamnus) schotti*):
Neopomphale sp. nov.
From *Aleuropleurocelus* sp. on lomboy (*Jatropha cinerea*):
Neopomphale sp. nov.
Eretmocerus sp. nov.
Encarsia meritoria
From *Aleuropleurocelus* sp. on desert lavender (*Hyptis emoryi*):
Encarsia sp. nov.

[a]Whitefly hosts were identified by Ray Gill of California Department of Food and Agriculture. Refer to text for discussion of species identification.

18.3.6.2 Urban Communities: Ornamentals

Ball and Weddle (1993) began surveys on various ornamental host plants, including hibiscus (*Hibiscus rosa-sinensis* L.), orchid tree (*Bauhinia variegata* L.), and snail vine (*Vigna caracalla* L.), in southern California deserts in 1991 and continued their survey in 1992–1993 in urban areas of Imperial County. They recorded three species of parasitoids attacking *B. tabaci* of which 94% were *Eret. eremicus* (reported as *Eret. californicus*) with the remainder being *Enc. luteola* and *Enc. meritoria.*

These same three species were found in surveys continued by APHIS-PPQ during the next several years. *Encarsia coquilletti,* sporadically reared from *B. tabaci* on desert host plants in low numbers, was not found in samples from urban areas. Later surveys included periodic collections of indigenous whiteflies on several ornamental shrubs and trees (Table 18.3). None of the parasitoids reared from *B. tabaci* were reared from these whiteflies (with the exception of *Enc. meritoria* or near *meritoria,* which was reared from an *Aleuropleurocelus* species on lomboy in samples from the Living Desert Museum, Palm Desert, California), nor were any of their parasitoids reared from *B. tabaci*. None of the introduced species of *Eretmocerus* or *Encarsia* were reared from any of the indigenous whiteflies collected in urban areas by 2000.

18.3.6.3 Agricultural Crops

Pre- and post-release surveys conducted in Imperial County, California and Maricopa and LaPaz Cos., Arizona resulted in many thousands of parasitoids being reared from *B. tabaci* on seasonal crops including cotton, cantaloupe, alfalfa, broccoli (*Brassica oleracea italica* group) and cauliflower (*Brassica oleracea* L. *botrytis* group). The same three species (*Eret. eremicus, Enc. luteola,* and *Enc. meritoria*) recorded from *B. tabaci* on ornamental and desert vegetation were found on these crops. *Encarsia coquilletti* was found rarely in *B. tabaci* on okra grown organically in Bard/Winterhaven, California, adjacent to the Colorado River, but was not found in other Imperial Co. okra samples. Proportional species composition varied depending on time of year and host crop, but *Eret. eremicus* was the most abundant species overall. Among the *Encarsia* species, *Enc. luteola* was more abundant than *Enc. meritoria*. The proportion of species represented among 3,550 parasitoids reared from *B. tabaci* during the first California and Arizona desert surveys in 1992–1993 was generally representative of the overall species composition obtained in later surveys (72% *Eret. eremicus*, 19% *Enc. luteola* females, 3% *Enc. meritoria* females, and 6% male *Enc. luteola* or *meritoria* indistinguishable by recognized characters). In pre-release surveys no specimens of exotic *Enc. sophia* or *Eret. mundus* were found that had been released against *B. tabaci* biotype A several years earlier, consistent with reports that establishment had not occurred from those releases.

Post-release surveys indicated the increasing presence of several introduced *Eretmocerus* species (*emiratus* Zolnerowich and Rose and nr. *emiratus, hayati* Zolnerowich and Rose, and *mundus*) and *Enc. sophia* in samples of *B. tabaci* (Chapter 15).

18.3.7 California: San Joaquin Valley

Crop (melons, cotton, alfalfa, sweet potato, and tomato) and weed (smooth pigweed, velvetleaf (*Abutilon theophrasti* Medic.), black nightshade (*Solanum americanum* P. Mill.) and morning glory (*Ipomoea* sp.)) hosts were sampled using timed searches in five counties in 1994 and about 20,000 nymphs were examined to determine levels of parasitism (Godfrey et al. 1995). In Kern Co. (southern San Joaquin Valley) parasitism levels were very low, averaging 1.5%, but increased from the 0.2% level previously found in 1993. Two thirds of the parasitism was due to unidentified *Eretmocerus* species; of the remaining parasitized nymphs about 80% was due to unidentified *Encarsia* species and 20% was unidentified due to mortality before emergence. In northern San Joaquin Valley counties overall parasitism was 0.5% in 1994 (up slightly from 0.1% in 1993); approximately 40% of this was due to *Eretmocerus* spp., 20% to *Encarsia* spp., and the rest was unidentified.

18.4 Hawaii

Although no releases of exotic parasitoids were made in Hawaii, several indigenous enemies of *B. tabaci* biotype B were noted. During a study in eggplant conducted in 1994 in Hawaii in which *Enc. formosa* was released augmentatively, no *Enc. formosa* were recovered in samples, but low numbers of *Enc. sophia, Enc. nigricephala,* and an unidentified *Eretmocerus* species were found. Several predators were also noted as possibly influencing *B. tabaci* numbers, including the coccinellid *Curinus coeruleus* Mulsant, and the phytoseiid mites *Phytoseiulus hawaiiensis* Prasad and *Amblyseius* sp. nr. *tetranychivorus* (Johnson 1995).

18.5 Conclusions

The greatest diversity of native parasitoid species was reported from surveys in Florida. This may be a result of the very high diversity of whitefly species in Florida, which is frequently invaded by exotic whiteflies. Some of these species may be the primary hosts of parasitoids that were reared in low numbers from *B. tabaci*. The surveys showed that only two or three species were responsible for most of parasitism of *B. tabaci* within any given region of the USA.

A certain amount of caution concerning rearing methods should be mentioned when comparing relative species compositions. In some surveys large quantities of parasitized whiteflies were held together in the same container until emergence was complete. This method of handling allows early emerging, unmated female *Encarsia* to hyperparasitize other parasitized whitefly and produce male offspring, potentially skewing the resulting emergence totals in favor of *Encarsia* at the expense of preferred hosts for hyperparasitism, such as *Eretmocerus*.

It is clear from the surveys in Texas and the southern California and Arizona deserts that *T. abutiloneus* serves as a natural reservoir for most of the indigenous parasitoids that attack *B. tabaci* (either biotype) in these areas. However, it is encouraging that as of the latest surveys in 2001 (California) and 2003 (Texas), the exotic species that were introduced have remained limited to their intended target, *B. tabaci*. Continued monitoring will be needed to demonstrate whether this remains the case over the long term.

References

Ball, J.C. and D. Weddle. 1993. Untitled report *in* Section D: Biocontrol. In T.J. Henneberry, N.C. Toscano, R.M. Faust, and J.R. Coppedge, eds., Sweetpotato Whitefly: 1993 Supplement to the 5-Year National Research and Action Plan - First Annual Review, Tempe, AZ, January 18–21, 1993, p. 93. US Department of Agriculture, ARS-112.

Bellows, T.S. Jr., T.M. Perring, and D.H. Headrick. 1994. Parasitism of silverleaf whitefly in California crops and weeds. In T.J. Henneberry, N.C. Toscano, R.M. Faust, and J.R. Coppedge, eds., Silverleaf Whitefly: 1994 Supplement to the 5-Year National Research and Action Plan – Second Annual Review, Orlando, FL, January 24–27, 1994, p. 122. US Department of Agriculture, ARS-125.

Brewster, C.C., J.C. Allen, D.J. Schuster, and P.A. Stansly. 1997. Simulating the dynamics of *Bemisia argentifolii* (Homoptera: Aleyrodidae) in an organic cropping system with a spatio-temporal model. Environmental Entomology 26: 603–616.

Ciomperlik, M.A., J.M. Rodriguez, and L.E. Wendel. 1995. Assessments of sweetpotato whitefly (*Bemisia tabaci,* biotype 'B') and indigenous parasite populations in agroecosystems of the Lower Rio Grande Valley, Texas. In T.J. Henneberry, N.C. Toscano, R.M. Faust, and J.R. Coppedge, eds., Silverleaf Whitefly: 1995 Supplement to the 5-Year National Research and Action Plan – Third Annual Review, San Diego, CA, January 28–31, 1995, p. 103. US Department of Agriculture, ARS-1995-2.

Gerling, D. 1966. Studies with whitefly parasites of Southern California. II. *Eretmocerus californicus* Howard (Hymenoptera: Aphelinidae). Canadian Entomologist 98: 1316–1329.

Gerling, D. 1967. Bionomics of the whitefly-parasite complex associated with cotton in Southern California (Homoptera: Aleurodidae; Hymenoptera: Aphelinidae). Annals of the Entomological Society of America 60: 1306–1321.

Godfrey, L.D., P.B. Goodell, T.M. Perring, T.S. Bellows, C.G. Summers, W.J. Bentley, T. Prather, and R. Covielli. 1995. Incidence of parasitism of silverleaf whitefly in the San Joaquin Valley. In T.J. Henneberry, N.C. Toscano, R.M. Faust, and J.R. Coppedge, eds., Silverleaf Whitefly: 1995 Supplement to the 5-Year National Research and Action Plan - Third Annual Review, San Diego, CA, January 28–31, 1995, p. 105. US Department of Agriculture, ARS-1995-2.

Goolsby, J.A., M.A. Ciomperlik, B.C. Legaspi, Jr., J.C. Legaspi, and L.E. Wendel. 1998. Laboratory and field evaluation of parasitoids of exotic parasitoids of *Bemisia tabaci* (Gennadius) (Biotype "B") (Homoptera: Aleyrodidae) in the Lower Rio Grande Valley of Texas. Biological Control 12: 127–135.

Hamon, A.B. and V. Salguero. 1987. *Bemisia tabaci*, sweetpotato whitefly, in Florida (Homoptera: Aleyrodidae: Aleyrodinae). Florida Department of Agriculture & Consumer Services, Entomology Circular 292.

Hennessey, R.D. 1993. Untitled report *in* Section D: Biocontrol. In T. J. Henneberry, N.C. Toscano, R. M. Faust, and J.R. Coppedge, eds., Sweetpotato Whitefly: 1993 Supplement to the 5-Year National Research and Action Plan – First Annual Review, Tempe, AZ, January 18–21, 1993, p. 99. US Department of Agriculture, ARS-112.

Heraty, J.M. and A. Polaszek. 2000. Morphometric analysis and descriptions of selected species in the *Encarsia strenua* group (Hymenoptera: Aphelinidae). Journal of Hymenoptera Research 9: 142–169.

Hoelmer, K.A. and G. Culver. 1997. Survey of desert host plants for whiteflies and parasitoids. In T.J. Henneberry, N.C. Toscano, T.M. Perring, and R.M. Faust, eds., Silverleaf Whitefly: 1997

Supplement to the 5-Year National Research and Action Plan - Progress Review, Technology Transfer, and New Research and Action Plan, 1997–2001, Fifth Annual Review, San Diego, CA, January 28–30, 1997, p. 149. US Department of Agriculture, ARS-1997-02.

Johnson, M. 1995. Natural enemies associated with silverleaf whitefly in Hawaii. In T.J. Henneberry, N.C. Toscano, R.M. Faust, and J.R. Coppedge, eds., Silverleaf Whitefly: 1995 Supplement to the 5-Year National Research and Action Plan – Third Annual Review, San Diego, CA, January 28–31, 1995, p. 116. US Department of Agriculture, ARS-1995-2.

Johnson, M.W., N.C. Toscano, H.T. Reynolds, E.S. Sylvester, K. Kido, and E.T. Natwick. 1982. Whiteflies cause problems for southern California growers. California Agriculture 36(9/10): 24–26.

Jones, W.A. 1993. Untitled report *in* Section D: Biocontrol. In T.J. Henneberry, N.C. Toscano, R.M. Faust, and J.R. Coppedge, eds., Sweetpotato Whitefly: 1993 Supplement to the 5-Year National Research and Action Plan – First Annual Review, Tempe, AZ, January 18–21, 1993, p. 101. US Department of Agriculture, ARS-112.

McAuslane, H.J., F.A. Johnson, D.A. Knauft, and D.L. Colvin. 1993. Seasonal abundance and within-plant distribution of parasitoids of *Bemisia tabaci* (Homoptera: Aleyrodidae) in peanuts. Environmental Entomology 22: 1043–1050.

McAuslane, H.J., F.A. Johnson, and D.A. Knauft. 1994. Population levels and parasitism of *Bemisia tabaci* (Gennadius) (Homoptera: Aleyrodidae) on peanut cultivars. Environmental Entomology 23: 1203–1210.

Moomaw, C., J. Woolley, M. Rose, and D. Riley. 1994. Evaluation of naturally occurring parasitic Hymenoptera attacking *Bemisia tabaci* (Genn.) in Texas. In T.J. Henneberry, N.C. Toscano, R.M. Faust, and J.R. Coppedge, eds., Silverleaf Whitefly: 1994 Supplement to the 5-Year National Research and Action Plan - Second Annual Review, Orlando, FL, January 24–27, 1994, p. 143. US Department of Agriculture, ARS-125.

Mound, L.A. and S.H. Halsey. 1978. Whitefly of the World: A Systematic Catalogue of the Aleyrodidae (Homoptera) with Host Plant and Natural Enemy Data. British Museum (Natural History) and Wiley, Chichester, UK.

Nuessly, G.S. 1990. Biological control of the sweetpotato whitefly in California. In R.K. Yokomi, K.R. Narayanan, and D.J. Schuster, eds., Sweetpotato Whitefly-Mediated Vegetable Disorders in Florida. Proceedings, University of Florida Institute of Food and Agricultural Science Workshop, Homestead, FL, February 1–2, 1990, pp. 81–82. University of Florida, Gainesville, FL.

Perring, T.M., A. Cooper, D.J. Kazmer, C. Shields, and J. Shields. 1991. New strain of sweetpotato whitefly invades California vegetables. California Agriculture 45(6): 10–12.

Perring, T.M., A.D. Cooper, R.J. Rodriquez, C.A. Farrar, and T.S. Bellows Jr. 1993. Identification of a whitefly species by genomic and behavior studies. Science 259: 74–77.

Polaszek, A.G., G.A. Evans, and F.D. Bennett. 1992. *Encarsia* parasitoids of *Bemisia tabaci* (Hymenoptera: Aphelinidae, Homoptera: Aleyrodidae): a preliminary guide to identification. Bulletin of Entomological Research 82: 375–392.

Riley, D.G. and A.N. Sparks Jr. 1993. Management of the sweetpotato whitefly in the Lower Rio Grande Valley of Texas. Texas A&M Extension Bulletin B-5082.

Riley, D.G. and M.A. Ciomperlik. 1997. Regional population dynamics of whitefly (Homoptera: Aleyrodidae) and associated parasitoids (Hymenoptera: Aphelinidae). Environmental Entomology 26: 1049–1055.

Rose, M. and G. Zolnerowich. 1997. *Eretmocerus* Haldeman (Hymenoptera: Aphelinidae) in the US, with descriptions of new species attacking *Bemisia* (*tabaci* complex) (Homoptera: Aleyrodidae). Proceedings of the Entomological Society of Washington 99: 1–27.

Schauff, M.E., G.A. Evans, and J.M. Heraty. 1996. A pictorial guide to the species of *Encarisia* (Hymenoptera: Aphelinidae) parasitic on whiteflies (Homoptera: Aleyrodidae) in North America. Proceedings of the Entomological Society of Washington 98: 1–35.

Schuster, D.J., J.F. Price, J.B. Kring, and P.H. Everett. 1989. Integrated management of the sweetpotato whitefly on commercial tomato. University of Florida, Institute of Food and Agricultural Science, Bradenton Gulf Coast Research and Education Center Research Report. BRA-1989.

Schuster, D.J., J.E. Polston, and J.F. Price. 1992. Reservoirs of the sweetpotato whitefly for tomatoes in west-central Florida. Proceedings of the Florida State Horticultural Society 105: 311–314.

Schuster, D.J., G.A. Evans, F.D. Bennett, P.A. Stansly, R.K. Jansson, G.L. Leibee, and S.E. Webb. 1998. A survey of parasitoids of *Bemisia* spp. whiteflies in Florida, the Caribbean, and Central and South America. International Journal of Pest Management 44: 255–260.

Simmons, A.M. 1993. Untitled report *in* Section D: Biocontrol. In T.J. Henneberry, N.C. Toscano, R.M. Faust, and J.R. Coppedge, eds., Sweetpotato Whitefly: 1993 Supplement to the 5-Year National Research and Action Plan – First Annual Review, Tempe, AZ, January 18–21, 1993, p. 110. US Department of Agriculture, ARS-112.

Simmons, A.M. 1995. Abundance of parasitoids of *Bemisia argentifolii* in sweetpotato. In T.J. Henneberry, N.C. Toscano, R.M. Faust, and J.R. Coppedge, eds., Silverleaf Whitefly: 1995 Supplement to the 5-Year National Research and Action Plan - Third Annual Review, San Diego, CA, January 28–31, 1995, p. 137. US Department of Agriculture, ARS-1995-2.

Simmons, A.M. and K.D. Elsey. 1994. Seasonal parasitism in South Carolina. In T.J. Henneberry, N.C. Toscano, R.M. Faust, and J.R. Coppedge, eds., Silverleaf Whitefly: 1994 Supplement to the 5-Year National Research and Action Plan – Second Annual Review, Orlando, FL, January 24–27, 1994, p. 150. US Department of Agriculture, ARS-125.

Smith, M.T., M. Allred, D. Fieselman, M. Ciomperlik, B. Layton, J. Jarratt, P. Harris, A. Milling, and R. Snyder. 1997. Survey and identification of the whitefly and associated parasitoid species in Mississippi. In T.J. Henneberry, N.C. Toscano, T.M. Perring, and R.M. Faust, eds., Silverleaf Whitefly: 1997 Supplement to the 5-Year National Research and Action Plan – Progress Review, Technology Transfer, and New Research and Action Plan, 1997–2001, Fifth Annual Review, San Diego, CA, January 28–30, 1997, p. 172. US Department of Agriculture, ARS-1997-02.

Stansly, P.A., D.J. Schuster, and T.-X. Liu. 1997. Apparent parasitism of *Bemisia argentifolii* (Homoptera: Aleyrodidae) by Aphelinidae (Hymenoptera) on vegetable crops and associated weeds in south Florida. Biological Control 9: 49–57.

US Department of Agriculture, Animal and Plant Health Inspection Service. 1995a. Field releases of nonindigenous parasitic wasps in the genus *Eretmocerus* (Hymenoptera: Aphelinidae) for biological control of whitefly pests (Homoptera: Aleyrodidae). Environmental Assessment. Riverdale, MD.

US Department of Agriculture, Animal and Plant Health Inspection Service. 1995b. Field releases of nonindigenous parasitic wasps in the genus *Encarsia* (Hymenoptera: Aphelinidae) for biological control of whitefly pests (Homoptera: Aleyrodidae). Environmental Assessment. Riverdale, MD.

US Department of Agriculture. 2005. Natural Resources Conservation Service Database. http://plants.usda.gov/.

van Lenteren, J.C. and L.P.J.J. Noldus. 1990. Whitefly-plant relationships: behavioral and ecological aspects. In D. Gerling, ed., Whiteflies: their Bionomics, Pest Status and Management, pp. 147–185. Intercept, Andover, UK.

Viggiani, G. 1989. Notes on some nearctic and neotropical *Encarsia* foerster (Hymenoptera: Aphelinidae). Bollettino del Laboratorio di Entomologia Agraria "Filippo Silvestri", Portici 43: 59–78.

Wendel, L., R. Weddle, and G. Culver. 1995. Survey of *Bemisia tabaci* on the Colorado Desert, Imperial Valley, California. In T.J. Henneberry, N.C. Toscano, R.M. Faust, and J.R. Coppedge, eds., Silverleaf Whitefly: 1995 Supplement to the 5-Year National Research and Action Plan – Third Annual Review, San Diego, CA, January 28–31, 1995, p. 20. US Department of Agriculture, ARS-1995-2.

Yokomi, R.K., K.A. Hoelmer, and L.S. Osborne. 1990. Relationships between the sweetpotato whitefly and the squash silverleaf disorder. Phytopathology 80: 895–900.

Epilogue

Controlling pests with natural enemies reduces pesticide use, stabilizes food supplies and ultimately benefits the public worldwide. The *Bemisia* program shared these objectives. Classical biological control practitioners often distribute natural enemies to researchers in other countries, subject to export and import regulations. Rather than reinventing the wheel, programs in one country can learn from the experiences of scientists in other countries, and biological control is often obtained with the same agents in various parts of the world. Foreign explorers from the USA were granted access to many countries throughout the world to collect natural enemies of *Bemisia*. In turn, natural enemies that performed well in the USA were shared with researchers in other nations affected by *Bemisia*.

Table 1 lists the dates, countries, cooperators, and natural enemies shipped. It is not yet known which species permanently established in most of the receiving countries, with the exception of Australia. Paul DeBarro (CSIRO Entomology), Indooroopilly, Queensland, Australia, conducted extensive research to determine if imported natural enemies were needed for control of *B. tabaci* biotype B that invaded Australia, causing problems starting in 1997. He concluded that *B. tabaci* did not have effective parasitoids despite the presence of parasitoids adapted to the indigenous *B. tabaci* 'Australasian' biotype. He then collaborated with US researchers to determine which single species should be the first to be imported and tested. The single release approach is required in Australia due to the extensive regulatory requirements for both insect and weed biological control agents. The decision was to import *Eretmocerus hayati*, originally from Pakistan, from field populations that had established in Texas. This was based on the fact that *Eret. hayati* had performed very well in the US program and was the most common exotic parasitoid in southern Texas, which is climatically similar to affected areas in Queensland, Australia. *Eretmocerus hayati* was imported in 2001. Host range testing of numerous related whitefly species that were indigenous to Australia showed that *Eret. hayati* posed little or no risk of unintended nontarget impacts. *Eretmocerus hayati* was released in Bundaberg, Queensland in late 2004 and has spread rapidly from the release sites. As of January 2005 it had been recovered far from release sites 300km to the south in the Lockyear Valley near Brisbane. Cotton and vegetable growers reported that 2004–2005 were the first years since 1995 that *B. tabaci* was not a significant pest in eastern Queensland (P. Debarro, personal communication).

Table 1 Shipments of *B. tabaci* parasitoids and predators to cooperators overseas.

Year	Country	Receiving scientist and institution	Species	M#	Origin	#'s released
1995	Honduras	Ron Cave	*Eret. mundus*	M92014	Spain	1,000
		Zamarano College	*Eret. mundus*	M92019	India	10,000
1996	Dominican Republic	Colmar Serra	*Enc. sophia*	M93003	Spain	2,000
		DR Department of Agriculture				
	Italy	Genaro Viggiani	*Enc. sophia*	M93003	Spain	100
		Silvestri Institute				
	Puerto Rico	Leyinska Wiscovitch	*Enc. sophia*	M93003	Spain	18,650
		USDA-APHIS-PPQ	*Enc. formosa*	M92030	Egypt	23,480
			Eret. mundus	M92014	Spain	42,644
			Serangium parcestosum	M93008	India	100
1997	Israel	Dan Gerling	*Enc. sophia*	M95107	Pakistan	800
		Tel Aviv University				
	Mexico	Jesus Vargas Camplis	*Eret. hayati*	M95012	Pakistan	33,000
		National Institute of Forestry, Agricultural	*Eret. mundus*	M92014	Spain	33,000
		and Animal Research	*Eret. emiratus*	M95104	U.A.E.	33,000
	Puerto Rico	Leyinska Wiscovitch	*Eret. hayati*	M95012	Spain	128,508
		USDA-APHIS-PPQ	*Eret. mundus*	M92014	Spain	11,000
			Eret. emiratus	M95104	U.A.E.	10,000
			Enc. sophia	M93003	Spain	8,000
			Enc. formosa	M92030	Egypt	10,000
2001	Australia	Paul DeBarro	*Eret. hayati*	M95012	Pakistan	5,000
		CSIRO Entomology				

Plans are presently being made to export *Eret. hayati* to China and *Eret. melanoscutus* to Colima, Mexico. Both countries have been affected by outbreaks of *B. tabaci* biotype B.

The recent discovery of *B. tabaci* biotype Q in Arizona and California has prompted calls for additional screenings of exotic *B. tabaci* parasitoids on this biotype. However, this may not be necessary, since many of the *Eretmocerus* spp. imported and established in the US program were obtained from collections of *B. tabaci* biotypes other than B. At least three native North American *Eretmocerus* species in Florida, Texas, Arizona and California readily adapted from *B. tabaci* biotype A to attack biotype B. It is likely that the already established native and introduced natural enemies will readily attack the Q biotype, as has been demonstrated for Spanish populations of *Eret. mundus* (Urbaneja and Stansly 2004).

Reference

Urbaneja, A. and P.A. Stansly. 2004. Host suitability of different instars of the whitefly *Bemisia tabaci* 'biotype Q' for *Eretmocerus mundus*. BioControl 49: 153–161.

Summary

The classical biological control program directed against *Bemisia tabaci* (Gennadius) biotype B (= *B. argentifolii* Bellows and Perring) in the 1990s was one the largest and most comprehensive programs in the history of biological control. A team of scientists from USDA-APHIS, USDA-ARS, CDFA, and several universities contributed to the discovery, importation, evaluation, release, and colonization of a suite of imported natural enemies. Field entomologists were stationed in Arizona, California and Texas to carry out the research where the infestations were the most damaging. To support the foreign exploration for natural enemies for *B. tabaci* and other biological control projects, a new APHIS quarantine facility was built in Mission, Texas, which eventually coordinated with mass rearing facilities at the field locations to maximize the release efforts for the biological control program. The substantial level of commitment by USDA matched the level of damage caused to a wide range of crops by this 'super pest'. The national program leadership of ARS and APHIS produced a remarkable achievement in redirecting research programs, obtaining additional research and implementation funding from Congress, and developing a 5-Year National Research and Action Plan for Development of Management and Control Methodology for the Sweetpotato Whitefly (1992–1996), which was followed by the Silverleaf Whitefly 5-Year National Research, Action, and Technology Transfer Plan (1997–2001) to help organize research and track progress.

The foreign exploration program for natural enemies of *B. tabaci* was comprehensive, covering 28 countries, and more than 130 shipments of natural enemies were sent to quarantine facilities in the USA between 1991 and 1998. Climate matching was used to compare the affected areas in the USA with locations within the native distribution of *B. tabaci* and thereby prioritize foreign exploration. The ARS European Biological Control Laboratory in Montpellier, France was extremely valuable to the biological control program; its staff engaged in nearly year-round exploration, which led to the discovery of many parasitoids, predators and pathogens for evaluation by US researchers.

In the USDA APHIS Mission Biological Control Laboratory quarantine, 55 populations of natural enemies were established in culture, including 13 species of *Eretmocerus* and *Encarsia* parasitoids. Only parasitoids reared from *B. tabaci* (any biotype) and with a primary or autoparasitic biology were considered for release.

Predictive, prerelease studies were conducted in quarantine and in field cages to determine which species showed the most potential to impact whitefly populations. This information was used to prioritize species for mass rearing and field release. Low-performing parasitoid species were also released in substantial numbers at selected locations to validate the predictions.

The sweetpotato whitefly program was the first large-scale biological control program to use molecular genetic methods to characterize the imported natural enemies prior to their release. RAPD-PCR was used to screen the natural enemies reared in quarantine to identify cryptic species and maximize available genetic diversity from the exploration efforts. This was critical because many of the most valuable *Eretmocerus* species were extremely similar morphologically, but had unique biological traits such as host-plant preferences and climatic adaptations. The molecular methods were also used to assure colony purity and to identify exotic specimens recovered in the field programs. Taxonomic keys were developed to identify and describe the native and imported *Eretmocerus* and *Encarsia* parasitoids of *B. tabaci* in North America.

Mass rearing facilities were established in Tucson, Arizona; Imperial and Sacramento, California; and Mission, Texas. Hundreds of millions of *Eretmocerus* and *Encarsia* species were mass reared for several years for release and evaluation in the areas affected by sweetpotato whitefly, which included the subtropical agricultural areas of the USA and Mexico. Mass rearing techniques improved dramatically over the course of the program, beginning with laboratory rearing in environmental chambers on whitefly-infested hibiscus plants, to heated, outdoor field cages with large pots of kale and eggplant, to highly managed greenhouses that used large-leaf eggplants and mechanical removal of parasitoid pupae. The substantial number of parasitoids available for release enabled a large-scale field evaluation of biological control as an integrated component of management programs.

Field evaluation programs were conducted in Phoenix, Arizona; Brawley and Sacramento, California; and Mission, Texas. Candidate natural enemies were tested in field cages on multiple crops, including alfalfa, broccoli, cotton, and melons. The results showed strong tri-trophic interactions and the significance of climatic adaptation. Four species of Palearctic *Eretmocerus* that established in the western USA were morphologically similar, representing a group of closely related taxa that appear to be specialist parasitoids of the *B. tabaci* complex that includes several known biotypes. Their ability to readily attack whiteflies in the *B. tabaci* complex may have given them an advantage in the field versus the native *Eret. tejanus* Rose and Zolnerowich and *Eret. eremicus* Rose and Zolnerowich, which have a broader host range that includes *Trialeurodes* species. To date no nontarget impacts have been detected from the exotic parasitoids, which is likely due to their narrow host range.

The climate in the native range of each of the four imported *Eretmocerus* spp. closely matched the climate in the areas of the USA where they established: *Eret. mundus* Mercet from Mediterranean Europe to San Joaquin Valley, California; *Eret. emiratus* Zolnerowich and Rose from the hot desert of the Arabian Peninsula to Imperial Valley, California; *Eret.* sp. nr. *emiratus* from the tropical desert of Ethiopia in equatorial Africa to Yuma and Phoenix, Arizona; and *Eret. hayati*

Zolnerowich and Rose from the subtropical desert of the Indus River Valley in Pakistan to the Rio Grande Valley, Texas. The exotic autoparasitoid *Enc. sophia* (Girault and Dodd) also established in the Imperial Valley of California, although several endemic autoparasitioids were previously present in this agroecosystem.

Several release methods were developed to enhance the likelihood of establishment of the parasitoids in the annual cropping systems. Refuge strips, home gardens, and commercial nurseries were used as release sites because they had stable year-round populations of *B. tabaci* and were free of broad spectrum pesticide use. In addition to inoculative release methods, a more efficient method for augmentation of parasitoids using seedling transplants as 'banker plants' was developed for cucurbit crops.

Field efforts to control *B. tabaci* were aided by the development of several narrow spectrum insecticides that were effective against *B. tabaci* while allowing substantial parasitoid activity. Biological control-intensive IPM strategies were developed to take advantage of the new selective insecticides. Banker plants were transplanted into imidacloprid treated fields, which reduced the cost of release and demonstrated how the BC-IPM could be integrated with current farming practices.

Trials with the entomopathogens *Beauveria bassiana* and *Paecilomyces fumosoroseus* showed they caused significant mortality to *B. tabaci* nymphs under high humidity conditions. Specialized spray equipment was developed and tested to apply the pathogens under commercial field conditions. Field trials showed that the biopesticides were effective during the early season and compatible with natural enemies. Two different formulations of *B. bassiana* (Mycotrol) and a formulation of *Paecilomyces fumosoroseus* were later commercialized.

The sweetpotato whitefly biological control program clearly demonstrated the potential benefits of classical biological control in annual row-crop agriculture. The program also demonstrated the utility of predictive evaluations, which showed that a multiple species release strategy was needed due the varied climates and crops that were impacted by *B. tabaci*. This strategy should be considered for future biological control programs directed at polyphagous invasive pests that become widely distributed in the USA.

Following their establishment, detailed quantitative studies are now needed to document the impact of the introduced species, which appeared to be substantial in the years immediately following establishment in areas such as the Lower Rio Grande Valley. Life table studies similar to those conducted for survivorship of *B. tabaci* on cotton in Maricopa, Arizona (Naranjo and Ellsworth 2005) are needed for all areas where the exotic parasitoids have become established. Only with such evaluations will we truly be able to accurately measure the impact and significance of benefits derived from the interagency sweetpotato whitefly biological control program.

Reference

Naranjo, S.E. and P.C. Ellsworth. 2005. Mortality dynamics and population regulation in *Bemisia tabaci*. Entomologia Experimentalis et Applicata 116: 93–108.

Index

Q

R

MIX
Papier aus verantwortungsvollen Quellen
Paper from responsible sources
FSC® C105338

If you have any concerns about our products,
you can contact us on
ProductSafety@springernature.com

In case Publisher is established outside the EU,
the EU authorized representative is:
Springer Nature Customer Service Center GmbH
Europaplatz 3, 69115 Heidelberg, Germany

Printed by Libri Plureos GmbH
in Hamburg, Germany